Lecture Notes in Artificial Intelligence   5790

Edited by R. Goebel, J. Siekmann, and W. Wahlster

Subseries of Lecture Notes in Computer Science

Emiel Krahmer   Mariët Theune (Eds.)

Empirical Methods in Natural Language Generation

Data-Oriented Methods and Empirical Evaluation

 Springer

Series Editors

Randy Goebel, University of Alberta, Edmonton, Canada
Jörg Siekmann, University of Saarland, Saarbrücken, Germany
Wolfgang Wahlster, DFKI and University of Saarland, Saarbrücken, Germany

Volume Editors

Emiel Krahmer
Tilburg University, Tilburg Center for Cognition and Communication (TiCC)
Faculty of Humanities
Department of Communication and Information Sciences (DCI)
P.O.Box 90153, 5000 LE Tilburg, The Netherlands
E-mail: e.j.krahmer@uvt.nl

Mariët Theune
University of Twente, Human Media Interaction (HMI)
Faculty of Electrical Engineering, Mathematics
and Computer Science (EEMCS)
P.O. Box 217, 7500 AE Enschede, The Netherlands
E-mail: m.theune@ewi.utwente.nl

Library of Congress Control Number: 2010933310

CR Subject Classification (1998): I.2, H.3, H.4, H.2, H.5, J.1

LNCS Sublibrary: SL 7 – Artificial Intelligence

ISSN   0302-9743
ISBN-10   3-642-15572-3 Springer Berlin Heidelberg New York
ISBN-13   978-3-642-15572-7 Springer Berlin Heidelberg New York

springer.com

© Springer-Verlag Berlin Heidelberg 2010

Typesetting: Camera-ready by author, data conversion by Scientific Publishing Services, Chennai, India
Printed on acid-free paper   06/3180

Preface

Natural language generation (NLG) is a subfield of natural language processing (NLP) that is often characterized as the study of automatically converting non-linguistic representations (e.g., from databases or other knowledge sources) into coherent natural language text. NLG is useful for many practical applications, ranging from automatically generated weather forecasts to summarizing medical information in a patient-friendly way, but is also interesting from a theoretical perspective, as it offers new, computational insights into the process of human language production in general. Sometimes, NLG is framed as the mirror image of natural language understanding (NLU), but in fact the respective problems and solutions are rather dissimilar: while NLU is basically a disambiguation problem, where ambiguous natural language inputs are mapped onto unambiguous representations, NLG is more like a choice problem, where it has to be decided which words and sentences best express certain specific concepts.

Arguably the most comprehensive currently available text book on NLG is Reiter and Dale's [7]. This book offers an excellent overview of the different subfields of NLG and contains many practical insights on how to build an NLG application. However, in recent years the field has evolved substantially, and as a result it is fair to say that the book is no longer fully representative of the research currently done in the area of NLG. Perhaps the most important new development is the current emphasis on data-oriented methods and empirical evaluation. In 2000, data-oriented methods to NLG were virtually non-existent and researchers were just starting to think about how experimental evaluations of NLG systems should be conducted, even though many other areas of NLP already placed a strong emphasis on data and experimentation. Now the situation has changed to such an extent that *all* chapters in this book crucially rely on empirical methods in one way or another.

Three reasons can be given for this important shift in attention, and it is instructive spelling them out here. First of all, progress in related areas of NLP such as machine translation, dialogue system design and automatic text summarization created more awareness of the importance of language generation, even prompting the organization of a series of multi-disciplinary workshops on Using Corpora for Natural Language Generation (UCNLG). In statistical machine translation, for example, special techniques are required to improve the grammaticality of the translated sentence in the target language. N-gram models can be used to filter out improbable sequences of words, but as Kevin Knight put it succinctly "automated language translation needs generation help badly" [6]. To give a second example, automatic summarizers which go beyond mere sentence extraction would benefit from techniques to combine and compress sentences. Basically, this requires NLG techniques which do not take non-linguistic information as input, but rather (possibly ungrammatical) linguistic information (phrases or text fragments), and as a result this approach to NLG is sometimes referred to as text-to-text generation. It bears a strong conceptual resemblance to text revision, an area of NLG which received some scholarly attention in the 1980s and 1990s (e.g., [8, 9]). It has turned out that text-to-text generation lends itself well for data-oriented approaches, in part because textual training and evaluation material are easy to come by.

In contrast, text corpora are of relatively limited value for "full" NLG tasks which are about converting concepts into natural language. For this purpose, one would prefer to have so-called semantically transparent corpora [4], which contain both information about the available concepts as well as human-produced realizations of these concepts. Consider, for instance, the case of referring expression generation, a core task of many end-to-end NLG systems. A corpus of human-produced referring expressions is only useful if it contains complete information about the target object (what properties does it have?) and the other objects in the domain (the distractors). Clearly this kind of information is typically not available in traditional text corpora consisting of Web documents, newspaper articles or comparable collections of data. In recent years various researchers have started collecting semantically transparent corpora (e.g., [5, 10]), and this has given an important boost to NLG research. For instance, in the area of referring expression generation, the availability of semantically transparent corpora has made it possible for the first time to seriously evaluate traditional algorithms and to develop new, empirically motivated ones.

The availability of suitable corpora also made it feasible to organize shared tasks for NLG, where different teams of researchers develop and evaluate their algorithms on a shared, held out data set. These kinds of shared tasks, including the availability of benchmark data sets and standardized evaluation procedures, have proven to be an important impetus on developments in other areas of NLP, and already a similar effect can be observed for the various NLG shared tasks ("generation challenges") for referring expression generation [1], for generation of references to named entities in text [2] and for instruction giving in virtual environments [3]. These generation challenges not only resulted in new-generation research, but also in a better understanding of evaluation and evaluation metrics for generation algorithms.

Taken together these three developments (progress in related areas, availability of suitable corpora, organization of shared tasks) have had a considerable impact on the field, and this book offers the first comprehensive overview of recent empirically oriented NLG research. It brings together many of the key researchers and describes the state of the art in text-to-text generation (with chapters on modeling text structure, statistical sentence generation and sentence compression), in NLG for interactive applications (with chapters on learning how to generate appropriate system responses, on developing NLG tools that automatically adapt to their conversation partner, and on NLG as planning under uncertainty, as applied to spoken dialogue systems), in referring expression generation (with chapters on generating vague geographic descriptions, on realization of modifier orderings, and on individual variation), and in evaluation (with chapters dedicated to comparing different automatic and hand-crafted generation systems for data-to-text generation, and on evaluation of surface realization, linguistic quality and affective NLG). In addition, this book also contains extended chapters on each one of the generation challenges organized so far, giving an overview of what has been achieved and providing insights into the lessons learned.

The selected chapters are mostly thoroughly revised and extended versions of original research that was presented at the 12th European Workshop on Natural Language Generation (ENLG 2009) or the 12th Conference of the European Association for Computational Linguistics (EACL 2009), both organized in Athens, Greece, between March 30 and April 3, 2009. Both ENLG 2009 and EACL 2009 were preceded by the usual extensive reviewing procedures and we thank Regina Barzilay, John Bateman, Anja Belz, Stephan Busemann, Charles Callaway, Roger Evans, Leo Ferres, Mary-Ellen Foster, Claire Gardent, Albert Gatt, John Kelleher, Geert-Jan Kruijff, David McDonald, Jon Oberlander, Paul Piwek, Richard Powers, Ehud Reiter, David Reitter, Graeme Ritchie, Matthew Stone, Takenobu Tokunaga, Kees van Deemter, Manfred Stede, Ielka van der Sluis, Jette Viethen and Michael White for their efforts.

April 2010

Emiel Krahmer
Mariët Theune

References

1. Belz, A., Gatt, A.: The attribute selection for GRE challenge: Overview and evaluation results. In: Proceedings of UCNLG+MT: Language Generation and Machine Translation, Copenhagen, Denmark, pp. 75–83 (2007)
2. Belz, A., Kow, E., Viethen, J., Gatt, A.: The GREC challenge 2008: Overview and evaluation results. In: Proceedings of the Fifth International Natural Language Generation Conference (INLG 2008), Salt Fork, OH, USA, pp. 183–191 (2008)
3. Byron, D., Koller, A., Striegnitz, K., Cassell, J., Dale, R., Moore, J., Oberlander, J.: Report on the first NLG challenge on generating instructions in virtual environments (GIVE). In: Proceedings of the 12th European Workshop on Natural Language Generation (ENLG 2009), Athens, Greece, pp. 165–173 (2009)
4. van Deemter, K., van der Sluis, I., Gatt, A.: Building a semantically transparent corpus for the generation of referring expressions. In: Proceedings of the 4th International Conference on Natural Language Generation (INLG 2006), Sydney, Australia, pp. 130–132 (2006)
5. Gatt, A., van der Sluis, I., van Deemter, K.: Evaluating algorithms for the generation of referring expressions using a balanced corpus. In: Proceedings of the 11th European Workshop on Natural Language Generation (ENLG 2007), Saarbrücken, Germany, pp. 49–56 (2007)
6. Knight, K.: Automatic language translation generation help needs badly. Or: "Can a computer compress a text file without knowing what a verb is?" In: Proceedings of UCNLG+MT: Language Generation and Machine Translation, Copenhagen, Denmark, pp. 1–4 (2007)

7. Reiter, E., Dale, R.: Building Natural Language Generation Systems. Cambridge University Press, Cambridge (2000)
8. Robin, J.: A revision-based generation architecture for reporting facts in their historical context. In: Horacek, H., Zock, M. (eds.) New Concepts in Natural Language generation: Planning, Realization and Systems. Frances Pinter, London (1993)
9. Vaughan, M.M., McDonald, D.D.: A model of revision in natural language generation. In: Proceedings of the 24th Annual Meeting of the Association for Computational Linguistics (ACL 1986), New York, NY, USA, pp. 90–96 (1986)
10. Viethen, J., Dale, R.: Algorithms for generating referring expressions: Do they do what people do? In: Proceedings of the 4th International Conference on Natural Language Generation (INLG 2006), Sydney, Australia, pp. 63–70 (2006)

Table of Contents

Shared Task Challenges for NLG

Probabilistic Approaches for Modeling Text Structure and Their Application to Text-to-Text Generation

Regina Barzilay

Computer Science and Artificial Intelligence Laboratory
Massachusetts Institute of Technology
77 Massachusetts Avenue, Cambridge MA 02139

Abstract. Since the early days of generation research, it has been acknowledged that modeling the global structure of a document is crucial for producing coherent, readable output. However, traditional knowledge-intensive approaches have been of limited utility in addressing this problem since they cannot be effectively scaled to operate in domain-independent, large-scale applications. Due to this difficulty, existing text-to-text generation systems rarely rely on such structural information when producing an output text. Consequently, texts generated by these methods do not match the quality of those written by humans – they are often fraught with severe coherence violations and disfluencies.

In this chapter,[1] I will present probabilistic models of document structure that can be effectively learned from raw document collections. This feature distinguishes these new models from traditional knowledge intensive approaches used in symbolic concept-to-text generation. Our results demonstrate that these probabilistic models can be directly applied to content organization, and suggest that these models can prove useful in an even broader range of text-to-text applications than we have considered here.

1   Introduction

Text-to-text generation aims to produce a coherent text by extracting, combining, and rewriting information given in input texts. Summarization, answer fusion in question-answering, and text simplification are all examples of text-to-text generation tasks. At first glance, these tasks seem much easier than the traditional generation setup, where the input consists of a non-linguistic representation. Research in summarization over the last decade has proven that texts generated for these tasks rarely match the quality of those written by humans. One of the key reasons is the lack of coherence in the generated text. As the

---

[1] This chapter is based on the invited talk given by the author at the 2009 European Natural Language Generation workshop. It provides an overview of the models developed by the author and her colleagues, rather than giving a comprehensive survey of the field. The original papers [2,3,4,26] provide the complete technical details of the presented models.

E. Krahmer, M. Theune (Eds.): Empirical Methods in NLG, LNAI 5790, pp. 1–12, 2010.

Table 1. An example of a summary generated by a multidocument summarization system participating in the 2003 Document Understanding Conference (DUC)

He said Cathay Pacific was still studying PAL's financial records.

Ailing Philippine Airlines and prospective investor Cathay Pacific Airways have clashed over the laying off of PAL workers, prompting PAL to revive talks with another foreign airline, an official said Tuesday.

PAL resumed domestic flights Oct. 7 and started restoring international flights last month after settling its labor problems.

PAL officials say Singapore Airlines is also interested in a possible investment. "As much as PAL is the flag carrier, we should see to it that PAL will always fly. But Philippine officials said Cathay and PAL had run into difficulties in two areas: who would manage PAL and how many workers would lose their jobs.

---

automatically-generated summary in Table 1 illustrates, coherence violations can drastically reduce the information value of the system output.

To illustrate current text-to-text generation methods, let us consider the task of automatic text summarization. Most current summarizers select content based on low-level features such as sentence length and word occurrence. The inability of summarizers to consider higher level contextual features leads to decreased performance and a need for greater amounts of training data. Similarly, when these systems render the selected content as a summary, they operate with no structural considerations. Therefore, they are ill-equipped to answer the questions of how the selected items fit together, what their optimal ordering is, or, in short, how to ensure that a fluent text is produced rather than a heap of sentences glued together randomly. What we are missing is an automatic mechanism that can score the well-formedness of texts and that can select the most fluent alternative among various text rendering strategies. The ability to automatically analyze the topical structure of actual and potential documents should also enable the development of more effective content selection algorithms that can operate on a more abstract and global level.

This current state of affairs is rather surprising, given that NLP researchers have developed elaborate discourse models that capture various facets of text structure. These models encompass textual aspects ranging from the morphological [27] to the intentional [9], and include a characterization of documents in terms of domain-independent rhetorical elements, such as schema items [22] or rhetorical relations [19,20]. In fact, for concept-to-text generation systems, encoding these structural theories has contributed significantly to output quality, making them almost indistinguishable from human writing. In contrast, for text-to-text generation systems, these theories are hard to incorporate as they stand: they rely on handcrafted rules, are valid only for limited domains, and have no guarantee of scalability or portability. This difficulty motivates the development of novel approaches for document organization that can rely exclusively on information available in textual input.

In this chapter, I will present models of document structure that can be effectively used to guide content selection in text-to-text generation. First, I will focus on unsupervised learning of domain-specific content models. These models capture the topics addressed in a text and the order in which these topics appear; they are similar in their functionality to the content planners traditionally used in concept-to-text generation. I will present an effective method for learning content models from unannotated domain-specific documents utilizing hierarchical Bayesian methods. Incorporation of these models into information ordering and summarization applications yields substantial improvements over previously proposed methods.

Next, I will present a method for assessing the coherence of a generated text. The key premise of this method is that the distribution of entities in coherent texts exhibits certain regularities. The models I will be presenting operate over an automatically-computed representation that reflects distributional, syntactic, and referential information about discourse entities. This representation allows us to induce the properties of coherent texts from a given corpus, without recourse to manual annotation or a predefined knowledge base. I will show how these models can be effectively employed for content organization in text-to-text applications.

2 Content Models

A content model is a structural representation that describes the main topics and their organization within a given domain of discourse. Modeling content structure is particularly relevant for domains that exhibit recurrent patterns in content organization, such as news and encyclopedia articles. These models aim to induce, for example, that articles about cities typically contain information about History, Economy, and Transportation, and that descriptions of History usually precede those of Transportation. Note that computing content models for an arbitrary domain is a challenging task due to the lack of explicit, unambiguous structural markers (e.g., subsections with the corresponding titles). Moreover, texts within the same domain may exhibit some variability in topic selection and ordering; this variability further complicates the discovery of content structure.

Since most concept-to-text generation systems operate in limited domains, using a set of deterministic rules is a feasible way to capture patterns in content organization [15,22,25]. Despite substantial effort involved in such induction, these rules proved to be essential for successful content planning – they govern the selection of relevant material, its grouping into topically homogeneous units, and its subsequent ordering. The success of this architecture in concept-to-text generation systems motivates exploring it in the context of text-to-text generation. Clearly, the first step in this direction is automating the computation of content models, as manual crafting is infeasible in large, complex domains that a text-to-text generation system should handle.

2.1 Computational Modeling

Fortunately, recent research has demonstrated the feasibility of acquiring content models automatically from raw texts. These approaches adapt a *distributional*

Table 2. A sample of Wikipedia sentences assigned the same section heading by the editors

Addis Ababa also has a railway connection with Djibouti City, with a picturesque French style railway station.
The rail network, connecting the suburbs in the tri-state region to the city, consists of the Long Island Rail Road, Metro-North Railroad and New Jersey Transit.
The most important project in the next decade is the Spanish high speed rail network, Alta Velocidad Española AVE.
Rail is the primary mode of transportation in Tokyo, which has the most extensive urban railway network in the world and an equally extensive network of surface lines.
The backbone of the city's transport, the Mumbai Suburban Railway, consists of three separate networks: Central, Western, and Harbour Line, running the length of the city, in a north-south direction.

view, learning content models via analysis of word-distribution patterns across texts within the same domain. This idea dates back at least to Harris [11], who claimed that "various types of [word] recurrence patterns seem to characterize various types of discourse."

The success of automatic induction greatly depends on how much variability is exhibited by the texts in the underlying domain. In formulaic domains, a simple word-based clustering can be sufficient for computing content models [26]. More commonly, however, the same topic can be conveyed using very different wordings. In such cases, word distribution on its own may not be a sufficient predictor of a segment topic. For instance, consider sentences that represent the *Railway Transportation* topic in several Wikipedia articles about cities (see Table 2). The first two sentences clearly discuss the same topic, but they do not share any content words in common. To properly handle such cases, more elaborate algorithms for learning content models are needed.

One of the early instances of such approaches is an algorithm for learning content models using Hidden Markov Models (HMMs) [3]. In these models, states correspond to types of information characteristic of the domain of interest (e.g., earthquake magnitude or previous earthquake occurrences) and state transitions capture possible information-presentation orderings within that domain. Like clustering algorithms, these models capture the intuition that sentences that belong to the same topic use similar vocabulary. In addition, HMM-based models can exploit regularities in ordering to further refine the induction process.

Our recent work on content model induction has focused on modeling *global* constraints on discourse organization which cannot be easily expressed in Markovian models. An example of such a global constraint is *topic continuity* – it posits that each document follows a progression of coherent, nonrecurring topics [10]. Following the example above, this constraint captures the notion that a single topic, such as History, is expressed in a contiguous block within the document, rather than spread over disconnected sections. Another global constraint concerns similarity in *global ordering*. This constraint guides the model toward selecting sequences with similar topic *ordering*, such as placing History before Transportation.

To effectively capture these global constraints, a content model posits a single distribution over the *entirety* of a document's content ordering [4]. Specifically, the model represents content structure as a *permutation* over topics. This naturally enforces the first constraint since a permutation does not allow topic repetition. Despite apparent intractability, this distribution over permutations can be effectively learned using the *Generalized Mallows Model* (GMM) [6].

By design, GMMs concentrate the most probability mass on a single permutation, the *canonical permutation*. While other permutations are plausible, their likelihood decays exponentially with their distance from the the canonical permutation. In comparison to HMMs, GMMs greatly restrict a set of possible topic orderings predicted for a given domain as permutations drawn from this distribution are likely to be similar. However this restriction actually enables the compact parametrization of GMMs, supporting effective inference of its parameters in a Bayesian framework.

We position the GMM within a larger hierarchical Bayesian model that explains how a set of related documents is generated. Figure 1 pictorially summarizes the steps of the generative process. At a high level, the model first selects how frequently each topic is expressed in the document, and how the topics are ordered. These topics then determine the selection of words for each paragraph. More specifically, for each document, the model posits that a topic ordering is drawn from the GMM, and that a set of topic frequencies is drawn from a multinomial distribution. Together, these draws specify the document's entire content structure, in the form of topic assignments for each textual unit. As with traditional topic models, words are then drawn from language models indexed by topic. Model parameters are estimated using Gibbs sampling.

2.2   Applications to Text-to-Text Generation

One of the important advantages of the automatically induced content models is that they can easily be integrated into existing generation applications. For instance, consider the task of information ordering, where the goal is to determine the sequence in which a pre-selected set of items is presented to the user. This is an essential step in concept-to-text generation, multi-document summarization, and other text-synthesis problems.

To apply a content model to this task, we assume we are provided with well structured documents from a single domain as training examples; once trained, the model is used to induce orderings of previously unseen collections of paragraphs from the same domain. The implementation of the ordering algorithms depends on the underlying content model. An HMM-based model searches the space of possible orderings based on the likelihood of a sequence as predicted by the model. Typically, A* and other heuristic search algorithms are used. The GMM-based model takes a very different approach for computing an ordering: it predicts the most likely topic for each paragraph independently using

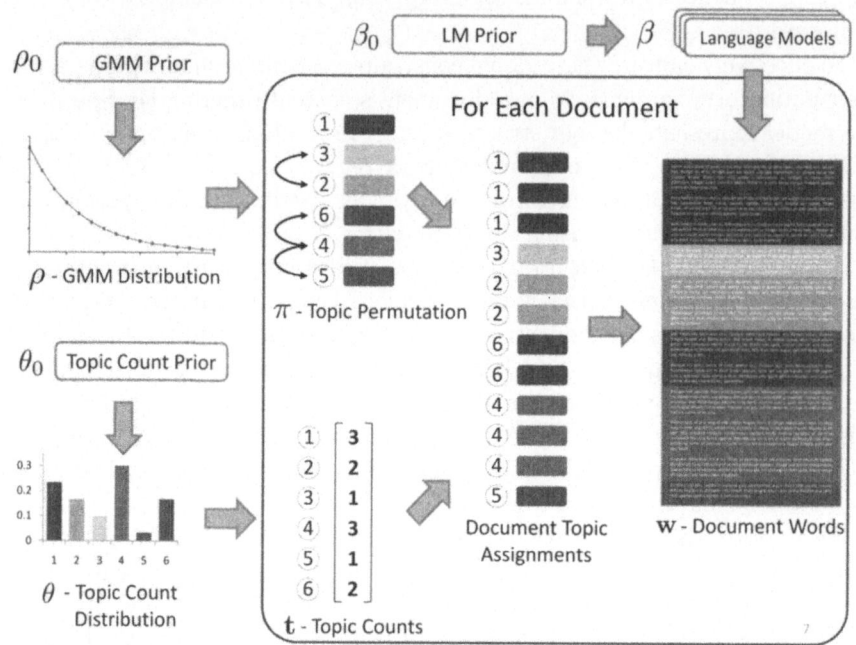

Fig. 1. The generative process for permutation-based content models

topic-specific language models. Because the GMM distribution concentrates probability mass around one known topic ordering, these topic assignments determine the best ordering of the paragraphs.

These content modeling algorithms have been tested on multiple domains, ranging from product reviews to Wikipedia articles. The results consistently indicate that the ordering retrieved by the model are close to the ordering of a human writer. For instance, the GMM-based algorithm achieves *Kendall's*[2] τ of 0.678 on ordering documents in the domain of cell phone reviews [4]. Moreover, the empirical results demonstrate the advantages of encoding global structural constraints into probabilistic content models [4]. In fact, the difference between the HMM-based and GMM-based content models is substantial: the former achieves *Kendall's* τ of only 0.256, when tested on the same domain.

In our recent work, we have considered a new application of content models – automatic generation of overview articles. These multi-paragraph texts are comprehensive surveys of a subject, generated by composing information drawn from the Internet [26]. Examples of such overviews include actor biographies

---

[2] Kendall's τ measures how much an ordering differs from the reference order. Specifically, for a permutation π of the sections in an N-section document, $\tau(\pi)$ is computed as $\tau(\pi) = 1 - 2\frac{d(\pi,\sigma)}{\binom{N}{2}}$, where $d(\pi,\sigma)$ is the number of swaps of adjacent textual units necessary to rearrange π into the reference order. The metric ranges from -1 (inverse orders) to 1 (identical orders).

Diagnosis . . . No laboratories offering molecular genetic testing for prenatal diagnosis of 3-M syndrome are listed in the GeneTests Laboratory Directory. However, prenatal testing may be available for families in which the disease-causing mutations have been identified in an affected family member in a research or clinical laboratory.

Causes Three M syndrome is thought to be inherited as an autosomal recessive genetic trait. Human traits, including the classic genetic diseases, are the product of the interaction of two genes, one received from the father and one from the mother. In recessive disorders, the condition does not occur unless an individual inherits the same defective gene for the same trait from each parent. . . .

Symptoms . . . Many of the symptoms and physical features associated with the disorder are apparent at birth (congenital). In some cases, individuals who carry a single copy of the disease gene (heterozygotes) may exhibit mild symptoms associated with Three M syndrome.

Treatment . . . Genetic counseling will be of benefit for affected individuals and their families. Family members of affected individuals should also receive regular clinical evaluations to detect any symptoms and physical characteristics that may be potentially associated with Three M syndrome or heterozygosity for the disorder. Other treatment for Three M syndrome is symptomatic and supportive.

Fig. 2. A fragment from the automatically created article for 3-M Syndrome

from IMDB and disease synopses from Wikipedia. An example of our system's output[3] is shown in Figure 2.

In this application, content models are employed for both selecting and ordering the material in a generated article. Given a corpus of human-authored texts with a corresponding content model, the algorithm learns a content extractor targeted for each topic of the content model, such as *diagnosis* and *treatment*. This extractor specifies a query for selecting candidate material from the web and provides a ranker that assigns a relevance score for each retrieved passage. To produce a new article, the algorithm employs these extractors for each topic of the content model and then jointly selects the best passages based on local and global constraints.

This method is related to content selection methods developed for extractive summarization. In fact, the design of individual extractors is similar to supervised methods for sentence extraction [18]. The difference, however, is in the extraction criteria. Traditional summarization methods filter sentences based on the generic notion of "importance," while our selection criteria is more focused: it is driven by the topic of the candidate passage. This architecture ensures that the overview article will have the breadth expected in a comprehensive summary, with content drawn from a wide variety of Internet sources.

We evaluate the quality of the generated articles by comparing them with the corresponding human-authored articles in Wikipedia. System performance

---

[3] This system output was added to Wikipedia at http://en.wikipedia.org/wiki/ 3-M_syndrome on June 26, 2008. The page's history provides examples of changes performed by human editors to articles created by our system.

is measured using ROUGE-1, a standard measure used in the summarization community [17]. We employ the system to generate articles in two domains – American Film Actors and Diseases. For each domain, we randomly select 90% of Wikipedia articles in the corresponding categories for training and test on the remaining 10%. To measure the impact of the content model-based architecture for content selection, we consider a baseline that does not use a template to specify desired topics. Instead, we train a single classifier that learns to extract excerpts that are likely to appear in the Wikipedia article, without explicitly capturing the topic they convey. The results convincingly show the advantages of the content model-based approach: in both domains, this approach outperforms the baseline by a statistically significant margin. For instance, in the Disease domain the full model achieves an F-measure of 0.37, while the baseline yields an F-measure of 0.28.

3   Coherence Models

While a content model is a powerful abstraction of document structure, by definition this representation is domain-specific. As many text-to-text applications are domain-independent, we need a model that can operate in such a context. In this section, we introduce a *coherence model* that captures text relatedness at the level of sentence-to-sentence transitions [1,7,14,16,21].

The key premise of this approach is that the distribution of entities in locally coherent texts exhibits certain regularities. This assumption is not arbitrary – some of these regularities have been recognized in Centering Theory [8] and other entity-based theories of discourse. Previous research has demonstrated that direct translation of these linguistic theories into a practical coherence metric is difficult: one has to determine ways of combining the effects of various constraints and to instantiate parameters of the theory that are often left unspecified [12,13,23,24].

3.1   Computational Modeling

The model is based on an expressive linguistic representation called the *entity-grid*, a two-dimensional array that captures the distribution of discourse entities across text sentences. The rows of the grid correspond to sentences, while the columns correspond to discourse entities. By *discourse entity* we mean a class of coreferent noun phrases. For each occurrence of a discourse entity in the text, the corresponding grid cell contains information about its presence or absence in the sentences. In addition, for entities present in a given sentence, grid cells contain information about their syntactic role reflecting whether the corresponding entity is a subject (**s**), object (**o**), or neither (**x**). Entities absent from a sentence are signaled by gaps (–). Table 4 illustrates a fragment of an entity grid constructed for the text in Table 3.[4] Grid representation can be automatically computed

---

[4] These two tables are borrowed from the Computational Linguistics article that introduced the grid representation [3].

Table 3. Summary augmented with syntactic annotations for grid computation

---

1 [The Justice Department]$_s$ is conducting an [anti-trust trial]$_o$ against [Microsoft Corp.]$_x$ with [evidence]$_x$ that [the company]$_s$ is increasingly attempting to crush [competitors]$_o$.

2 [Microsoft]$_o$ is accused of trying to forcefully buy into [markets]$_x$ where [its own products]$_s$ are not competitive enough to unseat [established brands]$_o$.

3 [The case]$_s$ revolves around [evidence]$_o$ of [Microsoft]$_s$ aggressively pressuring [Netscape]$_o$ into merging [browser software]$_o$.

4 [Microsoft]$_s$ claims [its tactics]$_s$ are commonplace and good economically.

5 [The government]$_s$ may file [a civil suit]$_o$ ruling that [conspiracy]$_s$ to curb [competition]$_o$ through [collusion]$_x$ is [a violation of the Sherman Act]$_{|o}$.

6 [Microsoft]$_s$ continues to show [increased earnings]$_o$ despite [the trial]$_x$.

---

using standard text processing algorithms such as a syntactic parser and a tool for coreference resolution.

Our analysis revolves around patterns of local entity transitions. These transitions, encoded as continuous subsequences of a grid column, represent entity occurrences and their syntactic roles in n adjacent sentences. According to Centering Theory, in a coherent text some transitions are more likely than others. For instance, grids of coherent texts are likely to have some dense columns (i.e., columns with just a few gaps such as *Microsoft* in Table 4) and many sparse columns which will consist mostly of gaps (see *markets, earnings* in Table 4). One would further expect that entities corresponding to dense columns are more often subjects or objects. These characteristics will be less pronounced in low-coherence texts.

Therefore, by looking at the likelihood of various transition types in a text, we can assess the degree of its coherence. To automatically uncover predictive patterns, we represent a text by a fixed set of transition sequences using a feature vector notation. Given a collection of coherent texts and texts with coherence violations, we can employ standard supervised learning methods to uncover entity distribution patterns relevant for coherence assessment.

3.2   Application to Text-to-Text Generation

The main utility of a coherence model is to automatically assess the quality of a generated output. When a text-to-text generation system is equipped with such a measure, it can select the most fluent candidate among possible output realizations. These possible realizations may correspond to different orders of the output sentences, to different sets of selected sentences, or to different ways entities are realized within each sentence.

To validate the usefulness of the coherence model, we consider two evaluation scenarios. In the first scenario, the model is given a pair of texts consisting of a well-formed document and a random permutation of its sentences. The task is to select the more coherent document, which in this case corresponds to the

Table 4. A fragment of the entity grid. Noun phrases are represented by their head nouns

	Department	Trial	Microsoft	Evidence	Competitors	Markets	Products	Brands	Case	Netscape	Software	Tactics	Government	Suit	Earnings	
1	s	o	s	x	o	–	–	–	–	–	–	–	–	–	–	1
2	–	–	o	–	–	x	s	o	–	–	–	–	–	–	–	2
3	–	–	s	o	–	–	–	–	s	o	o	–	–	–	–	3
4	–	–	s	–	–	–	–	–	–	–	–	s	–	–	–	4
5	–	–	–	–	–	–	–	–	–	–	–	–	s	o	–	5
6	–	x	s	–	–	–	–	–	–	–	–	–	–	–	o	6

original document. While this evaluation setup allows us to generate a large-scale corpus for training and testing the method, it only partially approximates the degrees of coherence violation observed in the output of text-to-text generation systems. Therefore, in our second evaluation scenario, we apply the model to assess coherence of automatically generated summaries. In particular, we use summaries collected for the 2003 Document Understanding Conference. These summaries are generated by different multidocument summarization systems, and therefore they exhibit a range of disfluencies.

In both experiments, the grid model achieves notably high performance. For instance, the algorithm can distinguish a coherent ordering from a random permutation with an accuracy of 87.3% when applied to reports from the National Transportation Safety Board Accident Database, and 90.4% when applied to Associated Press articles on Earthquakes. The task of coherence assessment turned out to be more challenging: the best configuration of the model achieves an accuracy of 81.3%. The results also demonstrate that incorporating salience and syntactic features, sources of information featured prominently in discourse theories, leads to a consistent increase in accuracy. For example, eliminating syntactic information decreases the ordering performance of the model by 10%. We also compare the performance of coherence models against content models. While these two types of models capitalize on different sources of discourse information, they achieve comparable performances – content models yield an accuracy of 88.0% and 75.8% on the two datasets, compared with the accuracy of 87.3% and 90.4% obtained by coherence models. Recent work has demonstrated that further improvements in performance can be achieved by combining coherence and content models [5].

4   Conclusions

In this chapter, I demonstrated that automatically-induced models of text structure advance the state of the art in text-to-text generation. Our experiments show that incorporating discourse constraints leads to more effective information selection and increases fluency and coherence of the system output. The key strength of the proposed discourse models is that they can be derived with minimal annotation effort, in some cases learning from only the raw text. The performance of these models validates the long-standing hypothesis stated by Harris about the connection between high-level discourse properties and distributional patterns at the word level.

An important future direction lies in designing statistical models of text structure that match the representational power of traditional discourse models. Admittedly, the models described in this chapter constitute a relatively impoverished representation of discourse structure. While this contributes to the ease with which they can be learned, it limits their potential to improve the quality of text-to-text generation systems. For instance, the models described above do not capture the hierarchical structure of discourse, which has been shown to be important for content planning. Another limitation of the above models is that they capture just one aspect of discourse structure rather than modeling text structure in a holistic manner. I believe that recent advances in machine learning (e.g., discriminative structure prediction algorithms and unsupervised Bayesian methods) would enable us to further refine statistical discourse models and consequently improve the output of text-to-text generation systems.

Bibliographic Note and Acknowledgments. This paper builds on several previously published results [2,3,4,26]. I would like to thank my co-authors – Branavan, Harr Chen, David Karger, Mirella Lapata, Lillian Lee and Christina Sauper – for their generous permission to use this material in this chapter. I thank Branavan, Harr Chen, Emiel Krahmer and Yoong Keok Lee for their thoughtful comments and suggestions for improving this manuscript. This work was funded by the National Science Foundation (NSF) CAREER grant IIS-0448168 and the Microsoft Research New Faculty Fellowship. Any opinions, findings, conclusions or recommendations expressed in this article do not necessarily reflect the views of NSF or Microsoft.

References

1. Althaus, E., Karamanis, N., Koller, A.: Computing locally coherent discourses. In: Proceedings of the ACL, pp. 399–406 (2004)
2. Barzilay, R., Lapata, M.: Modeling local coherence: An entity-based approach. Computational Linguistics 34(1), 1–34 (2008)
3. Barzilay, R., Lee, L.: Catching the drift: Probabilistic content models, with applications to generation and summarization. In: HLT-NAACL, pp. 113–120 (2004)
4. Chen, H., Branavan, S., Barzilay, R., Karger, D.R.: Content modeling using latent permutations. JAIR, 129–163 (2009)
5. Elsner, M., Austerweil, J., Charniak, E.: A unified local and global model for discourse coherence. In: Proceedings of HLT-NAACL, pp. 436–443 (2007)
6. Fligner, M., Verducci, J.: Distance based ranking models. Journal of the Royal Statistical Society, Series B 48(3), 359–369 (1986)

7. Foltz, P.W., Kintsch, W., Landauer, T.K.: Textual coherence using latent semantic analysis. Discourse Processes 25(2&3), 285–307 (1998)
8. Grosz, B., Joshi, A.K., Weinstein, S.: Centering: A framework for modeling the local coherence of discourse. Computational Linguistics 21(2), 203–225 (1995)
9. Grosz, B.J., Sidner, C.L.: Attention, intentions, and the structure of discourse. Computational Linguistics 12(3), 175–204 (1986)
10. Halliday, M.A.K., Hasan, R.: Cohesion in English. Longman, London (1976)
11. Harris, Z.: Discourse and sublanguage. In: Kittredge, R., Lehrberger, J. (eds.) Sublanguage: Studies of Language in Restricted Semantic Domains, pp. 231–236. Walter de Gruyter, Berlin (1982)
12. Hasler, L.: An investigation into the use of centering transitions for summarisation. In: Proceedings of the 7th Annual CLUK Research Colloquium., pp. 100–107. University of Birmingham (2004)
13. Karamanis, N.: Exploring entity-based coherence. In: Proceedings of CLUK4, Sheffield, UK, pp. 18–26 (2001)
14. Karamanis, N., Poesio, M., Mellish, C., Oberlander, J.: Evaluating centering-based metrics of coherence for text structuring using a reliably annotated corpus. In: Proceedings of the ACL, pp. 391–398 (2004)
15. Kittredge, R., Korelsky, T., Rambow, O.: On the need for domain communication language. Computational Intelligence 7(4), 305–314 (1991)
16. Lapata, M.: Probabilistic text structuring: Experiments with sentence ordering. In: Proceedings of the ACL, pp. 545–552 (2003)
17. Lin, C.Y.: ROUGE: A package for automatic evaluation of summaries. In: Proceedings of ACL, pp. 74–81 (2004)
18. Mani, I., Maybury, M.T.: Advances in Automatic Text Summarization. The MIT Press, Cambridge (1999)
19. Mann, W.C., Thompson, S.A.: Rhetorical structure theory: Toward a functional theory of text organization. TEXT 8(3), 243–281 (1988)
20. Marcu, D.: The rhetorical parsing of natural language texts. In: Proceedings of the ACL/EACL, pp. 96–103 (1997)
21. Marcu, D.: The Theory and Practice of Discourse Parsing and Summarization. MIT Press, Cambridge (2000)
22. McKeown, K.R.: Text Generation: Using Discourse Strategies and Focus Constraints to Generate Natural Language Text. Cambridge University Press, Cambridge (1985)
23. Miltsakaki, E., Kukich, K.: The role of centering theory's rough-shift in the teaching and evaluation of writing skills. In: Proceedings of the ACL, pp. 408–415 (2000)
24. Poesio, M., Stevenson, R., Eugenio, B.D., Hitzeman, J.: Centering: a parametric theory and its instantiations. Computational Linguistics 30(3), 309–363 (2004)
25. Rambow, O.: Domain communication knowledge. In: Fifth International Workshop on Natural Language Generation, pp. 87–94 (1990)
26. Sauper, C., Barzilay, R.: Automatically generating wikipedia articles: A structure-aware approach. In: Proceedings of the ACL/IJCNLP, pp. 208–216 (2009)
27. Sproat, R.: Morphology and Computation. MIT Press, Cambridge (1993)

Spanning Tree Approaches for Statistical Sentence Generation

Stephen Wan[1,2], Mark Dras[2], Robert Dale[1], and Cécile Paris[2]

[1] Centre for Language Technology
Department of Computing
Macquarie University
Sydney, NSW 2113
[2] ICT Centre, CSIRO, Australia

Abstract. In abstractive summarisation, summaries can include novel sentences that are generated automatically. In order to improve the grammaticality of the generated sentences, we model a global (sentence) level syntactic structure. We couch statistical sentence generation as a spanning tree problem in order to search for the best dependency tree spanning a set of chosen words. We also introduce a new search algorithm for this task that models argument satisfaction to improve the linguistic validity of the generated tree. We treat the allocation of modifiers to heads as a weighted bipartite graph matching problem (also known as the assignment problem), a well studied problem in graph theory. Using BLEU to measure performance on a string regeneration task, we demonstrate an improvement over standard language model baselines, illustrating the benefit of the spanning tree approach incorporating an argument satisfaction model.

Keywords: Statistical Text-to-Text Generation, Spanning Tree Problem, Assignment Problem, Dependency Model, Argument Satisfaction.

1 Introduction

Research on statistical text-to-text generation has the potential to extend the current capabilities of automatic text summarisation technology to include the generation of novel sentences, thereby advancing the state-of-the-art from sentence extraction to abstractive summarisation. In many statistical text-to-text generation scenarios we regularly encounter the need for a mechanism to generate a novel summary sentence, given some pre-selected content. In this work, we describe a new algorithm that statistically generates sentences that include such content. The mechanisms proposed focus on encoding grammatical constraints to improve the grammaticality of statistically generated sentences beyond standard language model baselines. We describe this algorithm in the context of a string regeneration task, which was first proposed as a surrogate for a grammaticality test by Bangalore et al. [2].[1] The goal of the string regeneration task is to

---

[1] As an evaluation metric, the string regeneration task is actually overly strict. We discuss this point further in Section 5.1.

E. Krahmer, M. Theune (Eds.): Empirical Methods in NLG, LNAI 5790, pp. 13–44, 2010.

recover a sentence once its words have been randomly ordered. Typically, this is performed by scoring and ranking candidate sentences using a language model, and choosing the top-ranked candidate as the regenerated sentence. Similarly, in many statistical generation scenarios, the goal is to rank potential candidates based on a language model. As an evaluation task, string regeneration reflects the issues that challenge the sentence generation components of machine translation, paraphrase generation, and summarisation systems [29].

In addition to a language model, other constraints can be used to restrict the space of candidate sentences explored. For example, these other constraints can represent aspects of content selection. In work on generating novel sentences for abstractive summarisation, we explored methods for selecting topically-related content [31]. We have also explored a model for selecting auxiliary contextual content to include in the generated summary sentence [30]. In this work, however, we focus on issues of grammaticality, and measure performance on the string regeneration task. We thus begin with only with an unordered list of words as input, assuming that all input words are to be included in the generated sentence. Typically, ranking is performed using an n-gram language model which scores a candidate sentence based on the probability of word sub-sequences of size n. N-gram language models appear to do well at a *local* level when examining word sequences smaller than n. However, beyond this window size, these methods are unable to model grammaticality at the sentence level, leading to ungrammatical sequences. Typically, n-gram approaches have difficulty with long-distance dependencies between words. In practice, the lack of sufficient training data means that n is often smaller than the average sentence length. Even if sufficient data exists, increasing the size of n leads to an increase in the time taken to find the best word sequence, due to the computational complexity of the search algorithm.We focus on algorithms that search for the best word sequence in a way that attempts to model grammaticality at the sentence level. Mirroring the use of spanning tree algorithms in parsing [24], we present an approach to statistical sentence generation. Given a set of scrambled words, the approach searches for the most probable dependency tree, where this notion of probability is defined with respect to some corpus, such that it contains each word of the input set. The tree is then traversed to obtain the final word ordering. To perform this task, we present two spanning tree algorithms. We first adapt the Chu-Liu-Edmonds (CLE) algorithm [8,15], which has been used for dependency parsing [24]. Our adaptation includes a basic argument model, added to keep track of linear precedence between heads and modifiers. While our adapted version of the CLE algorithm finds an optimal spanning tree, this does not always correspond to a linguistically valid dependency tree, primarily because the algorithm does not attempt to ensure that the words in the tree have plausible numbers of arguments. We then propose an alternative dependency-spanning tree algorithm which uses a finer-grained argument model representing argument positions. To find the best modifiers for argument positions, we treat the attachment of edges to the spanning tree as an instance of the weighted bipartite graph matching problem (or the *assignment* problem), a standard problem in graph theory.

The remainder of this paper is structured as follows. Section 2 outlines the graph representation of the spanning tree problem and describes a standard spanning tree algorithm. In Section 3, we introduce a model for argument satisfaction. Section 4 presents a search algorithm designed to find a spanning tree that corresponds to a dependency tree, which it does by incorporating the argument satisfaction model. We experiment to determine whether a global dependency structure, as found by our algorithm and our argument model, improves performance on the string regeneration problem, presenting the results in Section 5. Related work is presented in Section 6. Section 7 concludes that an argument model improves the linguistic plausibility of the generated trees, thus improving grammaticality in text generation.

2   Generation via Spanning Trees

We explore the potential for dependency models to impose constraints within the search for an optimal word ordering. In this section, we describe our representation of dependency relations, which includes the relative ordering of a head and its modifier. This information is encoded within the constructed dependency tree, and can be used to constrain the word order obtained when traversing the tree, creating the summary sentence. We then describe a formalism for couching sentence generation as a spanning tree problem.

2.1   Preliminiaries

In treating statistical generation as a spanning tree problem, this work is a generation analog of the parsing work by MacDonald et al. [24]. Given a bag of words with no additional constraints, the aim is to produce a dependency tree containing the given words, thus imposing an ordering on the words. In considering all trees, dependency relations are possible between any pair of words. We can represent the set of dependencies between all pairs of input words as a graph, as noted by MacDonald et al. [24]. Our goal is to find a subset of the edges in this graph such that each vertex in the graph is visited once, thereby including each word once. The resulting subset corresponds to a spanning tree, an acyclic graph which spans all vertices. Spanning trees can be given a score, allowing all trees to be ranked. This leads to the notion of a best tree. In this work, scores correspond to costs, and so the best tree is the one with the lowest overall cost.

Scoring functions that take a whole tree as an argument are typically too complex algorithmically, and so we compute the cost of a tree based on its component edges. This method, known as *edge-based factorisation* [24], provides an overall cost for the dependency tree, calculated as the sum of the costs of the edges in the spanning tree. When searching through all possible trees, the spanning tree algorithm is designed to select the tree that minimises the overall tree cost.

Following Collins [12], we weight each edge with the probability of the dependency relation that it represents. In our work, we use the negative log of these probabilities so that edge weights can be interpreted as costs. Hence, our problem is the minimum spanning tree (MST) problem.

As input to the spanning tree algorithm, a set of tokens, $w = \{w_1 \ldots w_n\}$, is provided. Each token represents either a single word or a multi-word phrase. Multi-word phrases, such as named entities, are treated as single units which cannot be broken up into their component words. In the case where a token represents a multi-word phrase, the token included in w is the head word of that phrase.

We define a directed graph (digraph) in a standard way, $G = (V, E)$, where V is a set of vertices and $E \subseteq \{(u, v) | u, v \in V\}$ is a set of directed edges. Given w, we define the digraph $G_w = (V_w, E_w)$ where $V_w = \{w_0, w_1, \ldots, w_n\}$, with w_0 a dummy root vertex, and $E_w = \{(u, v) | u \in V_w, v \in V_w \setminus \{w_0\}\}$. The actual word or multi-word phrase represented by a vertex $v \in V_w$ is retrieved using a function $\rho : V_w \to \Sigma$, where Σ is a string in the target language of the generation process.

An example of such a graph is presented in Figure 1, illustrating all the possible dependency relations for the input word set: *people, remain, displaced, in, burdening, still, hosts, dili*. The graph is fully connected (except for the root vertex w_0, which is only fully connected inwards) and is a representation of all possible dependencies. For an edge (u, v), we refer to u as the head and v as the modifier. As we do not have a dedicated mechanism for handling punctuation, for clarity in the examples, we omit punctuation marks. Figure 2 presents a spanning tree for the graph in Figure 1 corresponding to the sentence: *People still remain displaced in Dili, burdening hosts*.

We extend the original graph formulation of McDonald et al. [24] by adding a simplified representation of *argument positions* for a word, providing points to attach modifiers. This is necessary because, in the parsing task of McDonald et al. , word order is fixed; in generation it is not, and it needs to be included as a feature of the dependency representation to be considered during the search process. Adopting an approach similar to Johnson et al. [18], we look at the direction (left or right) of the head with respect to a given modifier. We consequently define a set $\mathcal{D} = \{l, r\}$ to represent this. Set $\mathcal{D}$ represents the linear precedence of the words in the dependency relation. Aside from its use in simplifying tree linearisation, this feature has the added advantage of approximating the distinction between syntactic roles like *subject* and *object*. We use the terms leftward and rightward to refer to direction: leftward children occur on the left of the parent; conversely, rightward children occur on the right of the parent.

Each edge has a pair of associated weights, one for each direction, defined by the function $s : E \times \mathcal{D} \to \mathbb{R}$, based on a probabilistic model of dependency relations. We adapt a dependency model from the literature [12] to calculate edge weights. Our adaptation uses direction rather than relation type (represented in the original as triples of non-terminals). Given a corpus, for some edge $(u, v) \in E$ and direction $d \in \mathcal{D}$, we calculate the edge weight as:

$$s((u, v), d) = -\log \, p_d(u, v, d) \tag{1}$$

To account for the different parts of speech (PoS) that a word may have, we define the set of PoS tags $\mathcal{P}$ and a function $pos : V \to \mathcal{P}$, which maps vertices

w: people, remain displaced, in, burdening, still, hosts, Dili

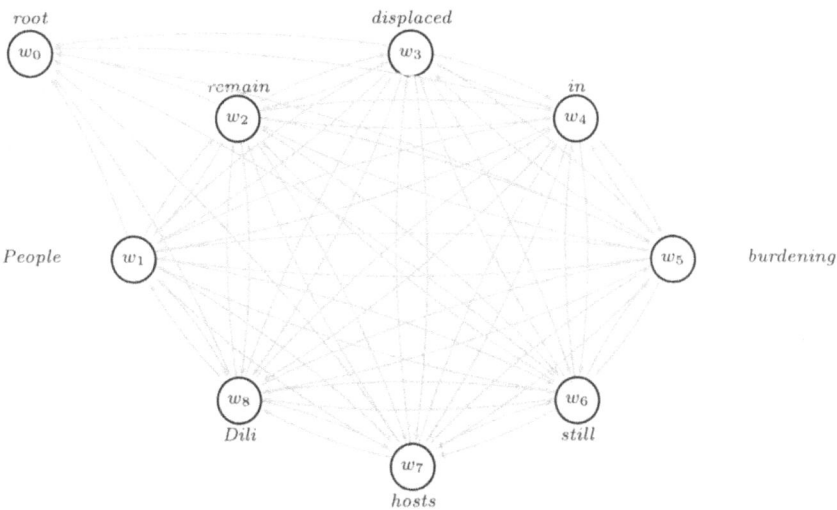

Fig. 1. A graph of all possible dependencies between words in the input set

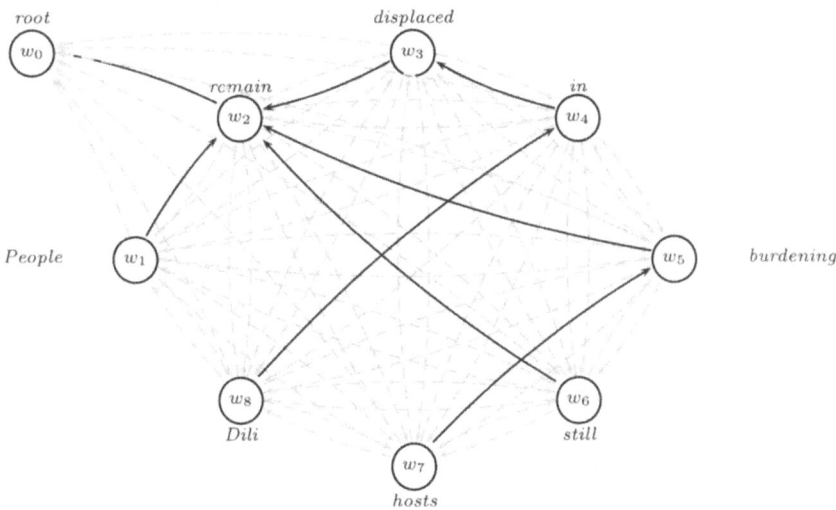

Fig. 2. A spanning tree (solid lines) for the ordering: *people still remain displaced in dili burdening hosts*

(representing words) to their PoS. For cases where the vertex, v, stands for a multi-word phrase, $pos(v)$ returns the PoS tag for the head word, since v represents the head of that phrase. To include PoS information in calculating the probability of a dependency relation, we define:

$$p_d(u, v, d) = \frac{cnt((u, pos(u)), (v, pos(v)), d)}{co\text{-}occurs((u, pos(u)), (v, pos(v)))} \tag{2}$$

where $cnt((u, pos(u)), (v, pos(v)), d)$ is the number of times where $(v, pos(v))$ and $(u, pos(u))$ are seen in a sentence in the training data, and $(v, pos(v))$ modifies $(u, pos(u))$ in direction d. The function $co\text{-}occurs((u, pos(u)), (v, pos(v)))$ returns the number of times that $(v, pos(v))$ and $(u, pos(u))$ are seen in a sentence in the training data.

We adopt the deleted interpolation smoothing strategy of Collins [12], which backs off to PoS for unseen dependency events. We base our implementation on the description of the model by Bikel [6]. This smoothing strategy relies on probability estimates of the dependency event that are based on a reduced set of features. Each reduced set of features essentially represents a more general dependency event for which sufficient training data may exist. In the extreme case, these reduced features, known as *contexts*, consist solely of part-of-speech information. Accordingly, the smoothed probability of a dependency event is defined using two levels of deleted interpolation, where p_{d1} represents the first level:

$$\begin{aligned}
p_{d1}(u, v, d) = \\
\lambda_1 \times p_d((u, pos(u)), (v, pos(v)), d) \\
(1 - \lambda_1) \times p_{d2}((u, pos(u)), (v, pos(v)), d)
\end{aligned} \tag{3}$$

where $p_{d2}((u, pos(u)), (v, pos(v)), d)$ represents the second level:

$$\begin{aligned}
p_{d2}((u, pos(u)), (v, pos(v)), d) = \\
\lambda_2 \times p_{d3}((u, pos(u)), (v, pos(v)), d) \\
(1 - \lambda_2) \times p_{d4}((u, pos(u)), (v, pos(v)), d)
\end{aligned} \tag{4}$$

The probability p_{d3} reduces the contextual features defining the dependency event by replacing either the head or modifier with its part-of-speech, one word at a time, as follows:

$$p_{d3}((u, pos(u)), (v, pos(v)), d) = \\
\frac{cnt(pos(u), (v, pos(v)), d) + cnt((u, pos(u)), pos(v), d)}{co\text{-}occurs(pos(u), (v, pos(v)), d) + co\text{-}occurs((u, pos(u)), pos(v), d)} \tag{5}$$

The next level of reduced features p_{d4} resorts to using only part-of-speech:

$$p_{d4}((u, pos(u)), (v, pos(v)), d) = \\
\frac{cnt(pos(u), pos(v), d)}{co\text{-}occurs(pos(u), pos(v), d)} \tag{6}$$

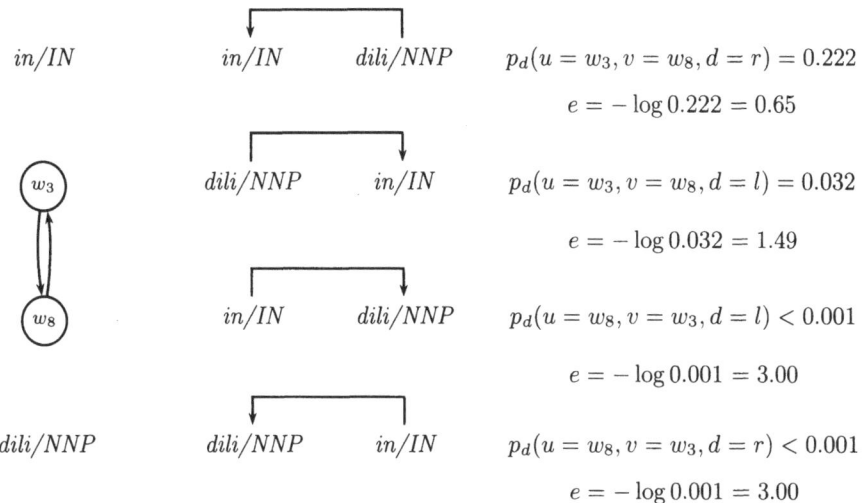

$p_d(u = w_3, v = w_8, d = r) = 0.222$

$e = -\log 0.222 = 0.65$

$p_d(u = w_3, v = w_8, d = l) = 0.032$

$e = -\log 0.032 = 1.49$

$p_d(u = w_8, v = w_3, d = l) < 0.001$

$e = -\log 0.001 = 3.00$

$p_d(u = w_8, v = w_3, d = r) < 0.001$

$e = -\log 0.001 = 3.00$

Fig. 3. Edge weights for two words, *in* and *dili*, that account for direction

In Equations (3) and (4), to gauge whether or not enough data exists to make the estimate reliable, λ weights based on a diversity measure are used to indicate the confidence of the estimated probabilities, p_d and p_{d3}, from both levels of deleted interpolation. Following Collins (as cited by Bikel [6]), λ is a weight of the form:

$$\lambda_i = \frac{c_i}{c_i + u_i} \tag{7}$$

where c_i is the size of the event space for which λ is a weight, and u_i represents its diversity. Thus, the larger c_i is, the more often we have seen the events needed to estimate its probability. We use $u_i = 1$ in this work, and λ_1 and λ_2 are calculated as follows:

$$\lambda_1 = \frac{co\text{-}occurs(u, v)}{co\text{-}occurs(u, v) + 1} \tag{8}$$

$$\lambda_2 = \frac{co\text{-}occurs(pos(u), (v, pos(v))) + co\text{-}occurs((u, pos(u)), pos(v))}{co\text{-}occurs(pos(u), (v, pos(v))) + co\text{-}occurs((u, pos(u)), pos(v)) + 1} \tag{9}$$

In Figure 3, we present an example of edge weights for the dependency relations between the two words *in/IN* and *dili/NNP*, using the Penn Treebank PoS tag set [22].[2] Because *dili/NNP* does not appear in the training corpus, the probabilities shown are for the case where *in/IN* is paired with any proper noun (*/NNP*).

2.2 The Chu-Liu/Edmonds Algorithm

To find the spanning tree with the lowest cost, we use the Chu-Liu/Edmonds (CLE) algorithm [8,15]. Given the graph $G_w = (V_w, E_w)$, the CLE algorithm

---

[2] We use the Penn Treebank notation, which separates words from their associated PoS tags using the "/" symbol. In the Penn Treebank PoS tag set, *IN* stands for a preposition, and *NNP* is a proper noun.

```
    /* initialisation */
1   Discard the edges exiting the w₀ if any. ;
2   Tw ← {} ;
    /* Chu-Liu/Edmonds Algorithm */
    input : Gw = (Ew, Vw)
    output: Tw ⊂ Ew
3   begin
4   |   Tw ← (u, v) ∈ Ew : ∀v∈V,d∈D arg min  s((u, v), d) ;
    |                                 (u,v)
5   |   while Mw = (Vw, Tw) has cycles do
6   |   |   forall the C ⊂ Tw : C is a cycle in Mw do
7   |   |   |   (e, d) ← arg min  s(e*, d*) : e ∈ C ;
    |   |   |           e*,d*
8   |   |   |   forall the c = (vh, vm) ∈ C and dc ∈ D do
9   |   |   |   |   forall the e' = (vi, vm) ∈ E and d' ∈ D do
10  |   |   |   |   |   s(e', d') ← s(e', d') − s(c, dc) − s(e, d) ;
11  |   |   |   |   end
12  |   |   |   end
13  |   |   |   s(e, d) ← s(e, d) + 1 ;
14  |   |   end
15  |   |   Tw ← (u, v) ∈ Ew : ∀v∈V,d∈D arg min  s((u, v), d) ;
    |   |                                 (u,v)
16  |   end
17  |   return Mw = (Vw, Tw) ;
18  end
```

Algorithm 1. The pseudo-code for the Chu-Liu Edmonds algorithm. Our adaptation includes the use of dependency edges which encode linear precedence.

finds a rooted directed spanning tree, specified by T_w, which is an acyclic set of edges in E_w minimising $\sum_{e \in T_w, d \in D} s(e, d)$. The algorithm is presented as Algorithm 1.[3]

There are two stages to the algorithm. The first stage finds, for each vertex, $v \in V_w$, the best edge connecting it to another vertex. To do this, all *outgoing* edges of v, that is edges where v is a modifier, are considered, and the one with the lowest edge weight (representing the smallest cost) is chosen. This *cost minimisation step* also ensures that each modifier has only one head.

If the chosen edges T_w produce a strongly connected subgraph $G_w^m = (V_w, T_w)$, then this is the Minimum Spanning Tree (MST). If they do not, a cycle exists amongst some subset of V_w that must be handled in the second stage. Essentially, one edge in the cycle is removed to produce a subtree. This is done by finding

---

[3] Adapted from McDonald et al. [24] and http://www.ce.rit.edu/~sjyeec/dmst.html. The difference from McDonald et al. concerns the direction of the edge and the edge weight function. We have also folded the function 'contract' in McDonald et al. into the main algorithm. Again following that work, we treat the function s as a data structure permitting storage of updated edge weights.

the best edge to join some vertex in the cycle to the main tree. This has the effect of finding an alternative head for some word in the cycle. The edge to the original head is discarded (to maintain one head per modifier), turning the cycle into a subtree. When all cycles have been handled, applying a greedy edge selection once more will yield the MST.

2.3   Generating a Word Sequence

Once the tree has been generated, all that remains is to obtain an ordering of words based upon it. Because the dependency relations encode the linear precedence of the head and modifier, it becomes relatively trivial to order child vertices with respect to a parent vertex. For a given head word, the only difficulty lies in finding a relative ordering for the leftward ($d \in \mathcal{D}, d = l$) children, and similarly for the rightward ($d \in \mathcal{D}, d = r$) children. We present an example of this scenario in Figure 4, based on a fragment of the spanning tree presented in Figure 2. The example illustrates an intermediate step in the traversal which determines the best ordering of the leftward and rightward children belonging to the vertex *remain*.

We traverse the spanning tree T_w using a greedy algorithm to order the siblings using the representation of linear precedence in the edges where possible. For siblings attached in the same direction with respect to the parent, we use an n-gram language model. We present this algorithm as pseudo-code in Algorithm 2, and refer to it as the *Language Model Ordering* (LMO) algorithm. The generated sentence is obtained by invoking the algorithm with the tree and its root, T_w and w_0, as parameters. The algorithm operates recursively if called on an inner node. If a vertex v is a leaf in the dependency tree, its string realisation $\rho(v)$ is returned. In Figure 5, we see an example of the realisation for a multi-word phrase, *Dili and districts*, being substituted for its head word, *dili*. Similarly, *hosts and families* is substituted for the head word *hosts*.

To find the relative ordering of siblings with the same linear precedence, we incrementally add each sibling into an ordered sequence, based on an n-gram language model. We keep track of the two resulting ordered sequences of children in the ordered sets, $R_d \subseteq w : d \in \mathcal{D}$, one for each direction. If the sibling set is leftwards, the ordered list R_l is initialised to be the singleton set containing

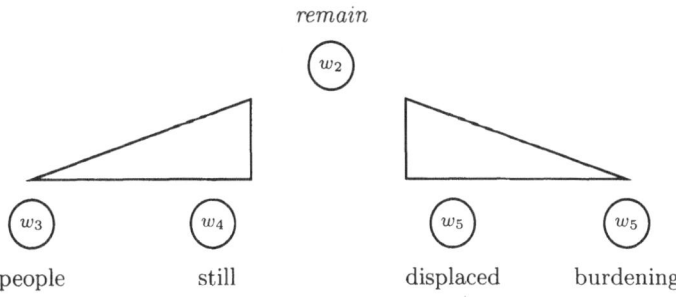

Fig. 4. The leftward and rightward children of the word *remain*

```
/* LMO Algorithm */
input  : v, T_w where v ∈ V_w
output: An ordering of V_w
1 begin
2    if isLeaf(v) then
3    |    return {ρ(v)}
4    end
5    else
6    |    C_l ← getLeftChildren(v, T_w) ;
7    |    C_r ← getRightChildren(v, T_w) ;
8    |    R_l ← (start) ;
9    |    R_r ← (ρ(v)) ;
10   |    while C_l ≠ {} do
11   |    |    c ← arg max pngram(R_l · LMO(c, T_w)) ;
     |    |         c∈C_l
12   |    |    R_l ← R_l · LMO(c, T_w) ;
13   |    |    C_l ← C_l \ {c} ;
14   |    end
15   |    while C_r ≠ {} do
16   |    |    c ← arg max pngram(R_r · LMO(c, T_w)) ;
     |    |         c∈C_r
17   |    |    R_r ← R_r · LMO(c, T_w) ;
18   |    |    C_r ← C_r \ {c} ;
19   |    end
20   |    return R_l · R_r ;
21   end
22 end
```

Algorithm 2. The Language Model Ordering algorithm for linearising an T_w. We denote concatenation with the operator: '·'.

a dummy start token with an empty realisation. If the sibling set is rightwards, then the ordered list R_r is initialised to be the realisation of the parent.

For a leftward or rightward sibling set, $C_d \subseteq V_w, d \in \mathcal{D}$, to be ordered the algorithm chooses the next vertex, $v \in C_d$, to append at the end of R_d. This is done by greedily maximising the probability of the string of words that would result if the string yield for the next vertex, $LMO(v, T_w)$, were concatenated with the string represented by realisations of the ordered elements in R_d.

The probability of the concatenation is calculated based on a window of words around the concatenation point. This window length is defined to be maximally $2 \times \text{floor}(n/2)$, for some n. In this work, we use a 4-gram language model, and so $n = 4$. The window size may be smaller if there are not enough words around the concatenation point. In such cases, the window is defined as the last $min(n - 1, |R_d|)$ of R_d for $d \in \mathcal{D}$ concatenated with the first $min(n - 1, |LMO(v, T_w)|)$ of $LMO(v, T_w)$, where R_d is a partially generated phrase and $LMO(v, T_w)$ represents a sequence of words that can be further appended. Figure 5 presents an example of the window that straddles the strings yielded by the two leftward

Let $R_l = \{start, displaced, in, dili, and, districts\}$

displaced   burdening

(w_2) (w_3)

displaced in <u>Dili and districts</u>   <u>burdening hosts</u> and families

$p_{ngram}(dili, and, districts, burdening, hosts, and) = 7.14 \times 10^{-13}$

Fig. 5. The n-gram window for concatenating two phrases. The two underlined phrases on either side of the concatenation point, each consisting of three words, are the two sides of the window. The overall probability for the underlined substring *dili and districts burdening hosts and* is used to rank this particular concatenation.

children of the parent node *remain*: these yields are obtained by applying the LMO algorithm to the subtrees rooted at *displaced* and *burdened*.

The probability of the window of words over the concatenation point, $w_0 \ldots w_j$, is given by the product of the probabilities of consecutive n-grams in that window:

$$p_{ngram}(w_0 \ldots w_j)$$
$$= \prod_{i=0}^{j-k} p_{mle}(w_i \ldots w_{i+k} | w_i \ldots w_{i+k-1}) \tag{10}$$

where $k = min(n-1, j)$, given an n-gram language model. The probability of each n-gram is computed using the Maximum Likelihood Estimate (MLE):

$$p_{mle}(w_0 \ldots w_{i+k} | w_0 \ldots w_{i+k-1})$$
$$= \frac{cnt(w_0 \ldots w_{i+k})}{cnt(w_0 \ldots w_{i+k-1})} \tag{11}$$

3   Argument Satisfaction and Spanning Trees

The CLE algorithm presented thus far utilises a simple model of argument positions based on an encoding of direction and is designed to find the optimal spanning tree.[4] However, the sentence represented by a spanning tree can fail to be a valid dependency tree by virtue of one or more words not having the required number of arguments. We refer to this requirement as *argument satisfaction*.

---

[4] In this work, we use the terms *argument* and *argument position* in a particular sense to broadly refer to the linguistic characteristic of a head word having a particular number of children. We do not use the terms with reference to existing linguistic definitions. We use the term *modifier* to refer to a word as it appears in a head-modifier relationship within a dependency tree.

In this section, we describe how a spanning tree produced by a standard spanning tree algorithm can differ from a dependency tree, motivating the need for a more sophisticated argument satisfaction model. The intuition behind the model is that a search algorithm should take into account the corresponding argument position when assigning a weight to an edge. In particular, when attaching an edge which connects a modifier to a head in the spanning tree, we count how many modifiers the head already has. We now introduce the argument satisfaction problem in more detail and present a model which enumerates argument positions.

3.1   The Argument Satisfaction Problem

A spanning tree that results from algorithms such as the CLE algorithm can fail in its argument satisfaction in two ways: it can be either under- or over-saturated in terms of its arguments. As an example of under-saturation, consider the case of intransitive, transitive and ditransitive verbs which require one, two and three arguments respectively, corresponding to the subject, direct object and indirect object. Any failure to provide the necessary arguments would result in an ungrammatical dependency structure. A preposition with two noun phrases attached as modifiers would constitute an example of over-saturation.

The case in Figure 6, which presents two head words competing for a modifier, illustrates the issues underlying modifier attachment. In general, the spanning tree algorithm makes no guarantee about which argument positions will be filled first, since attachment is based on the strength of the edge weight. In this example, modifiers are attached to heads one by one, and in this case, the word *dili* has already been attached to *in*. The figure shows potential edges as dashes, and demonstrates the effect of erroneously attaching the word *people* to the preposition. This would result in the under-saturation of the verb *remain*, which requires a subject, whilst also making the preposition *in* over-saturated.

A spanning tree algorithm like the CLE algorithm is prone to such errors as each attachment of a dependency edge to the spanning tree is based on the probability of the dependency tuple in isolation, without considering the number of arguments that are already attached to a word. In this case, the preposition *in* will effectively "steal away" the needed subject argument in the cost minimisation step of the CLE algorithm. Spanning tree algorithms that treat each edge independently will produce trees that are susceptible to this failing.

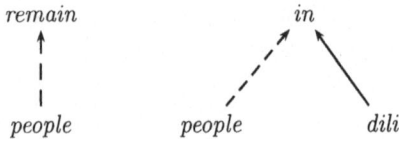

Fig. 6. Two competing attachment locations for the word *people*: the dashed edges represent the two proposed edges. The solid line indicates that the word *dili* has already been attached to *in*. In this example, one would want to restrict the preposition, *in*, from taking additional arguments.

3.2 Modelling Argument Satisfaction

One could rely on descriptions of English grammar to specify the minimum and maximum number of arguments allowed for each word, along with their types. For example, a transitive verb might be permitted only two arguments: the subject and direct object. Such constraints have been studied in related linguistic work in the form of verb subcategorisation rules [21], establishing the fact that words can have a canonical number of arguments. However, hand-coded grammatically resources have a number of limitations in the context of statistical text-to-text generation. For example, manual efforts to obtain a resource specifying the canonical arguments of words, for example a manually written grammar, may be hindered because of knowledge acquisition bottlenecks. Additionally, we observe that, while words like ditransitive verbs typically have set numbers of arguments, the corpus exhibits cases in which words can, under certain circumstances, accept a non-canonical number of arguments. For example, we can consider the phenomenon of transitive verb object drop in which one can say either *He ate the apple* or *He ate*. However, the related verb *devoured* seems to break this mold as the parallel constructions for are not so readily produced. The sentence **He devoured* seems less acceptable than *He devoured an apple*, illustrating that specifying the number of canonical arguments is a difficult task to do manually.

To allow for a non-canonical number of arguments, we use a representation of vertices that encodes argument positions. We describe this representation in the context of an iterative spanning tree algorithm, which constructs the spanning tree, T_w, incrementally across different time steps, denoted by t. We refer to the spanning tree at a particular time step as $T_{w,t}$.

To introduce this new representation, recall that we begin with the input set of words, w, which has a dependency graph $G_w = (V_w, E_w)$. For the purposes of defining an argument satisfaction model, we assume that we have the following at time t in some spanning tree algorithm, such as the CLE algorithm:

- $H_{w,t} \subseteq V_w$, the set of vertices in $T_{w,t}$, representing head words which can accept modifiers; and
- $M_{w,t} = V_w \backslash H_{w,t}$, the set of potential modifiers.

For example, at the start of the spanning tree algorithm, $H_{w,0}$ is empty. After adding the main verb of a sentence, $H_{w,1}$ would contain this verb. All remaining words would be stored in $M_{w,1}$.

For the potential heads, we want to define the set of argument positions in the spanning tree where the potential modifiers can attach. To talk about these attachment positions, we define the set of labels $\mathcal{L} = \{(d,j) \mid d \in \mathcal{D}, j \in \mathbf{Z}_+\}$, where an element (d,j) represents an attachment point in direction d, at position j. Valid attachment positions must be in sequential order and not missing any intermediate positions (e.g., if position 2 on the right is specified, position 1 must be also): so we define for some $i \in \mathbb{Z}_+$, and some direction $d \in \mathcal{D}$, a set $A_{d,i} \subseteq \mathcal{L}$ such that if the label $(d,j) \in A_{d,i}$ then the label $(d,k) \in A_{d,i}$ for $0 \leq k < j$. We also define an empty set, $A_{d,0} = \{\}$, to represent the case where no argument positions are available. Collecting these, we define $A = \{A_{r,i} \cup A_{l,j} \,|\, i,j \in \mathbb{N}\}$. For example, a partial enumeration of A is thus:

$$A = \left\{ \begin{array}{l} \{\}, \\ \{(l,1)\}, \\ \{(r,1)\}, \\ \{(l,1),(r,1)\}, \\ \{(l,1),(l,2)\}, \\ \{(r,1),(r,2)\}, \\ \ldots \end{array} \right\}$$

To map a potential head onto the set of attachment positions, we define a function $q : H_{w,t} \to A$. In talking about an individual attachment point $(d,j) \in q(v)$ for potential head v, we use the notation v_{dj}. For example, when referring to the second argument position on the right with respect to v, we use v_{r2}.

For the implementation of the algorithm, we have defined q to specify available attachment points as follows, given some $v \in H_{w,t}$:

$$q(v) = \left\{ \begin{array}{ll} \{v_{r1}\} & \text{if } v = w_0, \text{ the root} \\ \{v_{ri}, v_{lj} \,|\, i,j \in \mathbb{N}\} & \text{if pos}(v) \text{ is } VB \text{ (not } VBG) \text{ and } v \text{ is the child of } w_0 \\ \{v_{rj} \,|\, j \in \mathbb{N}\} & \text{otherwise} \end{array} \right.$$

Figure 7 presents some examples of attachment positions for some vertices. Defining q in such a way allows one to optionally incorporate linguistic information. For example, further constraints based on part-of-speech are possible.

We will also require a function to return the choice of the next available attachment point for a vertex v that can be filled in a particular tree $T_{w,t}$. We thus define the function $next : H_{w,t} \times 2^{E_w} \times \mathcal{D} \to H_{w,t} \times (\mathcal{L} \cup \{null\})$ that returns the position (d,j) with the smallest value of j for direction d for a particular vertex. This provides a pointer to the next available attachment point, for a vertex v, that can be filled in the tree $T_{w,t}$. The function $next$ calls $q(v)$ to obtain the set of attachment points for v. It then checks the spanning tree at time t, $T_{t,w}$, to see how many attachments points in $q(v)$ are already filled, returning the next available position in the direction d.

Computing the Probability of an Argument. We adopt a lexicalised corpus-based approach and propose an argument satisfaction model that provides the probability of a word having an arbitrary number of arguments. The

$$q(v_0 = root) = \{root_{r1}\}$$
$$q(v_1 = people/NN) = \{people_{r1}, people_{r2}, \ldots\}$$
$$q(v_2 = remain/VB) = \{remain_{r1}, remain_{r2}, \ldots, remain_{l1}, remain_{l2}, \ldots\}$$
$$q(v_3 = in/IN) = \{in_{r1}, in_{r2}, \ldots\}$$
$$q(v_5 = burdening/VBG) = \{burdening_{r1}, burdening_{r2}, \ldots\}$$

Fig. 7. Examples of attachment positions

model is able to encode the number of canonical arguments, while allowing words to take on additional arguments.

Given a word $v = (u, pos(u))$ and the direction d, and some step t in the creation of a spanning tree, the probability of attachment in the next available attachment point, $n = next(v, T_{w,t}, d)$, is defined as:

$$p_{arg}(v, n, d) = \frac{cnt(v, n, d)}{cnt(v)} \qquad (12)$$

Data smoothing can be performed using deleted interpolation, as in Collins [12]:

$$p_a(v, n, d) =$$
$$\lambda \times p_{arg}(v, n, d)$$
$$+ (1 - \lambda) \times p_{arg'}(v, n, d) \qquad (13)$$

where the probability $p_{arg'}((u, pos(u)), n, d)$ is calculated using a reduced context which ignores the lexical token, instead relying on just part-of-speech information. This is calculated as:

$$p_{arg'}(v, n, d) = \frac{cnt(pos(v), n, d)}{cnt(pos(v))} \qquad (14)$$

As in the dependency model, λ is defined in terms of the number of training instances seen:

$$\lambda = \frac{cnt(v)}{cnt(v) + 1} \qquad (15)$$

4   Assignment with an Argument Satisfaction Model

The model of argument satisfaction just described does not fit neatly into the CLE algorithm, for reasons we explain below. We therefore propose a novel algorithm to make use of the argument satisfaction model just described. With this model, the proposed algorithm finds not just the best spanning tree, but one that corresponds to a dependency tree. The proposed algorithm uses an argument satisfaction model representing argument positions. This new algorithm has been designed to focus on the problem of attaching modifiers to the appropriate head, optimising for argument satisfaction. Once more, we turn to graph theoretic approaches, recognising that this problem can be seen as an instance of the weighted bipartite graph matching problem, or the *assignment* problem (for an overview of the assignment problem, see Cormen et al. [13]).

4.1   Assigning Words to Argument Positions

One artifact of the argument satisfaction model presented in Section 3 is that the probability associated with an edge is now dependent on what other edges have already been attached to the spanning tree. In Figure 6, we saw a case where adding a second dependent child to a preposition would not be grammatically

1. Competing heads: Finding the best head attachment for a modifier

At some time, $t = i$, let:
$H_{w.i} = \{remain_{r1}, burdening_{l1}, remain_{l1}\}$
$M_{w.i} = \{\ldots, people, \ldots\}$

where we are searching for the best head for the word *people*. A cost minimisation step would consider the following dependency relations:

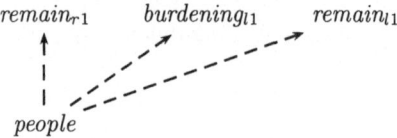

2. Competing modifiers: Finding the best modifier for a head

At some time, $t = j$, let:
$H_{w.j} = \{\ldots, burdening, \ldots\}$
$M_{w.j} = \{people, displaced, hosts\}$

where we are searching for the best modifier for the word *burdening*. A cost minimisation step would consider the following dependency relations:

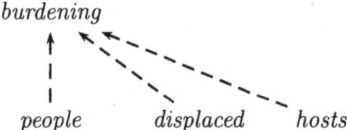

Fig. 8. Cost minimisations over one dimension: finding the best head for a modifier, or finding the best modifier for a head

correct. That is, the probability of that second attachment is dependent on the fact that there has already been a prior attachment of a modifier to the head. Accounting for argument positions makes an edge weight dynamic and dependent on surrounding tree context. This makes the search for an optimal tree an NP-hard problem [25], as all possible trees must be considered to find an optimal solution. For graphs with dynamically weighted edges, there are no known algorithms for finding a global optimum in polynomial time. So, in this section, we investigate the use of heuristic search algorithms which can incorporate the argument satisfaction model.

Greedy algorithms, like the CLE algorithm, iteratively select the next best dependency relation (an edge) using a cost minimisation step. For example, in Line 3 of Algorithm 1, we find the best head word to attach a modifier. However, it is hard to incorporate the argument satisfaction model in a straightforward cost minimisation step because edges, including dependency relations with a shared head, are independently attached in such algorithms. This limitation is

Searching for the best assignment of modifiers to argument positions. For this assignment, the following dependency relations would be considered:

At some time, $t = i$, let:
$H_{w.i} = \{remain_{r1}, burdening_{l1}, remain_{l1}\}$
$M_{w.i} = \{people, displaced, hosts\}$

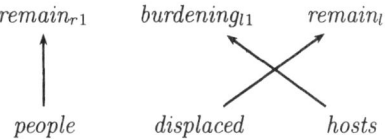

Fig. 9. Treating edge selection as an assignment problem. Potential assignments are shown with dashed lines, and desired assignments shown with solid lines.

exacerbated by the fact that, should a suboptimal path be followed, the attachment of a modifier cannot be undone; consequently, alternative attachment points are not considered. Figure 8 shows examples of cost minimisations over one dimension. The example word set, introduced in Figure 1, has been shortened to simplify the diagram.

An alternative treatment for choosing edges in the spanning tree is to consider minimisations over two dimensions. That is, each argument position competes for the unattached modifiers whilst, concurrently, each unattached modifier competes for attachment to a head (at a particular argument position). Thus, to consider both directions of competition, we design a new algorithm for constructing (dependency) spanning trees that casts edge selection as the assignment problem. The goal is to find a weighted alignment between objects of two distinct sets, where an object from one set is uniquely aligned to some object in the other set. For weighted alignments, as in the case of this work, the optimal alignment is one where the sum of alignment costs is minimal. The graph of all possible assignments is a weighted bipartite graph. In Figure 9, we present an example of edge selection as an assignment problem.

4.2   An Assignment-Based Spanning Tree Algorithm

We present an iterative spanning tree algorithm as Algorithm 3, and refer to this as the *Assignment-Based* (AB) algorithm. The AB algorithm operates on the dependency graph $G_w = (V_w, E_w)$, introduced in Section 2. Edges are incrementally selected until no more vertices remain unattached, or all edges in E_w have been considered (indicated by a failure to change the set of available argument positions). Thus, at some arbitrary iteration of the algorithm, t, we have the following:

- $T_{w,t} \subseteq E_w$, the set of edges in the spanning tree constructed so far;
- $H_{w,t} = \{u, v \mid (u, v) \in T_{w,t}\}$, the set of vertices in $T_{w,t}$, or "attached vertices", and therefore potential heads; and

- $M_{w,t} = V_w \backslash H_{w,t}$, the set of "unattached vertices", and therefore potential modifiers.

At each iteration, the edge selection is treated as an assignment problem. We use the Kuhn-Munkres (or Hungarian) algorithm [19], a standard polynomial-time solution to the assignment problem.[5] The use of the Kuhn-Munkres algorithm is embedded within an outer loop which repeatedly applies the algorithm until all modifiers are attached. The Kuhn-Munkres algorithm is $O(n^3)$ in the size of the bipartite sets, where $n < |V_w|$. The number of iterations of the outer loop, t, differs per generation case, but is also bounded by the number of input vertices. In the worst case, one word is attached to the growing spanning tree in each call to the Kuhn-Munkres algorithm, and $t = n$. Thus, the overall complexity for our approach is $O(n^4)$.

To apply the Kuhn-Munkres algorithm, we must map $H_{w,t}$ and $M_{w,t}$ into a bipartite graph representation. Here, to discuss bipartite graphs in general, we will extend our notation in a fairly standard way, to write $G^p = (H^p, M^p, E^p)$, where H^p, M^p are the disjoint sets of vertices and E^p the set of edges.

At time t, we consider the set of available modifiers, $M_{w,t}$, for assignment to the set of available argument positions. We define the set $H'_{w,t} = \{next(v, T_{w,t}, d) \,|\, v \in H_{w,t}, d \in D\}$, to keep track of the available attachment points for the vertices $v \in H_{w,t}$ in any direction $d \in D$, and given the current spanning tree $T_{w,t}$.

To describe the assignment problem at time t, we define the bipartite graph $G^p_{w,t} = (H^p_{w,t}, M^p_{w,t}, E^p_{w,t})$, in which $H'_{w,t} \subseteq H^p_{w,t}$ and $M_{w,t} \subseteq M^p_{w,t}$, respectively. Both $H^p_{w,t}$ and $M^p_{w,t}$ are padded with dummy vertices ϵ_i to make them equal in size. We also use extra padding to allow for null assignments to argument positions. Thus, $M^p_{w,t} = M_{w,t} \cup \{\epsilon_i\}$ for $1 \leq i \leq |H'_{w,t}|$. Similarly, $H^p_{w,t} = H'_{w,t} \cup \{\epsilon_i\}$ for $0 \leq i \leq |M_{w,t}|$.

Finally, $E^p_{w,t} = \{(u,v) \,|\, u \in H^p_{w,t}, v \in M^p_{w,t}\}$. In Figure 10, we present an example of edge selection as an assignment problem using epsilon tokens to pad out the bipartite graph.

The weight function is of the form $s_{ap} : E^p_{w,t} \rightarrow \mathbb{R}$ and incorporates a model of argument counts, as specified in Equation (13). We consider some $e \in E^p_{w,t}$: $e = (v^*, v)$ for some $v^* \in H^p_{w,t}, v \in M^p_{w,t}$; and $v^* = (u, (d,j))$ for some $u \in H_{w,t}, d \in D, j \in \mathbb{N}$. We then define s_{ap} as:

$$s_{ap}(e) = \begin{cases} c^\epsilon & if\, v = \epsilon_i, i \in \mathbb{N} \\ -\log(p_{as}(u,j,d) \times p_{d1}(u,v,d)) & if\, v \in M^p_{w,t} \end{cases} \qquad (16)$$

where c^ϵ is defined as a very large cost such that dummy leaves are only attached as a last resort.[6] Otherwise, the weight is based on both the probability

---

[5] To implement the Kuhn-Munkres algorithm, we use a graph theory library. The code can be found at:
http://sites.google.com/site/garybaker/hungarian-algorithm/assignment
[6] This is analogous to the use of a very large cost, 'M', with the use of artificial variables in linear programming to indicate a strongly dispreferred solution [32].

```
      /* initialisation */
 1    H_{w,t}, M_{w,t}, T_{w,t}, H'_{w,t}, M'_{w,t} ← {} for t ∈ N /* declare sets */
 2    H_{w,1} ← {w_0} ;
 3    M_{w,1} ← V_w ;
 4    H'_{w,1} ← {w_{0_{R1}}} ;
 5    H'_{w,0} ← {} ;

      /* The Assignment-based Algorithm */
      input  : H_{w,t}, M_{w,t}, T_{w,t}, H'_{w,t}, M'_{w,t}
      output: T_{w,t-1} ⊂ E_w
 6    begin
 7        t ← 1 ;
 8        while M_{w,t} ≠ {} and H'_{w,t-1} ≠ H'_{w,t} do
 9            ;
10            foreach ⟨(u,(d,j)),v⟩ ∈ Kuhn-Munkres(G^p_{w,t} = (H^p_{w,t}, M^p_{w,t}, E^p_{w,t})) do
11                if u ∈ H_{w,t} then
                      /* add to spanning tree */
12                    T_{w,t} ← T_{w,t} ∪ {(u,v)} ;

                      /* remove used argument positions */
13                    H'_{w,t+1} ← H'_{w,t} \ {(u,(d,j))} ;

                      /* add new argument positions for each direction */
14                    for d ∈ D do
15                        H'_{w,t+1} ← H'_{w,t+1} ∪ {next(u, T_{w,t}, d)} ;
16                        H'_{w,t+1} ← H'_{w,t+1} ∪ {next(v, T_{w,t}, d)} ;
17                    end

                      /* remove used modifier */
18                    M_{w,t+1} ← M_{w,t} \ {v} ;
19                    ;
                      /* add new head */
20                    H_{w,t+1} ← H_{w,t} ∪ {v} ;
21                end
22            end
23            t = t + 1 ;
24        end
25        return T_{w,t-1}
26    end
```

Algorithm 3. The Assignment-Based algorithm

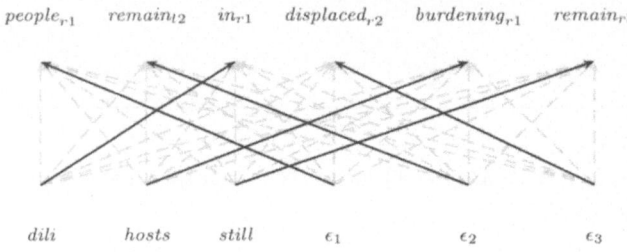

Fig. 10. Treating edge selection as an assignment problem

of the dependency relation and the probability of the argument position. The probability, $p_{d1}(u, v, d)$, is as defined in Equation 2.

The function $p_{as}(u, j, d)$ provides an estimate of the probability that a word u with j arguments assigned already can take on more arguments, and is defined as:

$$
p_{as}(u, j, d) \\
= \sum_{i=j+1}^{\infty} p_a(u, i, d) \\
= \frac{\sum_{i=j+1}^{\infty} cnt(u, i, d)}{cnt(u, d)} \tag{17}
$$

where $p_a(u, i, d)$ is defined in Equation (13) and $cnt(u, i, d)$ is the number of times word u has been seen with i arguments in direction d. The denominator $cnt(u, d) = \sum_{i \in \mathbb{N}} cnt(u, i, d)$. As the probability of argument positions beyond a certain value for i in a given direction will be extremely small, we approximate this sum by calculating the probability density up to a fixed maximum, in this case limiting argument positions to a maximum of 7, and assuming zero probability beyond that.

Greedy selection of the singleton child to w_0 may prematurely commit the generation process to the choice of a particular token $w_i \in w$ as the real root of the dependency tree. We can enumerate through each potential vertex in w in order to try it as the root. The Assignment-Based algorithm is then called with each enumerated root, and the tree based on the root with the mimimum cost is chosen. To iterate through all $w_i \in w$, we perform the initialisation specified in Algorithm 4.

4.3   An Example

Figure 11 presents a mock example for the generation of the sentence *People still remain displaced in Dili, burdening hosts*. In the first iteration of the algorithm, at time $t = 1$, a word is chosen as the singleton child of the dummy root node. In this case, it is the verb *remain* that has been chosen. We represent the null attachments with a single epsilon head to simplify the graph (there would actually be seven of these ϵ vertices in this case). In this example, we show a greedy selection of a single child for w_0.

```
/* Enumerating through possible roots: */
input  : G = (E_w, V_w)
output: T'_w ⊂ E_w
```

1 **begin**
2     $T'_w \leftarrow \{\}$;
3     $cost' \leftarrow$ INFINITY ;
4   **foreach** $w_i \in w$ **do**
       /* initialisation */
5        $H_{w,t}, M_{w,t}, T_{w,t}, H'_{w,t}, M'_{w,t} \leftarrow \{\}$ for $t \in \mathbf{N}$ /* declare sets */
6        $H_{w,1} \leftarrow \{w_0, w_i\}$;
7        $M_{w,1} \leftarrow V_w \setminus \{w_i\}$;
8        $H'_{w,1} \leftarrow \{w_{0_{R1}}, w_{i_{R1}}, w_{i_{L1}}\}$;
9        $H'_{w,0} \leftarrow \{\}$;
10        $T^i_w \leftarrow$ AB$(H_{w,t}, M_{w,t}, T_{w,t}, H'_{w,t}, M'_{w,t})$;
11        $cost = \sum_{e \in T^i_w} s(e)$;
12   **if** $cost' > cost$ **then**
13           $cost' \leftarrow cost$;
14           $T'_w \leftarrow T^i_w$;
15   **end**
16   **end**
17   **return** T'_w ;
18 **end**

Algorithm 4. Enumerating through all possible $w_i \in w$ as the true root of the dependency trees

In the second iteration at time $t = 2$, two attachment positions are available on the word *remain*. No other attachment positions are available from the dummy root, as defined by $q(w_0)$. Applying the Kuhn-Munkres algorithm at this stage assigns two words. The argument position *remain*$_{l1}$, which, in this case, corresponds to a syntactic subject, is filled by the word *people*. The position *remain*$_{r1}$ is filled by the verb *displaced*. All other modifiers have null attachments and are assigned to the epsilon token.

At time $t = 3$, there are more attachment points available. Importantly, the word *remain* now has one leftward and one rightward argument, and so the next available argument positions are *remain*$_{l2}$ and *remain*$_{r2}$. In addition, *people*$_{r1}$ and *displaced*$_{r1}$ are now available as argument positions. In this example, the edge selection assigns the preposition *in* to the argument position *displaced*$_{r1}$. The adverbial *still* is attached to *remain*$_{l2}$. The verb *burdening* is attached as a modifier to *remain*$_{r2}$.

Finally, at time $t = 4$, the only remaining modifiers are the words *dili* and *hosts*. Two heads are awaiting arguments, the preposition *in* and the verb *burdening*. Although the verbs *remain* and *displaced* are also possible heads for these modifiers, because they already have arguments attached, the available argument positions on these words have a lower associated probability. Applying

Iteration: $t = 1$
$H^p_{w.1} \leftarrow \{w_{0_{r1}}, \epsilon\}$
$H_{w.1} \leftarrow \{w_0\}$
$M_{w.1} \leftarrow \{\text{people, remain, displaced, still, in, Dili, hosts, burdening}\}$
$T_{w.1} \leftarrow \{\}$

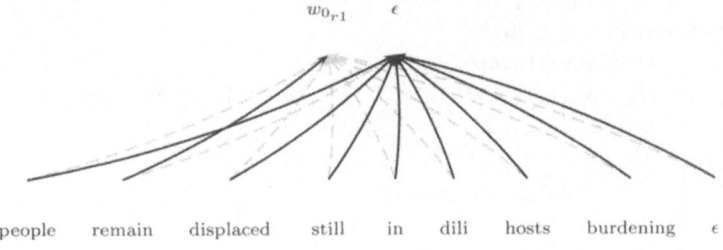

Iteration: $t = 2$
$H^p_{w.2} \leftarrow \{\text{remain}_{l1}, \text{remain}_{r1}, \epsilon\}$
$H_{w.2} \leftarrow \{w_0, \text{remain}\}$
$M_{w.2} \leftarrow \{\text{people, displaced, still, in, Dili, hosts, burdening}\}$
$T_{w.2} \leftarrow \{\langle w_{0_{r1}}, \text{remain}\rangle\}$

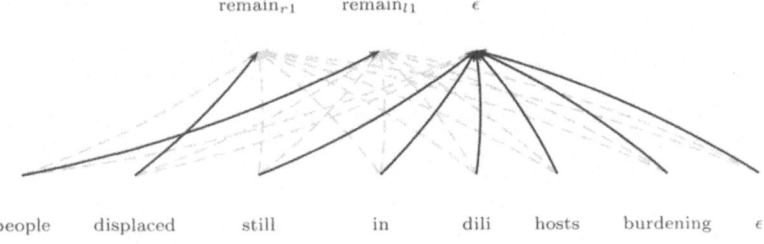

Iteration: $t = 3$
$H^p_{w.3} \leftarrow \{\text{remain}_{l2}, \text{remain}_{r2}, \text{people}_{r1}, \text{displaced}_{r1}, \epsilon\}$
$H_{w.3} \leftarrow \{w_0, \text{remain, people, displaced}\}$
$M_{w.3} \leftarrow \{\text{still, in, Dili, hosts, burdening}, \epsilon\}$
$T_{w.3} \leftarrow \{\langle w_{0_{r1}}, \text{remain}\rangle, \langle remain_{l1}, \text{people}\rangle, \langle remain_{r1}, \text{displaced}\rangle\}$

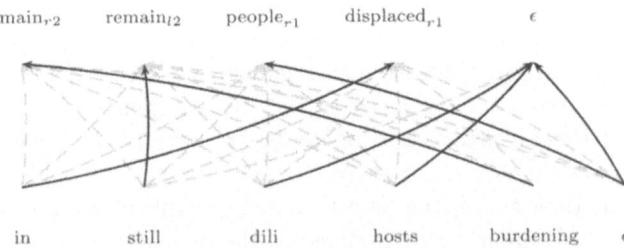

Fig. 11. An example of the generation of the sentence *People still remain displaced in Dili, burdening hosts*

Iteration: $t = 4$

$H^p_{w.4} \leftarrow \{\text{remain}_{l3}, \text{remain}_{r3}, \text{people}_{r1}, \text{displaced}_{r2}, \text{in}_{r1}, \text{burdening}_{r1}, \text{still}_{r1}\}$

$H_{w.4} \leftarrow \{w_0, \text{remain}, \text{people}, \text{displaced}, \text{in}, \text{burdening}, \text{still}\}$

$M_{w.4} \leftarrow \{\text{Dili}, \text{hosts}, \epsilon\}$

$T_{w.4} \leftarrow \{\langle w_{0,r1}, \text{remain}\rangle, \langle \text{remain}_{l1}, \text{people}\rangle, \langle \text{remain}_{r1}, \text{displaced}\rangle, \langle \text{displaced}_{r1}, \text{in}\rangle,$
$\langle \text{remain}_{r2}, \text{burdening}\rangle, \langle \text{remain}_{l2}, \text{still}\rangle\}$

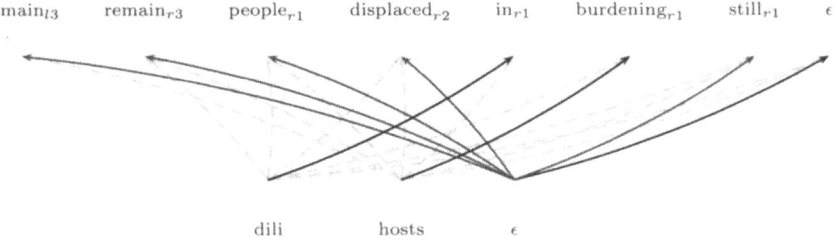

The spanning tree, $T_{w.4}$:

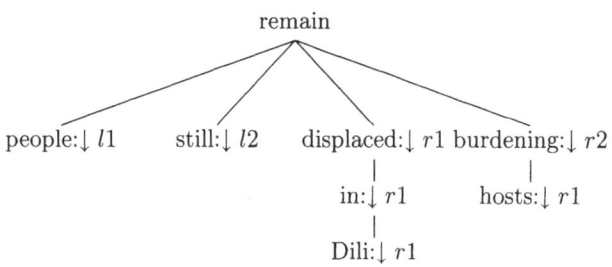

Fig. 12. Continuation of the example of the generation of the sentence *People still remain displaced in Dili, burdening hosts.* The dependency tree depicted in this figure places the argument position label on the right of the ↓ symbol.

the Kuhn-Munkres algorithm then favours the attachment positions in_{r1} and $burdening_{r1}$. The resulting tree is presented at the end of Figure 12. It is then linearised by the LMO algorithm described in Section 2.

5   Evaluation of Tree Building Algorithms

5.1   The String Regeneration Evaluation Framework

In research on statistical text generation, improvements in generation have previously been measured via the language modelling task of sentence regeneration, first proposed as an evaluation metric in this context by Bangalore et al. [2]. In this task, the goal is to regenerate a human-authored sentence given the words from that sentence in randomised order. The degree to which the sentence is regenerated is measured by a string comparison method. In our case, we used the BLEU metric [27], which has a range of 0 to 100. BLEU is a commonly used metric used in other language modelling related tasks. The BLEU score is applied in a standard way, on a set of test instances: that is, performance is measured across a set of regeneration test cases, removing the idiosyncracies of a particular test case as a confounding factor.

The string regeneration task is also a general generation task given that the only input data are the words. It reflects the issues that challenge the sentence generation components of machine translation, paraphrase generation, and summarisation systems [29]. The string regeneration task allows us to measure the success of statistical generation methods in producing the candidate sentences. Furthermore, content selection, as a factor, is held constant. Specifically, the probability of word selection is uniform for all words. Thus, the only factors that influence performance are the spanning tree algorithms and the language models.

As an evaluation metric, the degree to which a string is regenerated is overly strict. A set of input words and phrases could have more than one grammatical ordering. If we were to generate one of the alternative orderings, our score would be adversely affected. However, we assume that all algorithms and baselines compared suffer equally in this respect, and this issue is less problematic when averaging across multiple test cases.

Data Sets and Training Procedures. The approaches tested in this work require an n-gram language model, a dependency model and an argument satisfaction model. We use the Penn TreeBank corpus (PTB) as our training and testing data, as it contains human-annotated data on dependency relations and argument counts. We assume that the dependency model derived from the human annotations in PTB is as close to an ideal as practically possible, and thus differences in performance can be attributed predominantly to search errors made by the different spanning tree algorithms tested.

The corpus has about 3 million words of text from the Wall Street Journal (WSJ) with manual annotations of syntactic structures, from which the relevant models can be built. Dependency events were sourced from the events file of the Collins parser package [12], which contains the dependency events found in training Sections 2–22 of the corpus. Development was done on Section 00, and testing was performed on Section 23. A 4-gram language model (LM) was also obtained from the PTB training data, referred to as PTB-LM. N-gram models were smoothed using Katz's method, backing off to smaller values of n.

For this evaluation, tokenisation was based on that provided by the PTB data set. This data set also delimits base noun phrases (noun phrases without nested constituents). As described in Section 2, base noun phrases are multi-word phrases, and are treated here as single tokens. The rightmost word of the base-noun phrase is assumed to be the head. For the algorithms tested, the input set for any test case consisted of the single tokens identified by the PTB tokenisation, which includes the heads of base noun phrases. That is, we do not regenerate the base noun phrases. This would correspond to the use of a chunking algorithm or a named-entity recogniser to find noun phrases that could be re-used in sentence generation.

5.2   Algorithms and Baselines

We measure the performance of the Chu-Liu Edmonds (CLE) algorithm, defined in Section 2, to see if spanning tree algorithms do indeed improve upon conventional language modelling. We also evaluate the Assignment-Based (AB) algorithm, defined in Section 4, against the CLE algorithm to see if modelling argument assignments improves the resulting tree and thus the generated word sequence. Two baseline generators based on n-gram language models were used, representing approaches that optimise word sequences based on the local context of the n-grams.

The first baseline re-uses the *LMO* algorithm, defined in Section 2, on the same set of input words presented to the CLE and AB algorithms. As input to the LMO algorithm, we treat all input words in w as children of a dummy root vertex, each attached using a rightward relation. We apply the LMO algorithm on the dummy root vertex to order the rightward siblings. Ordering the siblings in this way greedily finds the best sequence beginning with a start-of-sentence token. Note that this baseline generator, like the two spanning tree algorithms, will score favourably using BLEU since, minimally, the word order of the base noun phrases will be correct when these are reinserted.

We test against a second language model baseline which uses a 4-gram language model to order individual words in the sentence, including those found in the base noun phrases. A Viterbi-style generator with a 4-gram model and a beam of 100 is used to generate the sequence. For this baseline, referred to as the Viterbi baseline, base noun phrases were separated into their constituent words and included in the input word set.

5.3   Evaluating the Assignment-Based Algorithm

We present the BLEU scores for the string regeneration task in Table 1. Significance was measured using the sign test and the sampling method described in [11]. The significant gain of 10 BLEU points by the LMO baseline over the Viterbi baseline shows the performance improvement that can be gained when reinserting the base noun phrases.

Table 1. String regeneration as measured in BLEU points using the WSJ language model. The systems compared are the standard Viterbi baseline, the dependency-based (DB) Viterbi baseline, the LMO baseline, the CLE algorithm, and the AB algorithm. BLEU scores are presented out of 100.

Algorithms	PTB-LM
Viterbi baseline	14.9
LMO baseline	24.3
CLE Algorithm	26.4
AB Algorithm	33.6

Table 2. String regeneration as measured in BLEU points using the BLLIP language model. The systems compared are the standard Viterbi baseline, the dependency-based (DB) Viterbi baseline, the LMO baseline, the CLE algorithm, and the AB algorithm. BLEU scores are presented out of 100.

Algorithms	BLLIP-LM
Viterbi baseline	18.0
LMO baseline	26.0
CLE	26.8
AB	33.7

Further gains are made by the CLE algorithm, significantly out-performing the LMO baseline by 2 BLEU points; from this, we conclude that incorporating a model for global syntactic structure and treating the search for a dependency tree as a spanning tree problem helps for novel sentence generation.

However, the real improvement can be seen in the performance of the AB system, which significantly out-performs all other methods, beating the CLE algorithm by 7 BLEU points. This illustrates the benefits of using a model of argument counts and of treating dependency tree building as an iterative application of the assignment problem for edge selection.

5.4   Comparing against Large N-Gram Models

One might reasonably ask if more n-gram data would narrow the gap between the tree algorithms and the baselines, which encode global and local information respectively. A new 4-gram language model was obtained from a subsection of the BLLIP'99 Corpus[7] containing three years of WSJ data from 1987 to 1989 [7]. As in Collins et al. [10], the 1987 portion of the BLLIP corpus containing 20 million words was also used to create a language model, referred to here as BLLIP-LM.

Our last evaluation repeats the same experiment as in Section 5.3. However, instead of the PTB-LM, we use the BLLIP-LM. The results are presented in Table 2 with significance again measured using the method outlined in Collins et al. [11]. All approaches improve significantly with the better language model. Unsurprisingly, the improvements are most evident in the baselines which rely heavily on the language model: the margin narrows between the CLE algorithm and the LMO baseline. However, the AB algorithm still significantly out-performs all other approaches by 7 BLEU points, highlighting the benefit of modelling dependency relations. Even with a language model that is one order of magnitude larger than the PTB-LM, the AB algorithm still maintains a sizeable lead in performance.

Figure 13 presents sample generated strings. In these sample sentences, the AB algorithm generates a word ordering that seems reasonable, aside from issues in punctuation, which are outside the scope of this work. In fact, it differs from

---

[7] LDC catalogue number: LDC2000T43.

Original:	at this point, the dow was down about 35 points
AB:	the dow at this point was down about 35 points
CLE:	was down about this point 35 points the dow at
LMO:	was this point about at down the dow 35 points
Viterbi:	the dow 35 points at was about this point down

Fig. 13. Example generated sentences using the BLLIP-LM

the original sentence in terms of the fronting of particular phrases, representing a novel sentence. In this case, the generated sentence fronts the reference *the dow*, whereas the original sentence fronts a time reference *at this point*. The sentence generated by the CLE algorithm is vaguely understandable, but has an undersatisfied preposition *at*.

The output of both baselines is noticeably poorer than the output of the spanning tree algorithms. The effects of constantly applying a 4-gram model can be seen in the output of the Viterbi generator. Interestingly, it manages to reconstruct the simple base noun phrases in this case.

6 Comparison to Related Work

6.1 Statistical Surface Realisers

This work is similar to research in statistical surface realisation (for example, [1,20]). These systems start with a semantic representation for which a specific rendering, an ordering of words, must be determined, often using pre-defined grammars to map semantic representations to candidate sentences. Candidate sentences are then typically ranked using n-gram language models.

The task presented in this chapter is different in that it is assumes a text-to-text scenario, which does not begin with a representation of semantics to be realised as a sentence. The similarity of our approach to surface realisation lies in the ranking of candidate sentences using statistical models. In our approach, we combine a dependency model and an argument satisfaction model with a standard n-gram language model to produce a dependency tree, which is then linearised using the LMO algorithm. These models heavily rely on word order statistics. As such, the utility of our approach is limited to human languages with minimal use of inflections, such as English. However, related corpus-based text-to-text approaches for inflectional languages, for example German, have been explored [16].

6.2 Text-to-Text Generation

As a text-to-text approach, our application of statistical generation is more similar to work on sentence fusion [5], a sub-problem in multi-document summarisation (see also [4] (this volume), [17], and [23] (this volume)). In work on sentence fusion, sentences presenting the same information, for example from multiple news articles describing the same event, are merged to form a single summary by aligning repeated words and phrases across sentences. The development of the AB algorithm was researched in the context of *sentence augmentation*, a related sentence fusion problem in which auxiliary information from additional sentences is merged into a key sentence (for more details on sentence augmentation, see Wan et al. [30]).

Other text-to-text approaches for generating novel sentences also aim to recycle sentence fragments where possible, as we do. Work on phrase-based statistical machine translation has been applied to paraphrase generation [3] and multi-sentence alignment in summarisation [14]. These approaches typically use n-gram models to find the best word sequence.

The WIDL formalism [29] was proposed to efficiently encode constraints, such as dependency information, that restrict possible word sequences. Explicitly representing each possible word sequence can lead to an inefficient use of computation and space resources. Our work here also avoids explicitly representing each possible word sequence, although it does not use a representation that is as easily parameterised as the WIDL formalism. The issue of an appropriate representation is orthogonal to the main focus of this work, and we assume that advances in representations can further improve our generation approach.

Our use of graph theoretic approaches and our treatment of edge selection in spanning tree construction as an assignment problem is related to statistical generation work which uses Integer Linear Programming (ILP). Global argument satisfaction constraints for sentence compression have previously been represented in an ILP framework [9]. The difference between our spanning tree approach and the sentence compression work is that we construct trees instead of pruning them.

For these text-to-text systems, the order of elements in the generated sentence is heavily based on the original order of words and phrases in the input sentences. Our approach has the benefit of considering all possible orderings of words, and thus allows a wide range of paraphrases unconstrained by the initial word orderings of the input text.

6.3   Parsing and Semantic Role Labelling

This text generation work is closely related to dependency parsing work [24]. Our work can be thought of as generating projective dependency trees (that is, without crossing dependencies).

The key difference between parsing and generation is that, in parsing, the word order is fixed, whereas for generation, this must be determined. In this chapter, we search across all possible tree structures whilst searching for the best word ordering. As a result, an argument model is needed to identify linguistically plausible spanning trees.

We treated the alignment of modifiers to head words as a bipartite graph matching problem. This method has previously been used in other language processing tasks, but has not previously been used for sentence generation.

Related assignment problem in the literature have modelled phenomena akin to argument satisfaction, such as in semantic role labelling [26]. The alignment of answers to question types as a semantic role labelling task can also employ the same method [28].

Our generation approach is also strongly related to work which constructs symbolic semantic structures via an assignment process in order to provide surface realisers with input [33]. Our approach differs in that we do not begin with a fixed set of semantic labels. Additionally, our end goal is a dependency tree that encodes word precedence order, bypassing the surface realisation stage.

7 Conclusions

We have presented a new use of spanning tree algorithms for generating sentences from an input set of words, a task common to many text-to-text scenarios. We explored the use of the CLE algorithm to find the best dependency tree for the purpose of improving the grammaticality of the generated string. We introduced the Assignment-Based algorithm, a novel generation algorithm that can improve the grammaticality of generated output by incorporating the use of an argument model. We model grammaticality at a global (sentence) level, integrating a dependency model into a spanning tree algorithm. Our algorithm also incorporates a model of argument satisfaction. We treat the attachment of modifiers to heads as an assignment problem, using the Kuhn-Munkres assignment algorithm. Although our representation of argument positions is relatively simple, we can nevertheless measure the effect of the basic principle embodied in the algorithm, which is to consider the attachment of modifiers to heads as an assignment problem.

We tested the performance of the approaches on the general task of string regeneration, a purely language modelling task. We found a significant improvement in performance when using a spanning tree algorithm instead of an n-gram based generator to generate sentences, as measured by BLEU. We conclude that global search algorithms, like the AB spanning tree algorithm, improve upon the grammaticality of the candidate sentence, as compared to search algorithms that optimise local constraints.

We also found that the Assignment-Based algorithm offered performance improvements over a standard spanning tree algorithm. We conclude that the argument satisfaction model used in our algorithm, and the treatment of edge selection in the spanning tree as an assignment problem, improves upon the grammaticality of the generated sentences. Furthermore, we found that the performance of the Assignment-Based algorithm was competitive even when we provide n-gram approaches with more data. We conclude that our new algorithm finds trees that are linguistically more valid than those found by standard spanning tree algorithms, thereby improving the grammaticality of the generated candidate sentences.

Acknowledgments. This work was funded by the CSIRO ICT Centre and Centre for Language Technology at Macquarie University. We would like to thank members of both groups for their helpful comments on this work. We would also like to thank the anonymous reviewers of EACL 2009, Erwin Marsi, and Emiel Krahmer for their feedback on earlier drafts of this chapter.

References

1. Bangalore, S., Rambow, O.: Exploiting a probabilistic hierarchical model for generation. In: Proceedings of the 18th Conference on Computational Linguistics (COLING 2000). Universität des Saarlandes, Saarbrücken (2000)
2. Bangalore, S., Rambow, O., Whittaker, S.: Evaluation metrics for generation. In: INLG 2000: Proceedings of the First International Conference on Natural Language Generation, Morristown, NJ, USA, pp. 1–8 (2000)
3. Bannard, C., Callison-Burch, C.: Paraphrasing with bilingual parallel corpora. In: Proceedings of the 43rd Annual Meeting of the Association for Computational Linguistics (ACL 2005), Ann Arbor, Michigan, pp. 597–604 (2005)
4. Barzilay, R.: Probabilistic approaches for modeling text structure and their application to text-to-text generation. In: Krahmer, E., Theune, M. (eds.) Empirical Methods in NLG. LNCS (LNAI), vol. 5790, pp. 1–12. Springer, Heidelberg (2010)
5. Barzilay, R., McKeown, K.R., Elhadad, M.: Information fusion in the context of multi-document summarization. In: Proceedings of the 37th Annual Meeting of the Association for Computational Linguistics (ACL 1999), Morristown, NJ, USA, pp. 550–557 (1999)
6. Bikel, D.M.: Intricacies of Collins' parsing model. Computional Linguistics 30(4), 479–511 (2004)
7. Charniak, E., Blaheta, D., Ge, N., Hall, K., Hale, J., Johnson, M.: Bllip 1987-89 WSJ Corpus Release 1. Tech. rep., Linguistic Data Consortium (1999)
8. Chu, Y.J., Liu, T.H.: On the shortest arborescence of a directed graph. Science Sinica 14, 1396–1400 (1965)
9. Clarke, J., Lapata, M.: Modelling compression with discourse constraints. In: Proceedings of the 2007 Joint Conference on Empirical Methods in Natural Language Processing and Computational Natural Language Learning (EMNLP-CoNLL), pp. 1–11 (2007)
10. Collins, C., Carpenter, B., Penn, G.: Head-driven parsing for word lattices. In: Proceedings of the 42nd Annual Meeting of the Association for Computational Linguistics (ACL 2004), Morristown, NJ, USA, p. 231 (2004)
11. Collins, M., Koehn, P., Kucerova, I.: Clause restructuring for statistical machine translation. In: Proceedings of the 43rd Annual Meeting of the Association for Computational Linguistics (ACL 2005), Morristown, NJ, USA, pp. 531–540 (2005)
12. Collins, M.J.: A new statistical parser based on bigram lexical dependencies. In: Proceedings of the 34th Annual Meeting of the Association for Computational Linguistics (ACL 1996), San Francisco (1996)
13. Cormen, T., Leiserson, C., Rivest, R.: Introduction to Algorithms. The MIT Press, Cambridge (1990)
14. Daumé III, H., Marcu, D.: A phrase-based HMM approach to document/abstract alignment. In: Lin, D., Wu, D. (eds.) Proceedings of Conference on Empirical Methods on Natural Language Processing, EMNLP 2004, Barcelona, Spain, pp. 119–126 (2004)

15. Edmonds, J.: Optimum branchings. Journal of Research of the National Bureau of Standards 71B, 233–240 (1967)
16. Filippova, K., Strube, M.: Generating constituent order in German clauses. In: Proceedings of the 45th Annual Meeting of the Association of Computational Linguistics (ACL 2007), Prague, Czech Republic, pp. 320–327 (June 2007)
17. Filippova, K., Strube, M.: Sentence fusion via dependency graph compression. In: Proceedings of the 2008 Conference on Empirical Methods in Natural Language Processing (EMNLP 2008), Honolulu, Hawaii, pp. 177–185 (2008)
18. Johnson, M.: Transforming projective bilexical dependency grammars into efficiently-parsable CFGs with unfold-fold. In: Proceedings of the Annual Meeting of the Association for Computational Linguistics (2007)
19. Kuhn, H.: The Hungarian method for the assignment problem. Naval Research Logistics Quarterly 219552, 83–97 (1955)
20. Langkilde, I., Knight, K.: The practical value of N-grams in derivation. In: Hovy, E. (ed.) Proceedings of the Ninth International Workshop on Natural Language Generation (INLG 1998), New Brunswick, New Jersey, pp. 248–255 (1998)
21. Levin, B.: English Verb Classes and Alternations: A Preliminary Investigation. University of Chicago Press, Chicago (1993)
22. Marcus, M.P., Santorini, B., Marcinkiewicz, M.A.: Building a large annotated corpus of English: The Penn Treebank. Computational Linguistics 19(2), 313–330 (1993)
23. Marsi, E., Krahmer, E., Hendrickx, I., Daelemans, W.: On the limits of sentence compression by deletion. In: Krahmer, E., Theune, M. (eds.) Empirical Methods in NLG. LNCS (LNAI), vol. 5790, pp. 45–66. Springer, Heidelberg (2010)
24. McDonald, R., Pereira, F., Ribarov, K., Hajic, J.: Non-projective dependency parsing using spanning tree algorithms. In: Proceedings of the Conference on Human Language Technology and Empirical Methods in Natural Language Processing (HLT 2005), Morristown, NJ, USA, pp. 523–530 (2005)
25. McDonald, R., Satta, G.: On the complexity of non-projective data-driven dependency parsing. In: Proceedings of the Tenth International Conference on Parsing Technologies, Prague, Czech Republic, pp. 121–132 (2007)
26. Padó, S., Lapata, M.: Optimal constituent alignment with edge covers for semantic projection. In: Proceedings of the 21st International Conference on Computational Linguistics and the 44th Annual Meeting of the Association for Computational Linguistics (COLING/ACL 2006), Morristown, NJ, USA, pp. 1161–1168 (2006)
27. Papineni, K., Roukos, S., Ward, T., Zhu, W.J.: BLEU: a method for automatic evaluation of machine translation. In: Proceedings of the 40th Annual Meeting of the Association for Computational Linguistics (ACL 2002), Philadelphia, pp. 311–318 (July 2002)
28. Shen, D., Lapata, M.: Using semantic roles to improve question answering. In: Proceedings of the 2007 Joint Conference on Empirical Methods in Natural Language Processing and Computational Natural Language Learning (EMNLP-CoNLL 2007), Prague, Czech Republic, pp. 12–21 (2007)
29. Soricut, R., Marcu, D.: Towards developing generation algorithms for text-to-text applications. In: Proceedings of the 43rd Annual Meeting of the Association for Computational Linguistics (ACL 2005), Ann Arbor, Michigan, pp. 66–74 (2005)

30. Wan, S., Dale, R., Dras, M., Paris, C.: Seed and grow: Augmenting statistically generated summary sentences using schematic word patterns. In: Proceedings of the 2008 Conference on Empirical Methods in Natural Language Processing (EMNLP 2008), Honolulu, Hawaii, pp. 543–552 (2008)
31. Wan, S., Dras, M., Paris, C., Dale, R.: Using thematic information in statistical headline generation. In: The Proceedings of the Workshop on Multilingual Summarization and Question Answering at ACL 2003. Sapporo, Japan (2003)
32. Winston, W.L.: Operations research: applications and algorithms. Duxbury Press, Belmont (1994)
33. Wong, Y.W., Mooney, R.: Generation by inverting a semantic parser that uses statistical machine translation. In: Proceedings of the Human Language Technology Conference of the North American Chapter of the Association for Computational Linguistics (NAACL/HLT 2007), Rochester, New York, pp. 172–179 (2007)

On the Limits of
Sentence Compression by Deletion

Erwin Marsi[1], Emiel Krahmer[1], Iris Hendrickx[2], and Walter Daelemans[2]

[1] Tilburg University
Tilburg, The Netherlands
{emarsi,ekrahmer}@uvt.nl
http://daeso.uvt.nl
[2] Antwerp University
Antwerpen, Belgium
{iris.hendrickx,walter.daelemans}@ua.ac.be

Abstract. Data-driven approaches to sentence compression define the task as dropping any subset of words from the input sentence while retaining important information and grammaticality. We show that only 16% of the observed compressed sentences in the domain of subtitling can be accounted for in this way. We argue that this is partly due to the lack of appropriate evaluation material and estimate that a deletion model is in fact compatible with approximately 55% of the observed data. We analyse the remaining cases in which deletion only failed to provide the required level of compression. We conclude that in those cases word order changes and paraphrasing are crucial. We therefore argue for more elaborate sentence compression models which include paraphrasing and word reordering. We report preliminary results of applying a recently proposed more powerful compression model in the context of subtitling for Dutch.

1 Introduction

The task of *sentence compression* (or *sentence reduction*) can be defined as summarizing a single sentence by removing information from it [17]. The compressed sentence should retain the most important information and remain grammatical. One of the applications is in automatic summarization in order to compress sentences extracted for the summary [17,20]. Other applications include automatic subtitling [9,27,28] and displaying text on devices with very small screens [8].

A more restricted version of the task defines sentence compression as dropping any subset of words from the input sentence while retaining important information and grammaticality [18]. This formulation of the task provided the basis for the noisy-channel and decision-tree based algorithms presented in [18], and for virtually all follow-up work on data-driven sentence compression [4,5,12,19,24,26,27,30] It makes two important assumptions: (1) only word deletions are allowed – no substitutions or insertions – and therefore no paraphrases; (2) the word order is fixed. In other words, the compressed sentence must be a

E. Krahmer, M. Theune (Eds.): Empirical Methods in NLG, LNAI 5790, pp. 45–66, 2010.

subsequence of the source sentence. We will call this *the subsequence constraint*, and refer to the corresponding compression models as *word deletion models*. Another implicit assumption in most work is that the scope of sentence compression is limited to isolated sentences and that the textual context is irrelevant.

Under this definition, sentence compression is reduced to a word deletion task. Although one may argue that even this counts as a form of text-to-text generation, and consequently an NLG task, the generation component is virtually non-existent. One can thus seriously doubt whether it really is an NLG task.

Things would become more interesting from an NLG perspective if we could show that sentence compression necessarily involves transformations beyond mere deletion of words, and that this requires linguistic knowledge and resources typical to NLG. The aim of this chapter is therefore to challenge the deletion model and the underlying subsequence constraint. To use an analogy, our aim is to show that sentence compression is less like carving something out of wood - where material can only be removed - and more like molding something out of clay - where the material can be thoroughly reshaped. In support of this claim we provide evidence that the coverage of deletion models is in fact rather limited and that word reordering and paraphrasing play an important role.

The remainder of this chapter is structured as follows. In Section 2, we introduce our text material which comes from the domain of subtitling in Dutch. We explain why not all material is equally well suited for studying sentence compression and motivate why we disregard certain parts of the data. We also describe the manual alignment procedure and the derivation of edit operations from it. In Section 3, an analysis of the number of deletions, insertions, substitutions, and reorderings in our data is presented. We determine how many of the compressed sentences actually satisfy the subsequence constraint, and how many of them could in principle be accounted for. That is, we consider alternatives with the same compression ratio which do not violate the subsequence constraint. Next is an analysis of the remaining problematic cases in which violation of the subsequence constraint is crucial to accomplish the observed compression ratio. We single out (1) word reordering after deletion and (2) paraphrasing as important factors. Given the importance of paraphrases, Section 4 discusses the perspectives for automatic extraction of paraphrase pairs from large text corpora, and tries to estimate how much text is required to obtain a reasonable coverage. Section 5 reports on a pilot experiment in which we apply a recently proposed and more expressive model for sentence compression [7] to the same Dutch data set. We identify a number of problems with the model, both when applied to Dutch and in general. We finish with a summary and discussion in Section 6.

2   Material

We study sentence compression in the context of subtitling. The basic problem of subtitling is that on average reading takes more time than listening, so subtitles can not be a verbatim transcription of the speech without increasingly lagging behind. Subtitles can be presented at a rate of 690 to 780 characters per minute,

Table 1. Degree of sentence alignment: shows the distribution of the number of other sentences (ranging from zero to four) that a given sentence is aligned to, for both autocue and subtitle sentences

Degree:	Autocue:	(%)	Subtitle:	(%)
0	3607	(20.74)	12542	(46.75)
1	12382	(71.19)	13340	(49.72)
2	1313	(7.55)	901	(3.36)
3	83	(0.48)	41	(0.15)
4	8	(0.05)	6	(0.02)

while the average speech rate is considerably higher [28]. Subtitles are therefore often a compressed representation of the original spoken text.

Our text material stems from the *NOS Journaal*, the daily news broadcast of the Dutch public television. It is parallel text with on source side the *autocue* sentences (aut), i.e. the text the news reader is reading, and on the target side the corresponding *subtitle* sentences (sub). It was originally collected and processed in two earlier research projects – ATRANOS and MUSA – on automatic subtitling [9,27,28]. All text was automatically tokenized and aligned at the sentence level, after which alignments were manually checked.

The same material was further annotated in a project called DAESO[1] (Detecting And Exploiting Semantic Overlap), in which the general goal is automatic detection of semantic overlap. All aligned sentences were first syntactically parsed using the Alpino parser for Dutch [3], after which their parse trees were manually aligned in more detail. Pairs of similar syntactic nodes – either words or phrases – were aligned and labeled according to a set of five semantic similarity relations [22,23]. For current purposes, only the alignment at the word level is used, ignoring phrasal alignments and relation labels.

Not all material in this corpus is equally well suited for studying sentence compression as defined in the introduction. As we will discuss in more detail below, this prompted us to disregard certain parts of the data.

Sentence deletion, splitting and merging. For a start, autocue and subtitle sentences are often not in a one-to-one alignment relation. Table 1 specifies the alignment degree (i.e. the number of other sentences that a sentence is aligned to) for autocue and subtitle sentences. The first thing to notice is that there is a large number of unaligned subtitles. These correspond to non-anchor text from, e.g., interviews or reporters abroad. More interesting is that about one in five autocue sentences is completely dropped. A small number of about 4 to 8 percent of the sentence pairs are not one-to-one aligned. A long autocue sentence may be split into several simpler subtitle sentences, each containing only a part of the semantic content of the autocue sentence. Conversely, one or more - usually short - autocue sentences may be merged into a single subtitle sentence.

---

[1] http://daeso.uvt.nl

These decisions of sentence deletion, splitting and merging are worthy research topics in the context of automatic subtitling, but they should not be confused with sentence compression, the scope of which is by definition limited to single sentences. Accordingly we disregarded all sentence pairs in which autocue and subtitle are not in a one-to-one relation with each other. This reduced the data set from 17393 to 12382 sentence pairs.

Word compression. A significant part of the reduction in subtitle characters is actually not obtained by deleting words but by lexical substitution of a shorter token. Examples of this include substitution by digits ("7" for "seven"), abbreviations or acronyms ("US" for "United States"), symbols (euro symbol for "Euro"), or reductions of compound words ("elections" for "state-elections"). We will call this *word compression.* Although an important part of subtitling, we prefer to abstract from word compression and focus here on sentence compression proper. Removing all sentence pairs containing a word compression has the disadvantage of further reducing the data set. Instead we choose to measure *compression ratio* (CR) in terms of tokens[2] rather than characters.

$$CR = \frac{\#tok_{sub}}{\#tok_{aut}} \tag{1}$$

This means that the majority of the word compressions do not affect the sentence CR.

Variability in compression ratio. The CR of subtitles is not constant, but varies depending (mainly) on the amount of provided autocue material in a given time frame. The histogram in Figure 1 shows the distribution of the CR (measured in tokens) over all sentence pairs (i.e. one-to-one aligned sentences). In fact, autocue sentences are most likely not to be compressed at all (thus belonging to the largest bin, from 1.00 to 1.09 in the histogram).[3] In order to obtain a proper set of compression examples, we retained only those sentence pairs with a compression ratio less than one.

Parsing failures. As mentioned earlier detailed alignment of autocue and subtitle sentences was carried out on their syntactic trees. However, for various reasons a small number of sentences (0.2%) failed to pass the Alpino parser and received no parse tree. As a consequence, their trees could not be aligned and there is no alignment at the word level available either. Variability in CR and parsing failures are together responsible for a further reduction down to 5233 sentence pairs, the final size of our data set, with an overall CR of 0.69. Other properties of this data set are summarized in Table 2.[4]

---

[2] Throughout this study we ignore punctuation and letter case.

[3] Some instances even show a CR larger than one, because occasionally there is sufficient time/space to provide a clarification, disambiguation, update, or stylistic enhancement.

[4] Notice that *Sum* is not meaningful for CR.

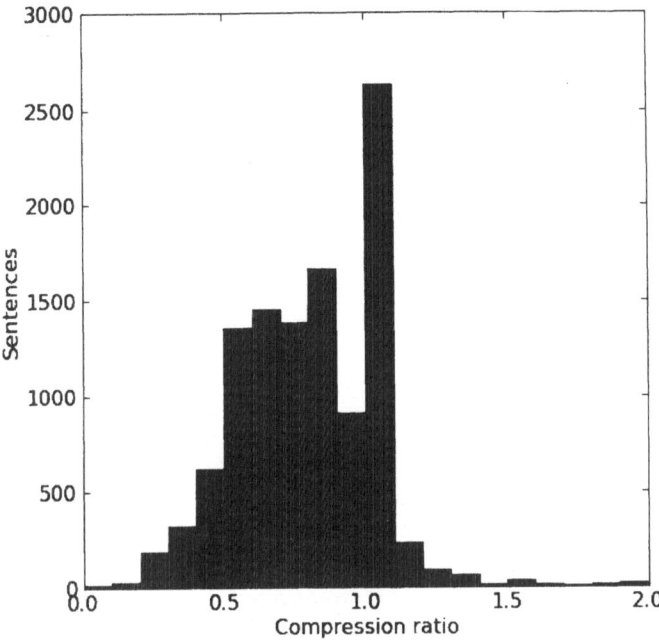

Fig. 1. Histogram of compression ratio: shows the distribution of the compression ratio (cf. equation 1) over all one-to-one aligned sentence pairs

Word deletions, insertions and substitutions Having a manual alignment of similar words in both sentences allows us to simply deduce word deletions, substitutions and insertions, as well as word order changes, in the following way:

- if an autocue word is not aligned to a subtitle word, then it is was deleted
- if a subtitle word is not aligned to an autocue word, then it was inserted
- if different autocue and subtitle words are aligned, then the former was substituted by the latter
- if alignments cross each other, then the word order was changed

The remaining option is that the aligned words are identical (ignoring differences in case).

Table 2. Properties of the final data set of 5233 pairs of autocue-subtitle sentences: minimum value, maximum value, total sum, mean and standard deviation for number of tokens per autocue/subtitle sentence and Compression Ratio

	Min:	Max:	Sum:	Mean:	SD:
aut-tokens	2	43	80651	15.41	5.48
sub-tokens	1	29	53691	10.26	3.72
CR	0.07	0.96		0.69	0.17

Without the word alignment, we would have to resort to automatically deriving the edit distance, i.e. the sum of the minimal number of insertions, deletions and substitutions required to transform one sentence into the other. However, this would result in different and often counter-intuitive sequences of edit operations. Our approach clearly distinguishes word order changes from the edit operations; the conventional edit distance, by contrast, can only account for changes in word order by sequences of the edit operations. Another problem with conventional edit distance is that substitution can also be accomplished by deletion and subsequent insertion, so we would either have to resort to assigning appropriate costs to the different operations or to abandon substitution altogether.

3   Analysis

3.1   Edit Operations

The observed deletions, insertions, substitutions and edit distances are shown in Table 3. For example, the minimum number of deletions observed in a single sentence pair (Min) is 1^5, whereas the maximum number of deletions observed in single sentence pair (Max) is 34. The total number of deletions over all sentence pairs is 34728, which amounts to a mean of 6.64 deletions per sentence pair and a standard deviation (SD) of 4.57. The rows for substitutions and insertions should be interpreted in a similar way. The edit distance is defined as the sum of all deletions, insertions and substitutions. On average, there are about 9 edit operations per sentence pair. As expected, deletion is the most frequent operation, with on average seven deletions per sentence. Insertion and substitutions are far less frequent. Note also that – even though the task is compression – insertions are somewhat more frequent than substitutions. Word order changes – which are not shown in the table – occur in 1688 cases, or 32.26% of all sentence pairs.

Another point of view is to look at the number of sentence pairs containing a certain edit operation. Here we find 5233 pairs (100.00%) with deletion, 2738 (52.32%) with substitution, 3263 (62.35%) with insertion, and 1688 (32.26%) with reordering.

Recall from the introduction that a subtitle is a *subsequence* of the autocue if there are no insertions, no substitutions, and no word order changes. In contrast, if any of these do occur, the subtitle is not a subsequence. The average CR for subsequences is 0.68 ($SD = 0.20$) versus 0.69 ($SD = 0.17$) for non-subsequences. A detailed inspection of the relation between the subsequence/non-subsequence ratio and CR revealed no clear correlation, so we did not find indications that non-subsequences occur more frequently at higher compression ratios.

3.2   Percentage of Subsequences

It turns out that only 843 (16.11%) subtitles are a subsequence, which is rather low. At first sight, this appears to be bad news for any deletion model, as it seems

---

[5] Every sentence pair must have at least one deletion, because by definition the CR must be less than one for all pairs in the data set.

Table 3. Observed word deletions, insertions, substitutions and edit distances

	Min:	Max:	Sum:	Mean:	SD:
del	1	34	34728	6.64	4.57
sub	0	6	4116	0.79	0.94
ins	0	17	7768	1.48	1.78
dist	1	46	46612	8.91	5.78

to imply that the model cannot account for close to 84% of the observed data. However, the important thing to keep in mind is that compression of a given sentence is a problem for which there are usually multiple solutions [2]. This is exactly what makes it so hard to perform automatic evaluation of NLG systems. There may very well exist semantically equivalent alternatives with the same CR which do satisfy the subsequence constraint. For this reason, a substantial part of the observed non-subsequences may have subsequence counterparts which can be accounted for by a deletion model. The question is: how many?

In order to address this question, we took a random sample of 200 non-subsequence sentence pairs. In each case we tried to come up with an alternative subsequence subtitle with the same meaning and the same CR (or when opportune, even a lower CR), but without compromising grammaticality. The task was carried out by one of the authors and subsequently checked by another author (both native speakers of Dutch), resulting in only a few minor improvements. Table 4 shows the distribution of the difference in tokens between the original non-subsequence subtitle and the manually-constructed equivalent subsequence subtitle. It demonstrates that 95 out of 200 (47%)

Table 4. Distribution of difference in tokens between original non-subsequence subtitle and equivalent subsequence subtitle

token-diff:	count:	(%:)
-2	4	2.00
-1	18	9.00
0	73	36.50
1	42	21.00
2	32	16.00
3	11	5.50
4	9	4.50
5	5	2.50
7	2	1.00
8	2	1.00
9	1	0.50
11	1	0.50

subsequence subtitles have the same (or even fewer) tokens, and thus the same (or an even lower) compression ratio. This suggests that the subsequence constraint is not as problematic as it seemed and that the coverage of a deletion model is in fact far better than it appeared to be. Recall that 16% of the original subtitles were already subsequences, so our analysis suggests that a deletion model can provide adequate output for 55% (16% plus 47% of 84%) of the data.

3.3 Problematic Non-subsequences

Another result of this exercise in rewriting subtitles is that it allows us to identify those cases in which the attempt to create a proper subsequence fails. In (1), we show one representative example of a problematic subtitle, for which the best equivalent subsequence we could obtain still has nine more tokens than the original non-subsequence.

(1)   **Aut**   *de bron   was een geriatrische patient die   zonder   het zelf te*
                 the source was a   geriatric   patient who without it   self to
                 merken uitzonderlijk veel   larven bij   zich bleek   te dragen
                 notice   exceptionally many larvae with him appeared to carry
                 en   een grote verspreiding veroorzaakte
                 and a   large spreading   caused
                 "the source was a geriatric patient who unknowingly carried exceptionally many larvae and caused a wide spreading"
     Sub   *een geriatrische patient met   larven heeft de   verspreiding*
                 a   geriatric   patient with larvae has   the spreading
                 veroorzaakt
                 caused
     Seq   *de bron   was een geriatrische patient die   veel   larven bij*
                 the source was a   geriatric   patient who many larvae with
                 zich bleek   te dragen en   een verspreiding veroorzaakte
                 him appeared to carry   and a   spreading   caused

These problematic non-subsequences reveal where insertion, substitution and/or word reordering are essential to obtain a subtitle with a sufficient CR (i.e. the CR observed in the real subtitles). At least three different types of phenomena were observed.

Word order. In some cases deletion of a constituent necessitates a change in word order to obtain a grammatical sentence. In example (2), the autocue sentence has the PP modifier *in verband met de lawineramp in galür* in its topic position (first sentence position).

(2)   **Aut**   *in verband met de   lawineramp   in galür   hebben de*
              in relation to   the avalanche-disaster in Galtür have   the
              politieke partijen in tirol   gezamenlijk besloten de
              political parties in Tirol together   decided the
              verkiezingscampagne voor het regionale parlement op te
              election-campaign   for   the regional parliament up to
              schorten
              postpone
              "Due to the avalanche disaster in Galür, political parties in Tirol
              have decided to postpone the elections for the regional parlia-
              ment."

      Sub   *de politieke partijen in tirol   hebben besloten de   verkiezingen*
              the political parties in Tirol have   decided the elections
              op te schorten
              up to postpone

Deleting this modifier, as is done in the subtitle, results in a sentence that starts
with the verb *hebben*, which is interpreted as a yes-no question. For a declarative
interpretation, we have to move the subject *de politieke partijen* to the first position, as in the subtitle. Incidentally, this indicates that it is instructive to apply
sentence compression models to multiple languages, as a word order problem
like this never arises in English.

Similar problems arise whenever an embedded clause is promoted to a main
clause, which requires a change in the position of the finite verb in Dutch. In
total, a word order problem occurred in 24 out 200 sentences.

Referring expressions. Referring expressions are on many occasions replaced by
shorter ones – usually a little less precise. For example, *de belgische overheid* 'the
Belgian authorities' is replaced by *belgie* 'Belgium'. Extreme cases of this occur
where a long NP such as *deze tweede impeachment-procedure in de amerikaanse
geschiedenis* 'this second impeachment-procedure in the American history' is
replaced by an anaphor such as *het* 'it'.

Since a referring expression or anaphor must be appropriate in the given
context, substitutions such as these transcend the domain of a single sentence
and require taking the preceding textual context into account. This is especially
clear in examples such as (3) in which 'many of them' is replaced by the 'refugees'.

(3)   **Aut**   *velen van hen   worden door de serviërs in volgeladen treinen*
              many of   them are   by   the Serbs   in crammed trains
              gedeporteerd
              deported
              "Many of them are deported by Serbs in overcrowded trains."

      Sub   *vluchtelingen worden per trein gedeporteerd*
              refugees   are   by train deported

It is questionable whether these types of substitutions belong to the task of
sentence compression. We prefer to regard rewriting of referring expressions as

one of the additional tasks in automatic subtitling, apart from compression. As expected the challenge of generating appropriate referring expressions is also relevant for automatic subtitling.

Paraphrasing. Apart from the reduced referring expressions, there are nominal paraphrases reducing noun phrases such as *medewerkers van banken* 'employees of banks' to compound words such as *bankmedewerkers* 'bank-employees'. Likewise, there are adverbial paraphrases such as *sinds een paar jaar* 'since a few years' to *tegenwoordig* 'nowadays', and *van de afgelopen tijd* 'of the past time' to *recent* 'recent'. However, the majority of the paraphrasing concerns verbs as in the three examples below.

(4)   **Aut**   *X zijn doorgegaan met hun stakingen*
                 X are continued   with their strikes
                 "X continued their strikes"
      Sub   *X staakten*
                 X striked

(5)   **Aut**   *X neemt het initiatief tot oprichting van Y*
                 X takes the initiative to founding of Y
      Sub   *X zet Y op*
                 X sets Y up

(6)   **Aut**   *X om zijn uitlevering vroeg maar Y die weigerde*
                 X for his extradition asked but   Y that refused
      Sub   *Y hem niet wilde   uitleveren aan X*
                 Y him not wanted extradite to   Y
                 "Y refused to extradite him to Y"

Even though not all paraphrases are actually shorter, it seems that at least some of them boost compression beyond what can be accomplished with only word deletion. In the next section, we look at the possibilities of automatic extraction of such paraphrases.

3.4   Semantic Relations between Aligned Phrases

The aligned phrases in our corpus were also manually labeled according to a set of five different semantic similarity relations. By way of example, we use the following pair of Dutch sentences:

(7)   a.   *Dagelijks koffie vermindert risico op Alzheimer en   Dementie.*
           Daily   coffee diminishes risk   on Alzheimer and Dementia
      b.   *Drie   koppen koffie per dag reduceert kans   op Parkinson en*
           Three cups   coffee a   day reduces   chance on Parkinson and
           Dementie.
           Dementia

The corresponding syntax trees and their (partial) alignment are shown in Figure 2. It should be noted that for expository reasons the alignment shown in

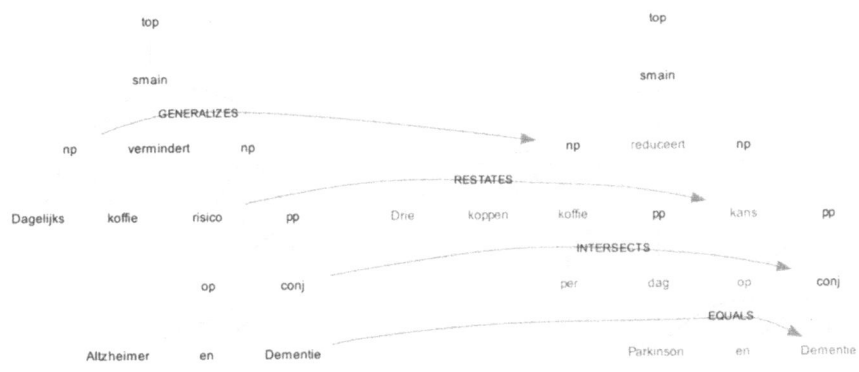

Fig. 2. Example of two (partially) aligned syntactic trees

the figure is not exhaustive. We distinguish the following five mutually exclusive similarity relations:

1. v **equals** v' iff STR(v) and STR(v') are literally identical (abstracting from case). Example: "Dementie" equals "Dementie";
2. v **restates** v' iff STR(v) is a paraphrase of STR(v') (same information content but different wording). Example: "risico" restates "kans";
3. v **generalizes** v' iff STR(v) is more general than STR(v'). Example: "dagelijks koffie" generalizes "drie koppen koffie per dag";
4. v **specifies** v' iff STR(v) is more specific than STR(v'). Example: "drie koppen koffie per dag" specifies "dagelijks koffie";
5. v **intersects** v' iff STR(v) and STR(v') share some informational content, but also each express some piece of information not expressed in the other. Example: "Alzheimer en Dementie" intersects "Parkinson en Dementie"

The distribution of the semantic relations in our autocue-subtitle corpus is shown in Table 5. The bulk of the alignments concerns Equals (67%). As is to be expected, the next most frequent class is Specifies (14%), where the information in the autocue is more specific than that in the compressed subtitle, followed by Restates (11%), where information is paraphrased. Only a small percentage

Table 5. Distribution of semantic relations between aligned phrases

	#Alignments:	%Alignment:
Equals	91609	67.46
Restates	15583	11.48
Generalizes	3506	2.58
Specifies	19171	14.12
Intersects	5929	4.37

of Generalizes (3%) and Intersects (4%) relations are present. These numbers confirm our intuition that paraphrasing and generalization are the most frequent operations in sentence compression.

4   Perspectives for Automatic Paraphrase Extraction

There is a growing amount of work on automatic extraction of paraphrases from text corpora [1,10,15,21]. One general prerequisite for learning a particular paraphrase pattern is that it must occur in the text corpus with a sufficiently high frequency, otherwise the chances of learning the pattern are proportionally small. In this section, we investigate to what extent the paraphrases encountered in our random sample of 200 pairs (cf. Section 3.2) can be retrieved from a reasonably large text corpus.

In a first step, we manually extracted all 106 paraphrase patterns observed in out data set. We filtered these patterns and excluded anaphoric expressions, general verb alternation patterns such as active/passive and continuous/non-continuous, as well as verbal patterns involving more than two arguments. After this filtering step, 59 pairs of paraphrases remained, including the examples shown in the preceding section.

The aim was to estimate how big our corpus has to be to cover the majority of these paraphrase pairs. We started with counting for each of the paraphrase pairs in our sample how often they occur in a corpus of Dutch news texts, the Twente

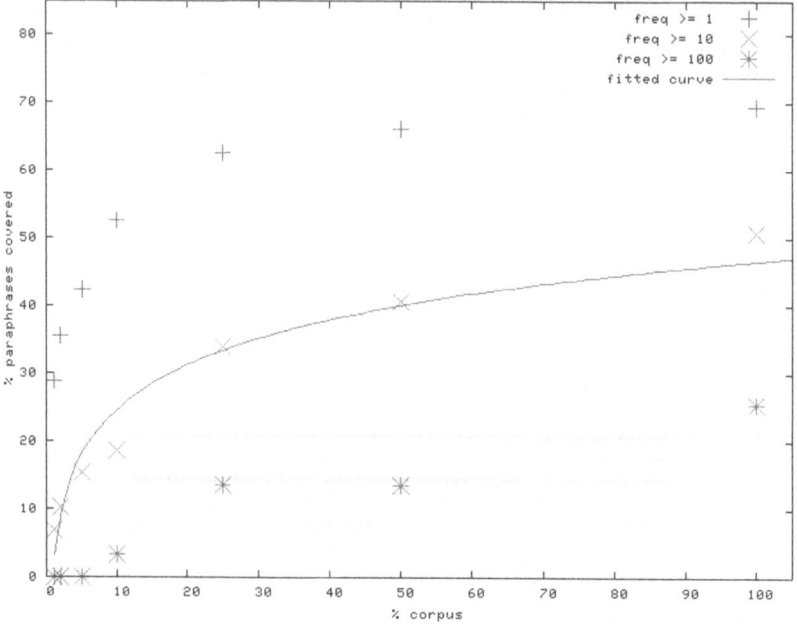

Fig. 3. Percentage of covered paraphrases as a function of the corpus size

News Corpus[6], which contains approximately 325M tokens and 20M sentences. We employed regular expressions to count the number of paraphrase pattern matches. The corpus turned out to contain 70% percent of all paraphrase pairs (i.e. both patterns in the pair occur at least once). We also counted how many pairs have frequencies of at least 10 or 100. To study the effect of corpus size on the percentage of covered paraphrases, we performed these counts on 1, 2, 5, 10, 25, 50 and 100% of the corpus. Figure 3 shows the percentage of covered paraphrases dependent on the corpus size. The most strict threshold that only counts pairs that occur at least 100 times in our corpus, does not retrieve any counts on 1% of the corpus (3M words). At 10% of the corpus size only 4% of the paraphrases is found, and on the full data set 25% of the pairs is found.

For 51% percent of the patterns we find substantial evidence (a frequency of at least 10) in our corpus of 325M tokens. We fitted a curve through our data points, and found a logarithmic line fit with adjusted R^2 value of .943 (which provides a measure between one and zero of how well future outcomes are likely to be predicted by the model). This suggests that in order to get 75% of the patterns, we would need a corpus that is 18 times bigger than our current one, which amounts to roughly 6 billion words. Although this seems like a lot of text, using the WWW as our corpus would easily give us these numbers. Today's estimate of the Index Dutch World Wide Web is 439 million pages[7]. If we assume that each page contains at least 100 tokens on average, this implies a corpus size of 43 billion tokens.

We are aware of the fact that these are very rough estimates and that the estimation method is to some extent questionable. For example, the extrapolation assumes that the relative distribution of words remains the same for a certain corpus and a superset of that corpus. Nevertheless we think that these estimates support the intuition that significantly more data is needed in order to extract the required paraphrases.

Not also that the patterns used here are word-based and in many cases express a particular verb tense or verb form (e.g. 3rd person singular) and word order. This implies that our estimations are the minimum number of matches one can find. For more abstract matching, we would need syntactically parsed data [21]. We expect that this would also positively affect the coverage.

5 Exploring Sentence Compression for Dutch

The preceding analysis of sentence compression in the context of subtitling provides evidence that deletion models are not sufficient for sentence compression. More elaborate models involving reordering and paraphrasing are therefore required. In recent work [7], Cohn & Lapata acknowledge the limitations of the deletion model and propose an interesting alternative which goes beyond word deletion. In this section, we describe a first attempt to apply this more powerful model to our Dutch subtitle data.

---

[6] http://www.vf.utwente.nl/~druid/TwNC/TwNC-main.html

[7] http://www.worldwidewebsize.com/index.php?lang=NL, as measured December 2009.

5.1   Sentence Compression as Tree Transduction

Cohn & Lapata [7] regard sentence compression as a tree-to-tree transduction process in which a source parse tree is transformed to a compressed parse tree. The formalization is based on Synchronous Tree Substitution Grammars (STSG) as proposed by [11], a formalism that allows local distortions of the tree structure and can therefore accommodate for substitutions, insertions and reordering. We refrain from a detailed formal description of their model, which can be found in [7], and instead provide an informal description by means of an (artificial) example.

The grammar rules of a STSG define aligned pairs of source and target tree fragments. For example, Rule 1 expresses the observation that the source NP *een nieuwe vakbond* 'a new union' can be rewritten to the compressed NP *een vakbond* 'a union' by deleting the adjective *nieuwe*.

Rule 1:

```
<NP, NP> ==> < (NP (Det een) (A nieuwe) (N vakbond)),
               (NP (Det) (N vakbond)) >
```

The tree fragments can be of arbitrary depth, allowing for an extended specification of syntactic context. Rule 2, for example, shows an example of substitution in which the source tree *medewerkers van banken* 'employees of banks' has a depth of four levels.

Rule 2:

```
<NP, NP> ==> < (NP (N medewerkers) (PP (P van) (NP (N banken)))),
               (NP (N bankmedewerkers)) >
```

To allow generalization, rules can contain variables. For instance, rule (3) has two NP slots, where the numeric indices define the alignment between the slots in the source and target tree (cf. example (5) for a gloss).

Rule 3:

```
<S, S> ==> < (S (NP [1]) (V nemen) (VP (NP (Det het)
             (N initiatief) (NP (P voor) (NP [2]))))),
             (S (NP [1]) (VP (V zetten) (NP [2]) (V_part op))) >
```

The variables are also the points of recursion in the transductive process. As in a normal context-free grammar, a source tree can be reproduced by top-down application of the left part of the synchronous grammar rules. Thus the source tree in the example below can be reproduced by first applying (the left part of) rule 3, followed by application of rules 1 and 2 to expand the two NP variables. Because of the aligned right part in the synchronous grammar rules, synchronous application of the aligned right parts produces the corresponding compressed tree.

Source tree:

```
(S  (NP (N medewerkers) (PP (P van) (NP (N banken))))) (V nemen)
(VP (NP (Det het) (N initiatief) (NP (P voor) (NP (Det een)
(A nieuwe) (N vakbond)) ))))
```

Target tree:

```
(S (NP (N bankmedewerkers))) (VP (V zetten) (NP (Det een)
(N vakbond) (V_part op)))
```

In order to automatically obtain a grammar, Cohn & Lapata rely on a parallel corpus of source and compressed sentences which are (automatically) aligned at the word level. From these word alignments, compatible constituent alignments are inferred. The resulting tree alignments are then input to an algorithm that derives the synchronous grammar rules.

Given an input source tree and a synchronous grammar, sentence compression amounts to finding the optimal target tree licensed by the grammar. One of the factors in scoring the output is an n-gram language model. Cohn & Lapata describe algorithms for training and decoding in this setting based on discriminative learning within a large margin framework.

5.2   Application to Dutch

Cohn & Lapata report state-of-the-art results for English, albeit on sentence compression by deletion only rather than the more general case which includes reordering, substitutions and insertions [7]. As they generously made their implementation publicly available as the Tree Transducer Toolkit [8] (T3), this paves the way to application to other corpora and languages. We think it is interesting to test their model in a different domain and for a new language. In the remainder of this section, we describe our experiences with a first attempt to apply this approach to Dutch.

The Dutch corpus has both advantages and disadvantages. An advantage is that it includes reordering, substitutions and insertions, and is therefore better suited to fully testing the claimed expressiveness of the model than the compression-by-deletion corpora used in [7]. A disadvantage is that it contains many sentences with a slightly ungrammatical word order. In order to understand the reason for this, it is necessary to know that the internal representation of a syntactic parse as used by the Alpino parser for Dutch is a dependency *graph* rather than a constituent tree, and may therefore contain crossing edges. Although the parser can output parse trees – which is what we use – crossing edges can only be resolved at the cost of moving some terminal nodes around to a different position in the tree. As a consequence, the yield of the parse tree, i.e., the sequence of terminals from left to right, is not always identical to the original input sentence. Hence the word order in the tree may differ slightly from that in the input sentence, and is in

---

[8] http://www.dcs.shef.ac.uk/~tcohn/t3/

fact often ungrammatical. A typical example of the difference between an input sentence and the yield of its parse tree is given in (8).

(8)   a.   *Met een flinke wind erbij   kan de sneeuw zich ophopen*
           with a   strong wind included can the snow   itself pile
           "In combination with a strong wind, snow can pile up"
      b.   kan de sneeuw Met een flinke wind erbij zich ophopen

What is needed is a reverse step which converts the constituent tree back to a dependency graph, thereby restoring the grammatical word order. This task of word (re)ordering is addressed in for example [13] and [29] (this volume), but is currently missing in our experimental setup.

We used the corpus of Dutch autocue and subtitle sentences as described in Section 2. We divided it into 3865 sentence pairs for training and 1354 sentence pairs for testing, in such way that all test material comes from another month of news broadcasts than the training material. Syntax trees were converted to the labeled bracket format as used in the Penn Tree Bank. Word alignments were obtained by removing alignments involving non-terminal nodes from our manual tree alignments. A trigram language model was trained on a background corpus of over 566M tokens of Dutch news text from the Twente News Corpus [25]. In all subsequent steps, we used settings identical to those in the sample script provided with the T3 distribution. Grammar rules were harvested from the training data, amounting to a total of over 124k unique rules.[9] Features were derived, the model was trained, and subsequently applied to decode the test material.

Given the preliminary nature of this exercise, we performed no formal evaluation experiment. Instead we first discuss a number of issues specific to our Dutch data set that we encountered upon inspecting the output. We also came across some general problems with the approach, which will be discussed in the next subsection.

Overall it seems that most acceptable compressions are the result of only deletion. Even though we did not inspect all 1354 test sentences, we were unable to find clear examples in which a combination of reordering, substitution or insertion resulted in a shorter paraphrase.

The most obvious problem in the output is ungrammatical word order. However, this can to a large extent be blamed on the training material containing ungrammatical word order, as explained above. It is therefore to be expected that the word order in the compressed output is also somewhat distorted.[10]

Apart from word order issues, two other problems – as far as grammaticality is concerned – are grammatical incongruence and missing complements. In

---

[9] This includes *epsilon rules* and *copy rules*, but no *deletion rules*, because a *head table* for Dutch, which specifies the head for each syntactic phrase, was not readily available. These rules will be explained in Section 5.3.

[10] This word order issue may also affect the n-gram language model, which plays a part in the scoring function for the target tree. The n-gram model is trained on Dutch sentences with a proper word order, whereas the yield of the source tree to be scored may have a distorted word order.

addition to subject-verb agreement, Dutch has determiner-noun and adjective-noun agreement. It turns out that these agreement constraints are often violated in the compressed output, presumably because the n-gram model has a limited capability for capturing these phenomena. Likewise, obligatory arguments, usually verbal complements, are frequently missing. A related issue concerns wrong/missing functions words such as determiners, complementizers, prepositions or verbal particles. Note that that this grammatical information is in principle available, as the edges of the parse tree are labeled with dependency relations such as *su* (subject) and *obj1* (verbal/prepositional complement) and *predc* (predicative complement). If we can force the model to take this dependency information into account – perhaps simply by concatenating constituent and dependency labels into a specialized label – this may have a positive impact on the grammaticality of the output

The remaining problems have to do with content selection where deletions or substitutions radically change the meaning of the compressed sentence or even render it nonsensical. One frequent error is the reduction of a source sentence to just its subject or object NP.

5.3   Some General Issues

We encountered a number of issues which we feel are general drawbacks of the model. The first of these is that it has a tendency to insert ungrounded lexical material. That is, the compressed output contains information that was in no form present in the source sentence. In (9), for example, the NP *de grootste in z'n soort in de wereld* is replaced by the completely unrelated phrase *Macedonie*. Likewise, (10) contains an extra piece of information in the form of the PP *van het concertgebouw*, for which there is no ground in the input.

(9)   a.   *De high tech campus wordt   de grootste in z'n soort in de*
             The high tech campus becomes the biggest in its sort in the
             wereld
             world

       b.   *De high tech campus wordt   Macedonie*
             The high tech campus becomes Macedonia

(10)   a.   *Tot   vorige week werkte ingenieur De Kwaadsteniet hier aan*
             until last   week worked engineer   De Kwaadsteniet here on
             mee
             with
             Until last week, engineer De Kwaadsteniet agreed to this

       b.   *van   vorige week werkte De Kwaadsteniet ingenieur van het*
             from last   week worked De Kwaadsteniet engineer of   the
             Concertgebouworkest   hier aan mee
             Concertgebouw-orchestra here on   with

The same problem is also observed in [24, p. 396] using automatically aligned comparable text as training material. One possible explanation is that the training material contains examples in which the source and compressed sentences are only *partly* overlapping in meaning. This would occur when multiple autocue sentences are compressed into a single subtitle sentence, so the subtitle may contain content that is not present in a least one of the autocue sentences. However, recall from Section 2 that we disregarded all cases of one-to-many and many-to-one sentence alignment. Just to check, we also ran an experiment in which we included these non-uniquely aligned sentences in the training material, and found that that indeed insertion of ungrounded material became a major problem, often giving rise to nonsensical output.

The second general problem is related to the compression ratio: it cannot be directly adapted to fit the level of compression desired by the user. As mentioned in [7], it can currently only be indirectly changed by modifying the loss function of the model. However, this means that different models must be trained for a discrete range of compression ratios, which seems impractical for real applications.

The third and final general problem has to do with coverage. The grammar rules induced from the development data are highly unlikely to cover all the lexical and syntactic variations in *unseen* data. This means that a source tree cannot be produced with the given rules, and consequently the transduction process will fail, resulting in no output. This problem seems unavoidable, even when training on massive amounts of data. The solution proposed by Cohn & Lapata is to add back-off rules. There are *epsilon rules* which delete unaligned constituents in the subtree, *deletion rules* which delete one or more non-head child nodes, and finally *copy rules* which simply copy a node and its child nodes from the source to the target tree (resulting in zero compression). However, the important thing to notice is that in the experiments reported, both here and in [6,7], it is implicit that these rules are derived not only from the development data, but also from the test data. While this is arguably a fair methodology in an experimental setting, it is problematic from a practical point of view. It means that for each unseen input sentence, we have to derive the epsilon, deletion and copy rules from the corresponding source tree, add them to the rule base, and retrain the model. Since training the support vector machine underlying the model takes considerable computing time and resources, this is clearly prohibitive in the case of online application.

To illustrate the point, we repeated our experiment for Dutch, with the difference that this time epsilon and copy rules were derived from the training data only, excluding the test data. In this setting, just 94 out of the total of 1354 test sentences (less than 7%) result in an output sentence.

To sum up, exploring sentence compression for Dutch with the tree transducer model from [7] gave results which are not immediately encouraging. However, these are preliminary results, and tuning of input and parameters may lead to significant improvements.

6   Discussion

In the first part of this chapter we performed an in depth analysis of sentence compression as observed in the context of subtitling for Dutch. We found that only 16.11% of 5233 subtitle sentences were proper subsequences of the corresponding autocue sentence, and therefore 84% can not be accounted for by a deletion model. One conclusion appears to be that the subsequence constraint greatly reduces the amount of available training material for any word deletion model. However, an attempt to rewrite non-subsequences to semantically equivalent sequences with the same CR suggests that a deletion model could in principle be adequate for 55% of the data. Moreover, in those cases where an application can tolerate a little slack in the CR, a deletion model might be sufficient. For instance, if we are willing to tolerate up to two more tokens, we can account for as much as 169 (84%) of the 200 non-subsequences in our sample, which amounts to 87% (16% plus 84% of 84%) of the total data.

It should be noted that we have been very strict regarding what counts as a semantically equivalent subtitle: every piece of information occurring in the non-subsequence subtitle must reoccur in the sequence subtitle.[11]. However, looking at our original data, it is clear that considerable liberty is taken as far as conserving semantic content is concerned: subtitles often drop substantial pieces of information. If we relax the notion of semantic equivalence a little, an even larger part of the non-subsequences can be rewritten as proper sequences.

The remaining problematic non-subsequences are those in which insertion, substitution and/or word reordering are essential to obtain a sufficient CR. One of the issues we identified is that deletion of certain constituents must be accompanied by a change in word order to prevent an ungrammatical sentence. Since changes in word order appear to require grammatical modeling or knowledge, this brings sentence compression closer to being an NLG task.

Nguyen and Horiguchi [19] describe an extension of the decision tree-based compression model [18] which allows for word order changes. The key to their approach is that dropped constituents are temporarily stored on a *deletion stack*, from which they can later be re-inserted in the tree where required. Although this provides an unlimited freedom for rearranging constituents, it also complicates the task of learning the parsing steps, which might explain why their evaluation results show marginal improvements at best.

In our data, most of the word order changes appear to be minor though, often only moving the verb to second position after deleting a constituent in the topic position. We believe that unrestricted word order changes are perhaps not necessary and that the vast majority of the word order problems can be solved by a fairly restricted way of reordering, in particular, a parser-based model with an additional swap operation that swaps the two topmost items on the stack. We expect that this is more feasible as a learning task than an unrestricted model with a deletion stack.

Apart from reordering, other problems for word deletion models are the insertions and substitutions as a result of paraphrasing. Within a decision tree-based model, paraphrasing of words or continuous phrases may be modeled by a combination of a paraphrase lexicon and an extra operation which replaces the n

---

[11] As far as reasonably possible, because in a few cases the substitute contains extra information that is simply not present in the autocue.

topmost elements on the stack by the corresponding paraphrase. However, paraphrases involving variable arguments, as typical for verbal paraphrases, cannot be accounted for in this way. More powerful compression models may draw on existing NLG methods for text revision [16] to accommodate full paraphrasing.

We also looked at the perspectives for automatic paraphrase extraction from large text corpora. About a quarter of the required paraphrase patterns was found at least a hundred times in our corpus of 325M tokens. Extrapolation suggests that using the web at its current size would give us a coverage of approximately ten counts for three quarters of the paraphrases.

In the second part of this chapter we explored sentence compression with the tree transducer model as proposed by Cohn & Lapata [7]. We reported preliminary results of applying this more powerful compression model to the task of subtitle compression for Dutch. In theory the proposed model looks very promising, because it can handle and learn reordering, substitution and insertion in an elegant way. In practice, the results were not immediately encouraging. We identified a number of problems with the model, both when applied to Dutch and in general. We might interpret these findings as support for choosing a hybrid approach to sentence compression – explicitly modeling linguistic knowledge – rather than a fully data-driven approach, at least if the goal is to model more complicated forms of compression beyond deletion.

Incidentally, we identified two other tasks in automatic subtitling which are closely related to NLG. First, splitting and merging of sentences [17], which seems related to content planning and aggregation. Second, generation of a shorter referring expression or an anaphoric expression, which is currently one of the main themes in data-driven NLG [14].

In conclusion, we have presented evidence that deletion models for sentence compression are not sufficient, at least not as far as concrete application in subtitle compression is concerned. More elaborate models involving reordering and paraphrasing are therefore required, which puts sentence compression in the field of NLG.

Acknowledgments. We would like to thank Nienke Eckhardt, Paul van Pelt, Hanneke Schoormans and Jurry de Vos for the corpus annotation work, and Erik Tjong Kim Sang and colleagues for the autocue-subtitle material from the ATRANOS project, Martijn Goudbeek for help with curve fitting, and Peter Berck for text material to train the Dutch n-gram model. This work was conducted within the DAESO project funded by the STEVIN program (De Nederlandse Taalunie).

References

1. Barzilay, R., Lee, L.: Learning to paraphrase: an unsupervised approach using multiple-sequence alignment. In: Proceedings of the 2003 Conference of the North American Chapter of the Association for Computational Linguistics on Human Language Technology, Morristown, NJ, USA, pp. 16–23 (2003)

2. Belz, A., Reiter, E.: Comparing automatic and human evaluation of NLG systems. In: Proceedings of the 11th Conference of the European Chapter of the Association for Computational Linguistics, pp. 313–320 (2006)

3. Bouma, G., van Noord, G., Malouf, R.: Alpino: Wide-coverage computational analysis of Dutch. In: Daelemans, W., Sima'an, K., Veenstra, J., Zavre, J., et al. (eds.) Computational Linguistics in the Netherlands 2000. Selected Papers from the Eleventh CLIN Meeting, Rodopi, Amsterdam, New York, pp. 45–59 (2001)

4. Clarke, J., Lapata, M.: Models for sentence compression: a comparison across domains, training requirements and evaluation measures. In: Proceedings of the 21st International Conference on Computational Linguistics and the 44th Annual Meeting of the Association for Computational Linguistics, Morristown, NJ, USA, pp. 377–384 (2006)

5. Clarke, J., Lapata, M.: Global inference for sentence compression an integer linear programming approach. Journal of Artificial Intelligence Research 31, 399–429 (2008)

6. Cohn, T., Lapata, M.: Sentence compression beyond word deletion. In: Proceedings of the 22nd International Conference on Computational Linguistics, vol. 1, pp. 137–144. Association for Computational Linguistics (2008)

7. Cohn, T., Lapata, M.: Sentence compression as tree transduction. J. Artif. Int. Res. 34(1), 637–674 (2009)

8. Corston-Oliver, S.: Text compaction for display on very small screens. In: Proceedings of the Workshop on Automatic Summarization (WAS 2001), Pittsburgh, PA, USA, pp. 89–98 (2001)

9. Daelemans, W., Höthker, A., Tjong Kim Sang, E.: Automatic sentence simplification for subtitling in Dutch and English. In: Proceedings of the 4th International Conference on Language Resources and Evaluation, pp. 1045–1048 (2004)

10. Dolan, B., Quirk, C., Brockett, C.: Unsupervised construction of large paraphrase corpora: Exploiting massively parallel news sources. In: Proceedings of the 20th International Conference on Computational Linguistics, Morristown, NJ, USA, pp. 350–356 (2004)

11. Eisner, J.: Learning non-isomorphic tree mappings for machine translation. In: Proceedings of 41st Annual Meeting of the Association for Computational Linguistics, Sapporo, Japan, pp. 205–208 (July 2003)

12. Filippova, K., Strube, M.: Sentence fusion via dependency graph compression. In: EMNLP 2008: Proceedings of the Conference on Empirical Methods in Natural Language Processing, pp. 177–185. Association for Computational Linguistics, Morristown (2008)

13. Filippova, K., Strube, M.: Tree linearization in English: improving language model based approaches. In: NAACL 2009: Proceedings of Human Language Technologies: The 2009 Annual Conference of the North American Chapter of the Association for Computational Linguistics, pp. 225–228. Association for Computational Linguistics, Morristown (2009) (Companion Volume: Short Papers)

14. Gatt, A., Belz, A.: Attribute selection for referring expression generation: New algorithms and evaluation methods. In: Proceedings of the Fifth International Natural Language Generation Conference, pp. 50–58. Association for Computational Linguistics, Columbus (2008)

15. Ibrahim, A., Katz, B., Lin, J.: Extracting structural paraphrases from aligned monolingual corpora. In: Proceedings of the 2nd International Workshop on Paraphrasing, Sapporo, Japan, vol. 16, pp. 57–64 (2003)

16. Inui, K., Tokunaga, T., Tanaka, H.: Text revision: A model and its implementation. In: Proceedings of the 6th International Workshop on Natural Language Generation: Aspects of Automated Natural Language Generation, pp. 215–230. Springer, London (1992)

17. Jing, H., McKeown, K.: Cut and paste based text summarization. In: Proceedings of the 1st Conference of the North American Chapter of the Association for Computational Linguistics, San Francisco, CA, USA, pp. 178–185 (2000)

18. Knight, K., Marcu, D.: Summarization beyond sentence extraction: A probabilistic approach to sentence compression. Artificial Intelligence 139(1), 91–107 (2002)

19. Le, N.M., Horiguchi, S.: A new sentence reduction based on decision tree model. In: Proceedings of the 17th Pacific Asia Conference on Language, Information and Computation, pp. 290–297 (2003)

20. Lin, C.Y.: Improving summarization performance by sentence compression - A pilot study. In: Proceedings of the Sixth International Workshop on Information Retrieval with Asian Languages, vol. 2003, pp. 1–9 (2003)

21. Lin, D., Pantel, P.: Discovery of inference rules for question answering. Natural Language Engineering 7(4), 343–360 (2001)

22. Marsi, E., Krahmer, E.: Annotating a parallel monolingual treebank with semantic similarity relations. In: Proceedings of the 6th International Workshop on Treebanks and Linguistic Theories, Bergen, Norway, pp. 85–96 (2007)

23. Marsi, E., Krahmer, E.: Detecting semantic overlap: A parallel monolingual treebank for Dutch. In: Verberne, S., van Halteren, H., Coppen, P.A. (eds.) Computational Linguistics in the Netherlands (CLIN 2007): Selected papers from the 18th meeting, Rodopi, Amsterdam, pp. 69–84 (2008)

24. Nomoto, T.: A Comparison of Model Free versus Model Intensive Approaches to Sentence Compression. In: Proceedings of the 2009 Conference on Empirical Methods in Natural Language Processing, Singapore, pp. 391–399 (2009)

25. Ordelman, R., de Jong, F., van Hessen, A., Hondorp, H.: Twnc: a multifaceted Dutch news corpus. ELRA Newsletter 12(3/4), 4–7 (2007)

26. Turner, J., Charniak, E.: Supervised and unsupervised learning for sentence compression. In: Proceedings of the 43rd Annual Meeting of the Association for Computational Linguistics, Ann Arbor, Michigan, pp. 290–297 (June 2005)

27. Vandeghinste, V., Pan, Y.: Sentence compression for automated subtitling: A hybrid approach. In: Proceedings of the ACL Workshop on Text Summarization, pp. 89–95 (2004)

28. Vandeghinste, V., Tjong Kim Sang, E.: Using a Parallel Transcript/Subtitle Corpus for Sentence Compression. In: Proceedings of LREC 2004 (2004)

29. Wan, S., Dras, M., Dale, R., Paris, C.: Spanning tree approaches for statistical sentence generation. In: Krahmer, E., Theune, M. (eds.) Empirical Methods in NLG. LNCS (LNAI), vol. 5790, pp. 13–44. Springer, Heidelberg (2010)

30. Zajic, D., Dorr, B.J., Lin, J., Schwartz, R.: Multi-candidate reduction: Sentence compression as a tool for document summarization tasks. Information Processing Management 43(6), 1549–1570 (2007)

Learning Adaptive Referring Expression Generation Policies for Spoken Dialogue Systems

Srinivasan Janarthanam[1] and Oliver Lemon[2]

[1] School of Informatics, University of Edinburgh
s.janarthanam@ed.ac.uk
[2] School of Mathematical and Computer Sciences, Heriot Watt University
o.lemon@hw.ac.uk
http://www.classic-project.org

Abstract. We address the problem that different users have different lexical knowledge about problem domains, so that automated dialogue systems need to adapt their generation choices online to the users' domain knowledge as it encounters them. We approach this problem using Reinforcement Learning in Markov Decision Processes (MDP). We present a reinforcement learning framework to learn adaptive referring expression generation (REG) policies that can adapt dynamically to users with different domain knowledge levels. In contrast to related work we also propose a new statistical user model which incorporates the lexical knowledge of different users. We evaluate this framework by showing that it allows us to learn dialogue policies that automatically adapt their choice of referring expressions online to different users, and that these policies are significantly better than hand-coded adaptive policies for this problem. The learned policies are consistently between 2 and 8 turns shorter than a range of different hand-coded but adaptive baseline REG policies.

Keywords: Reinforcement Learning, Referring Expression Generation, Spoken Dialogue System.

1 Introduction

We present a reinforcement learning framework for learning adaptive referring expression generation (REG) policies in interactive settings such as spoken dialogue systems. An adaptive REG policy allows a dialogue system to dynamically modify its utterances by choosing appropriate referring expressions in order to adapt to users' domain knowledge levels. For instance, in a technical support task, the dialogue agent could use technical jargon with experts, descriptive expressions with beginners, and a mixture of the two with intermediate users. Similarly, in a city navigation task, the dialogue agent could use proper names for landmarks with locals but descriptive expressions with foreign tourists. Issacs and Clark [17] show how two interlocutors adapt their language in a conversation by assessing each other's domain expertise during dialogue, by observing how

E. Krahmer, M. Theune (Eds.): Empirical Methods in NLG, LNAI 5790, pp. 67–84, 2010.

they react to each other's referring expression choices. This is called alignment through *Audience Design* [1,8,9]. Clark and Murphy [9] suggest that the dialogue partners adapt to each other by predicting each other's *community membership*. Inappropriate use of referring expressions in instructions has been identified as a serious problem affecting system's usability [27]. Similarly, Wittwer shows how under-estimating or over-estimating a layperson's domain knowledge impairs the knowledge acquisition process between experts and laymen [42].

In current *troubleshooting* spoken dialogue systems (SDS)[3,41] the major part of the conversation is directed by the system, while the user follows the system's instructions. Once the system decides what instruction to give to the user (at the dialogue management level), it faces several decisions to be made at the natural language generation (NLG) level. Although these include deciding which concepts and referring expressions (RE) to use in the utterance, in most systems the utterances are simply pre-scripted strings. However, it would be beneficial to choose appropriate referring expressions based on the user's domain expertise. It is a relatively simple problem to choose the right referring expressions when the system interacts with a known user, but it is impossible to accurately predict the expertise of an unknown user before a conversation starts, so we need adaptive REG policies which can estimate a new user's linguistic expertise based on the conversation so far.

One solution would be to learn the adaptive behaviour of a human expert to produce an adaptive REG policy. However, it has been shown in psychological studies that human experts in the healthcare domain do not present information using appropriate referring expressions. Although they lexically entrain (alignment due to priming) to a layperson's terms in their responses, they continue to use medical terms for the concepts never mentioned by the layperson [6]. Hinds notes that the cognitive heuristics of these experts interfere with their ability to predict a layperson's knowledge, which is termed as *the curse of expertise* [16]. Therefore, we shall learn adaptive strategies by interacting with different kinds of users using a reward based learning model, which rewards adaptive behaviour and penalises inappropriate use of referring expressions. Recently, van Deemter [12] proposed a similar idea of applying utility based decision making from Game Theory to solve NLG decision problems such as the appropriate use of vague expressions.

Using a reinforcement learning framework, we study to what extent a system could automatically align to the user's lexical knowledge by adapting its RE choices, by estimating the user's domain expertise dynamically during the course of the dialogue, and how this can be modelled and optimised within a computational learning paradigm [18,24,31]. We study this using a referential communication task in the context of a troubleshooting dialogue application – where users wish to mend their broken internet connection.

The example given below describes the kind of lexical alignment behaviour that we want the system to learn. The system chooses "small white box" instead of "ADSL filter" and "monitor symbol" instead of "network icon", because it has estimated that the user is a novice based on their clarification requests.

However, it switches to using technical terms like "browser", when it later estimates that the user is not a complete novice (since s/he verifies the description
for the network icon in move 4).

Example 1.
Sys 1>> Is your router connected to the computer?
Usr 1>> Uh. What's a router?
Sys 2>> It's the big black box.
Usr 2>> Ok.. yes.
Sys 3>> Do you see a small white box connected to the router?
Usr 3>> Yes.
Sys 4>> Ok. Is there a flashing monitor symbol at the bottom right
of the screen?
Usr 4>> the network icon?
Sys 5>> Yes. Is it flashing?
Usr 5>> Yes. It is flashing.
Sys 6>> Ok. Please open your browser.

In this paper, after a review of related work (Section 2), we present a reinforcement learning model of lexical alignment due to audience design (in Section 3).
In Sections 3.1, 3.2, 3.3, we present the finer details of the dialogue system architecture, the statistical user simulation and the environment simulation. In
Section 4, we show how the reward for reinforcement learning is computed. In
Section 5, we train the setup to learn adaptive REG policies that gets optimal
rewards for different kinds of users generated by the user simulation. Finally,
we compare policies learned in interaction with the User Simulation with hand-
coded policies, and present the results in Section 6.

2 Related Work

Several psycholinguistic studies have examined how humans use different referring expressions (proper names vs. descriptive expressions) for different users.
Issacs and Clark [17] show how interlocutors adapt their choice of referring expressions according to various users. Recently, Heller et al. [15] present how
human users keep track of shared and privileged ground when using different
referring expressions with different users. However, in contrast, we present a
framework to learn to align with the users automatically based on the user's
knowledge of the domain.

Similarly, there have been several studies of lexical alignment due to *priming*
[28]. This is also called *Lexical Entrainment* [29]. This refers to the act of using
the same referring expressions as used by the interlocutor during a conversation, although one might have used different referring expressions to begin with.
Buschmeier et al. [7] (this volume) present an alignment-capable microplanner
which aligns to the user due to priming effects. Such studies have also been
carried out on human-computer interaction, which show that human users align

lexically and syntactically with computers [4,5]. In contrast, we only study lexical alignment due to *Audience Design*.

Reinforcement Learning (RL) [40] has been successfully used for learning dialogue management policies [25]. The learned policies allow the dialogue manager to optimally choose appropriate dialogue acts such as instructions, confirmation requests, and so on, under uncertain noise or other environment conditions. Recently, Lemon [24] presented natural language generation in dialogue as a Reinforcement Learning or statistical planning problem. Rieser and Lemon [31,33] (this volume) presented a model to learn information presentation strategies using reinforcement learning. In contrast, we present a framework to learn to choose appropriate referring expressions based on a user's domain knowledge.

Referring expression generation has previously been treated as a content selection or attribute determination problem by several researchers [10,11,13,23,30,38], to determine the content of descriptive referring expressions. Later, it has also been extended to interactive situated dialogue contexts to take into account spatial context features [39]. Similarly, referring expression choices between descriptive expressions and pronouns in discourse contexts have been examined [2]. Reiter et al. discuss the idea of using a user model (and a corresponding UserKnows function) that can inform the NLG module on which attributes of the referent the user knows [30]. However, this model is static and predefined and can only be used for a known user. In contrast to the above perspectives, we examine referring expression generation in settings like conversations between dialogue systems and unknown users with differing levels of domain expertise.

Dialogue systems that can adapt to users based on their domain expertise or skills have been presented before. McKeown et al. [26] presented a multimodal dialogue system that adapts to the user's expertise level. The user model is either preset to an appropriate expertise level or is set after the system explicitly asks the user to choose. Boye [3] presented a system that has a fallback option to simplified instructions as soon as a complex instruction is not understood by the user. Komatani et al. [21,22] presented a system that provides users with appropriate amounts of bus travel information based on their domain knowledge. The user knowledge level is predicted using a trained decision tree. In contrast to these systems, we build a system that can dynamically adapt to different kinds of users based on a policy that can be automatically learned using reinforcement learning methods.

User simulations are a part of the reinforcement learning framework. Several statistical user simulation models that model a user's behaviour in a conversation have been proposed [14,34,35]. These models issue task specific dialogue acts such as informing their search constraints, confirming values, rejecting misrecognised values, and so on. Although these models have been successfully used in learning dialogue management policies, they are inadequate for learning adaptive REG policies for two reasons. First, they do not model a user population with varying domain expertise. Second, they are not sensitive to the system's referring expression choices and none of these models seek clarification at conceptual or lexical levels that occur naturally in conversations between real users. Janarthanam and Lemon [18] presented a user simulation model that simulates a variety of users with different domain knowledge profiles. Although this model incorporated clarification acts at the conceptual level, these simulated users ignore the issues concerning the user's understanding of the referring expressions used by the system. In this work, we present a user simulation model which explicitly encodes the user's lexical knowledge of the domain, is sensitive to system's referring expressions, and issues clarification requests at the lexical level.

3   Reinforcement Learning Environment

The Reinforcement Learning environment consists of the dialogue system, user simulation, and environment simulation as shown in Figure 1. In a fully developed system, modules like the speech recogniser, semantic decoder, and speech synthesizer will also be present. However, during learning, the dialogue system does not interact with the simulated user at the surface level and the learning environment is therefore simplified and is represented as a Markov Decision Process. At every turn (t), the dialogue system interacts with the simulated user by issuing a System Dialogue Act $(A_{s,t})$ along with a set of referring expressions, called the System Referring Expression Choices $(REC_{s,t})$. The list of system dialogue acts is given Table 1. $REC_{s,t}$ contains the expressions that refer to various domain objects in the dialogue act $A_{s,t}$. The user simulation responds to the system dialogue act and its choice of referring expressions with a user dialogue act $(A_{u,t})$. The user simulation also issues an environment action $(EA_{u,t})$ to either observe or manipulate the environment state $(S_{e,t})$ based on the system's instruction. The user simulation also rewards the dialogue system for its choice of referring expressions, giving more reward for choosing appropriate referring expressions. The user's response is updated to the system dialogue state $(S_{s,t+1})$

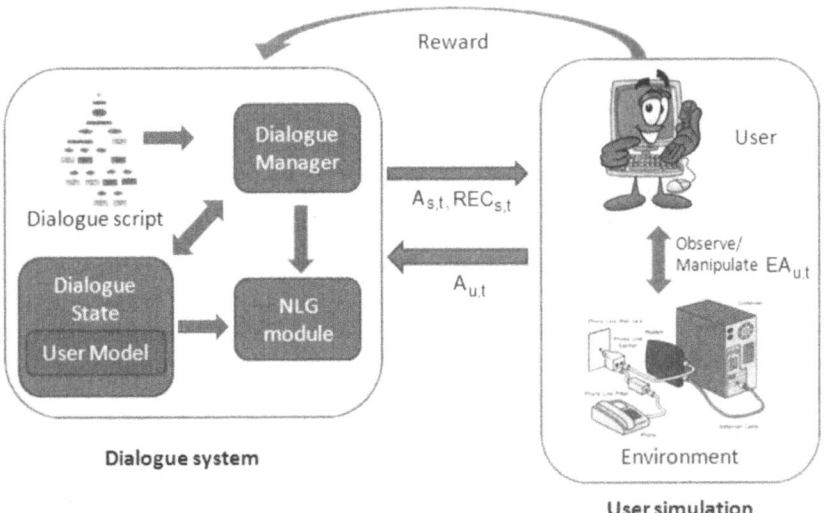

Fig. 1. Reinforcement Learning setup

Table 1. System Dialogue acts

greet_the_user
request_status(x)
request_action(x)
give_description(x)
accept_verification(x,y)
give_location(dobj)
give_procedure(daction)
close_dialogue

and the system goes back to choosing the next dialogue act $(A_{s,t+1})$ and referring expression choices $(REC_{s,t+1})$ based on the current state.

3.1 Dialogue System

The dialogue system plays the role of the domain expert who directs the conversation to first identify the problem with the user's broadband setup and then provide instructions to rectify it. The architecture of our system is shown in Figure 2. The dialogue manager decides what instruction to give the user at every dialogue turn. Besides providing instructions, the system also resolves different types of clarification. The dialogue act $(A_{s,t})$ is chosen based on the system dialogue state $(S_{s,t})$ as directed by the dialogue management policy (π_{DM}), which maps any given dialogue state to an optimal dialogue action. The list of all possible system dialogue acts is given in Table 1. Our objective is to build a system that is able to dynamically track the user's domain expertise and adapt its choices of referring expressions. We therefore introduce a *user model* as a part of the dialogue state and an *adaptive REG policy* as shown in Figure 2. The REG policy (π_{REC}) decides which referring expression to choose $(REC_{s,t})$ based on the referent and the user's domain knowledge given by the user model $(UM_{s,t})$.

Our objective is to learn an adaptive REG policy and not the dialogue management policy, so the dialogue management policy is coded in the form of a troubleshooting decision tree.[1] In order to learn an adaptive NLG policy (π_{REC}), we configure the NLG module as a reinforcement learning agent in a Markov Decision Process framework [24].

The system issues various repair moves when the users are unable to carry out the system's instructions due to ignorance, non-understanding or the ambiguous nature of the instructions. The *give_description* act is used to give the user a description of the domain object previously referred to using a technical term. It is also used when the user requests disambiguation. Similarly, *accept_verification* is given by the system when the user wants to verify whether the system is referring to a certain domain object y using the expression x (for example, "Yes, the black box is the router").

---

[1] The Troubleshooting decision tree was hand-built using guidelines from www.orange. co.uk and is similar to the one used by their Customer Support personnel.

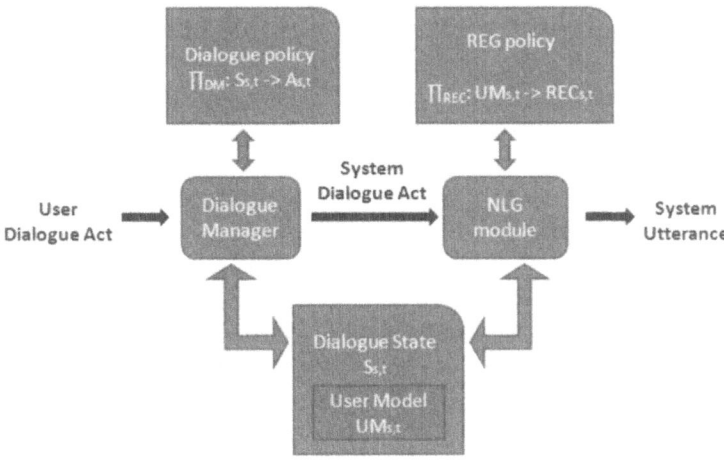

Fig. 2. Dialogue System with Adaptive NLG

After selecting the dialogue act $A_{s,t}$, a set of referring expressions must be chosen to refer to each of the domain objects/actions used in the dialogue act. For instance, the dialogue act *request_status(router_dsl_light)* requires references to be made to domain objects "router" and "DSL light". For each of these references, the system chooses a referring expression, creating the System REC $REC_{s,t}$. In this study, we have 7 domain objects and they can either be referred to using technical terms or descriptive expressions. For instance, the DSL light on the router can be descriptively referred to as the "second light on the panel" or using the technical term, "DSL light". Sometimes the system has to choose between a lesser known technical term and a well-known one. Some descriptive expressions may be underspecified and therefore can be ambiguous to the user (for example, "the black box"). Choosing inappropriate expressions can make the conversation longer with lots of clarification and repair episodes. This can lead to long frustrating dialogues, affecting the task success rate. Therefore, the dialogue system must learn to use appropriate referring expressions in its utterances. The referring expression choices available to the system are given in Table 2.

Table 2. System Referring Expression choices

1. router / black box / black box with lights
2. power light / first light on the panel
3. DSL light / second light on the panel
4. online light / third light on the panel
5. network icon / flashing computer symbol
6. network connections / earth with plug
7. WiFi / wireless

The system's referring expression choices are based on a user's dialogue behaviour. User's understanding of the technical terms are recorded as a part of the dialogue state called the *user model*. This is very similar to user knowledge base used in the Incremental Algorithm [30]. However, the difference is that the model is not predefined to the user's knowledge. These variables are initially set to *unknown (u)*. During the dialogue, they are updated to *user_knows (y)* or *user_doesnot_know (n)* states. The user's lexical knowledge is inferred during the course of the dialogue from their dialogue behaviour. For instance, if the user asks for clarification on a referring expression, his knowledge of the expression is set to n and when no clarifications are requested, it is set to y. In addition to the above, the dialogue state also records other information like the state of progress of the given task, the perceived state of the environment, and so on. Although these pieces of information are important for deciding the next dialogue act (by the dialogue manager), they are not important for NLG choices. A part of the dialogue state relevant to system's referring expression choices is given in Table 3.

Table 3. (Part of) Dialogue state for Lexical Alignment

Feature	Values
user_knows_router	y/n/u
user_knows_power_light	y/n/u
user_knows_dsl_light	y/n/u
user_knows_online_light	y/n/u
user_knows_network_icon	y/n/u
user_knows_network_connections	y/n/u
user_knows_wifi	y/n/u

The state can be extended to include other relevant information like the usage of various referring expressions by the user as well to enable alignment with the user through priming [28] and personal experience [8]. However they are not yet implemented in the present work.

3.2   User Simulation

The user simulation module simulates dialogue behaviour of different users, and interacts with the dialogue system by exchanging both dialogue acts and referring expressions. It produces users with different knowledge profiles. The user population produced by the simulation comprises a spectrum from complete novices to experts in the domain. Simulated users behave differently to one another because of differences in their knowledge profiles. Simulated users are also able to learn new referring expressions during interaction with the SDS. These new expressions are held in the user simulation's short term memory for later use in the conversation.

Domain Knowledge Model. Domain experts know most of the technical terms that are used to refer to domain objects whereas novice users can only reliably identify them when descriptive expressions are used. While in the earlier model [18], knowledge profiles were presented only at conceptual levels (e.g. does the user know what a modem is?), we present them in a more granular fashion. In this model, the user's domain knowledge profile is factored into lexical $(LK_{u,t})$, factual $(FK_{u,t})$ and procedural knowledge $(PK_{u,t})$ components. A user's lexical knowledge is encoded in the format:

$$vocab(referring\_expressions, domain\_object)$$

where *referring_expressions* can be a list of expressions that the user knows can be used to talk about each *domain_object*.

Whether the user knows facts like the location of the domain objects *(location(domain_object))* is encoded in the factual component. Similarly, the procedural component encodes the user's knowledge of how to find or manipulate domain objects *(procedure(domain_action))*. Table 4 shows an example user knowledge profile.

Table 4. Knowledge profile - Intermediate user

Lexical knowledge $LK_{u,t}$
vocab([modem, router], dobj1)
vocab([wireless, WiFi], dobj3)
vocab([modem power light], dobj7)
Factual knowledge $FK_{u,t}$
location(dobj1)
location(dobj7)
Procedural knowledge $PK_{u,t}$
procedure(replace_filter)
procedure(refresh_page)

In order to create a knowledge spectrum, a Bayesian knowledge model is used. The current model incorporates patterns of only the lexical knowledge among the users. For instance, people who know the word "router" most likely also know "DSL light" and "modem" and so on. These dependencies between referring expressions are encoded as conditional probabilities in the Bayesian model. Figure 3 shows the dependencies between knowledge of referring expressions.

Using this Bayesian model, we instantiate different knowledge profiles for different users. The current conditional probabilities were set by hand based on intuition. In future work, these values will be populated based on simple knowledge surveys performed on real users [20]. This method creates a spectrum of users from ones who have no knowledge of technical terms to ones who know all the technical jargon, though every profile will have a different frequency

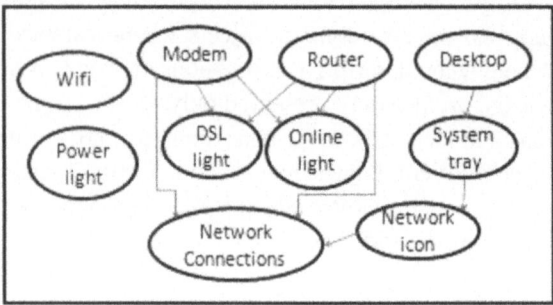

Fig. 3. Bayes Net for User Lexical Knowledge

of occurrence. This difference in frequency reflects that expert users are less common than novice users.

The user's domain knowledge can be dynamically updated. The new referring expressions (both technical and descriptive) presented by the system through clarification moves are stored in the user's short term memory. Exactly how long (in terms of dialogue turns) to retain the newly acquired knowledge is given by a retention index RI_u. At the end of RI_u turns, the lexical item is removed from user's short term memory.

User Dialogue Action Set. Apart from environment-directed acts, simulated users issue a number of dialogue acts. The list of dialogue actions that the user can perform in this model is given in Table 5. It consists of default moves like *provide_info* and *acknowledge* as well as some clarification moves. *Request_description* is issued when the SDS uses technical terms that the simulated user does not know, e.g. "What is a router?". *Request_verification* is issued when the SDS uses descriptive lexical items for domain objects that the user knows more technical terms for, e.g. System: "Is the black box plugged in?" User: "Do you mean the router?". *Request_disambiguation* is issued when the user faces an underspecified and ambiguous descriptive expression, e.g."User: Is have two black boxes here - one with lights and one without. Which one is it?". These

Table 5. User Dialogue Acts

report_problem
provide_info(dobj, info)
acknowledge
request_verification(x, y)
request_description(x)
request_disambiguation(x, [y1,y2])
request_location(dobj)
request_procedure(daction)
thank_system

clarification strategies have been modeled based on [36]. The user simulation also issues *request_location* and *request_procedure* dialogue acts, when it does not know the location of domain objects or how to manipulate them, respectively.

User Action Selection. The user's dialogue action behaviour is described in the action selection algorithm (Table 6). First, the user's knowledge of referring expressions used by the system ($REC_{s,t}$) to refer to different domain objects (*dobj*) and domain actions (*daction*) in the system instruction (step 1) is checked. In case of ambiguous or unknown expressions, the user raises an appropriate clarification request. Users identify the referents when the system uses descriptive expressions for those objects that they cannot identify using jargon (step 1a). However, they ask for clarification when the system uses jargon in such cases (step 1b). We assume that users who know jargon expressions for certain referents

Table 6. Algorithm: Simulated User Action Selection

Input:	System Dialogue Act $A_{s,t}$, System Referring Expressions Choice $REC_{s,t}$ and User State $S_{u,t}$: $LK_{u,t}$, $FK_{u,t}$, $PK_{u,t}$

Step 1.	$\forall$ x $\in REC_{s,t}$
Step 1a.	if (vocab(x, dobj)$\in LK_{u,t}$) then next x.
Step 1b.	else if (description(x, dobj) & $\exists$ j ((is_jargon(j) & vocab(j, dobj) $\notin LK_{u,t}$))) then next x.
Step 1c.	else if (is_jargon(x) & (vocab(x, dobj) $\notin LK_{u,t}$)) then return request_description(x).
Step 1d.	else if (is_ambiguous(x)) then return request_disambiguation(x).
Step 1e.	else if (description(x, dobj) & $\exists$ j ((is_jargon(j) & vocab(j, dobj) $\in LK_{u,t}$))) then return request_verification(x, j).

Step 2.	if ($\exists$dobj location(dobj) $\in FK_{prereq}$ & location(dobj) $\notin FK_{u,t}$) then return request_location(dobj).

Step 3.	else if ($\exists$daction procedure(daction) $\in PK_{prereq}$ & procedure(daction) $\notin PK_{u,t}$) then return request_procedure(daction).

Step 4.	else if ($A_{s,t} =$ request_status(dobj)) then observe_env(dobj, status), return provide_info(dobj, status)

Step 5.	else if ($A_{s,t} =$ request_action(daction)) then manipulate_env(daction), return acknowledge.

Step 6.	else if ($A_{s,t} =$ give_description(j, d) & description(d, dobj)) then add_to_short_term_memory(vocab(j, dobj)), return acknowledge.

Step 7.	else if ($A_{s,t} =$ give_location(dobj)) then add_to_short_term_memory(location(dobj)), return acknowledge.

Step 8.	else if ($A_{s,t} =$ give_procedure(daction)) then add_to_short_term_memory(procedure(daction)), return acknowledge.

prefer the system to use jargon expressions for those referents. So, they verify the referent when descriptive expressions are used instead (step 1d).

Next, the factual and procedural knowledge (FK_{prereq} and PK_{prereq}) required to comprehend the given instruction is checked against the user's knowledge. The user simulation issues an appropriate clarification request (steps 2 & 3), when the user does not know the prerequisite factual or procedural knowledge components. Once the knowledge requirements are satisfied, the user issues environment directed actions and responds to system instruction $A_{s,t}$ (steps 4 & 5) by either informing the status of domain objects (e.g. broadband light is flashing) or acknowledging an instruction to manipulate the state of the environment. When the system provides a clarification, it is added to the user's short term memory (steps 6-8). Although the action selection process is deterministic at this level, it is dependent on the users' diverse knowledge profiles, which ensures stochastic dialogue behaviour amongst different users created by the module.

3.3   Environment Simulation

The environment simulation includes both physical objects, such as the computer, modem, and ADSL filter, and virtual objects such as the browser, control panel, and so on in the user's environment. Physical and virtual connections between these objects are also simulated. At the start of every dialogue, the environment is initiated to a faulty condition. Following a system instruction or question, the user issues two kinds of environment acts ($EA_{u,t}$). It issues an observation act to observe the status of a domain object and a manipulation act to change the state of the environment ($S_{e,t}$). The simulation also includes task irrelevant objects in order to present the users with underspecified descriptive expressions. For instance, we simulate two domain objects that are black in colour - an external hard disk and a router. So, the users may get confused when the system uses the expression "black box".

4   Reward Function

The reward function calculates the reward awarded to the reinforcement learning agent at the end of each dialogue session. Successful task completion is rewarded with 1000 points. Dialogues running beyond 50 turns are deemed unsuccessful and are awarded 0 points. The number of turns in each dialogue varies according to the system's RE choices and the simulated user's response moves. Each turn costs 10 points. The final reward is calculated as follows:

$TaskCompletionReward(TCR) = 1000$
$TurnCost(TC) = 10$
$TotalTurnCost(TTC) = \#(Turns) * TC$
$FinalReward = TCR - TTC$

The reward function therefore gives high rewards when the system produces shorter dialogues, which is possible by adaptively using appropriate referring expressions for each user. In future work, the reward function will be set empirically, following the method of [32].

5 Learning REG Policies

The system was trained to learn an adaptive REG policy which can adapt to users with different lexical knowledge profiles. In this phase, the dialogue system is made to interact with the user simulation exchanging dialogue acts and referring expression choices. The user simulation rewards the system for choosing the referring expressions appropriate to the user's lexical knowledge. During these training interactions, the system explores various choices to learn the user model-REC combinations that score better rewards. After several interactions, the system learns to choose the most appropriate referring expressions based on the dynamically updated user model. Ideally, the system must interact with a number of different users in order to learn to align with them. However, with a large number of distinct Bayesian user profiles (there are 90 possible user profiles), the time taken for learning to converge is exorbitantly high. Hence the system was trained with selected profiles from the distribution. It was initially trained using two user profiles from the very extremes of the knowledge spectrum produced by the Bayesian model - complete experts and complete novices. In this study, we calibrated all users to know all the factual and procedural knowledge components, because the learning exercise was targeted only at the lexical level. With respect to the lexical knowledge, complete experts knew all the technical terms in the domain. Complete novices, on the other hand, knew only one: *power_light*. We set the RI_u to 10, so that the users do not forget newly learned lexical items for 10 subsequent turns. Ideally, we expected the system to learn to use technical terms with experts and to use descriptive expressions with novices and a mixture for intermediates. The system was trained using the SARSA reinforcement learning algorithm [37,40], with linear function approximation, for 50000 cycles. It produced around 1500 dialogues and produced an alignment policy (RL1) that adapted to users after the first turn which provides evidence about the kind of user the system is dealing with.

The system learns to get high reward by producing shorter dialogues. By learning to choose REs by adapting to the lexical knowledge of the user it avoids unnecessary clarification and repair episodes. It learns to choose descriptive expressions for novice users and jargon for expert users. It also learns to use technical terms when all users know them (for instance, "power_light"). Due to the user's high retention (10 turns), the system learned to use newly presented items later in the dialogue.

We also trained another alignment policy (RL2) with two other intermediate high frequency user lexical profiles. These profiles (Int1 and Int2) were chosen from either ends of the knowledge spectrum close to the extremes. Int1 is a knowledge profile that is close to the novice end. It only knows two technical terms: "power_light" and "WiFi". On the other hand, Int2 is profile that is close to the expert end and knows all technical terms except: "dsl_light" and "on-line_light" (which are the least well-known technical terms in the user population). With respect to the other knowledge components - factual and procedural, both users know every component equally. We trained the system for 50000 cycles following the same procedure as above. This produced an alignment policy (RL2) that learned to optimize the moves, similar to RL1, but with respect to the given distinct intermediate users.

Both policies RL1 and RL2, apart from learning to adapt to the users, also learned not to use ambiguous expressions. Ambiguous expressions lead to confusion and the system has to spend extra turns for clarification. Therefore both policies learnt to avoid using ambiguous expressions. Figure 4 and Figure 5 show the overall dialogue reward and dialogue length variation for the 2 policies during training.

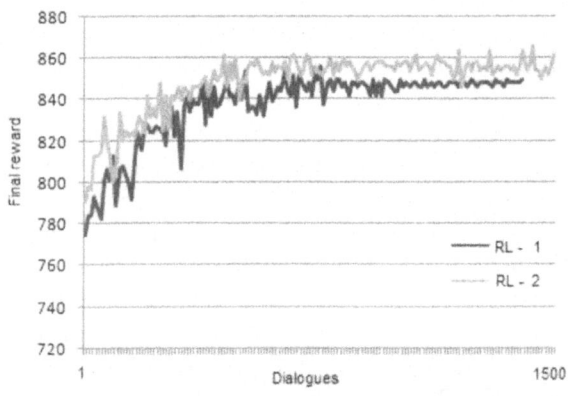

Fig. 4. Final reward for RL1 & RL2

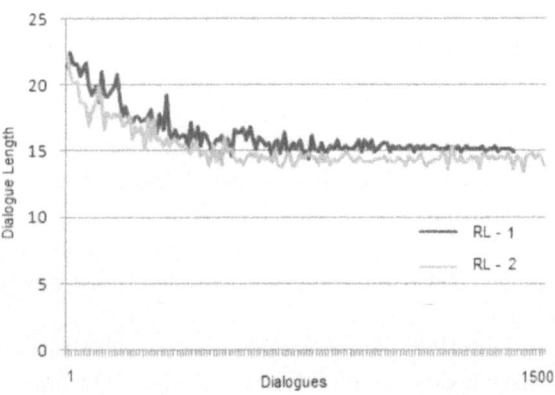

Fig. 5. Dialogue length for RL1 & RL2

6   Evaluation and Baselines

We evaluated both the learned policies using a testing simulation and compared the results to other baseline hand-coded policies. Unlike the training simulation, the testing simulation used the Bayesian knowledge model to produce all different kinds of user knowledge profiles. It produced around 90 different profiles in varying distribution, resembling a realistic user population. The tests were run over 250 simulated dialogues each.

Several rule-based baseline policies were manually created for the sake of comparison:

1. Random - Choose REs at random.
2. Descriptive only - Only choose descriptive expressions. If there is more than one descriptive expression it picks one randomly.
3. Jargon only - Chooses the technical terms.
4. Adaptive 1 - It starts with a descriptive expression. If the user asks for verification, it switches to technical terms for the rest of the dialogue.
5. Adaptive 2 - It starts with a technical term and switches to descriptive expressions if the user does not understand in the first turn.
6. Adaptive 3 - This rule-based policy adapts continuously based on the previous expression. For instance, if the user did not understand the technical reference to the current object, it uses a descriptive expression for the next object in the dialogue, and vice versa.

The first three policies (random, descriptive only and jargon only) are equivalent to policies learned using user simulations that are not sensitive to system's RE choices. In such cases, the learned policies will not have a well-defined strategy to choose REs based on user's lexical knowledge. Table 7 shows the comparative results for the different policies. RL (1 & 2) are significantly better than all the hand-coded policies. Also, RL2 is significantly better than RL1 $(p < 0.05)$.[2]

Table 7. Rewards and Dialogue Length

Policy	Avg. Reward	Avg. Length
RL2	830.4	16.98
RL1	812.3	18.77
Adaptive 1	809.6	19.04
Adaptive 2	792.1	20.79
Adaptive 3	780.2	21.98
Random	749.8	25.02
Desc only	796.6	20.34
Jargon only	762.0	23.8

---

[2] All significance testing was done using Wilcoxon signed-rank tests.

Ideally the system with complete knowledge of the user would be able to finish the dialogue in 13 turns. Similarly, if it got it wrong every time it would take 28 turns. From Table 7 we see that RL2 performs better than other policies, with an average dialogue length of around 17 turns. The learned policies were able to discover the hidden dependencies between lexical items that were encoded in the Bayesian knowledge model. Although trained only on two knowledge profiles, the learned policies adapt well to unseen users, due to the generalisation properties of the linear function approximation method. Many unseen states arise when interacting with users with new profiles and both the learned policies generalise well in such situations, whereas the baseline policies do not.

7 Conclusion

We have presented a reinforcement learning framework to learn adaptive REG policies. These policies adapt to a group of users with varying domain expertise and interact with them using appropriate referring expressions. We have presented a statistical User Simulation that is sensitive to referring expression choices so that we are able to learn REG policies that adapts based on audience design. We have also shown that adaptive REG policies learned with this type of simulation are better than a range of hand-coded policies.

Although REG policies could be hand-coded, the designers would need to invest significant resources every time the list of referring expressions is revised or the conditions of the dialogue change. Using reinforcement learning, near-optimal REG policies can be learned quickly and automatically. This model can be used in any task where interactions need to be tailored to different users' lexical knowledge of the domain.

7.1 Future Work

Adaptive REG policies, presented above, align with the users lexically due to Audience Design. Similary, lexical alignment in dialogue, which happens due to priming [28] and personal experience [8] can also be accounted for when choosing the referring expressions. Trade-offs in various conditions, like 'instruct' versus 'teach' and low versus high retention users can be examined. Using Wizard-of-Oz studies and knowledge surveys, data-driven and realistic user simulation models can be built for training and testing [19,20], and setting reward functions [32].

Acknowledgements. The research leading to these results has received funding from the European Community's Seventh Framework (FP7) under grant agreement no. 216594 (CLASSiC Project www.classic-project.org), EPSRC project nos. EP/E019501/1 and EP/G069840/1, and the British Council (UKIERI PhD Scholarships 2007-08).

References

1. Bell, A.: Language style as audience design. Language in Society 13(2), 145–204 (1984)
2. Belz, A., Varges, S.: Generation of repeated references to discourse entities. In: Proc. ENLG 2007 (2007)
3. Boye, J.: Dialogue management for automatic troubleshooting and other problem-solving applications. In: Proc. SIGDial 2007 (2007)
4. Branigan, H.P., Pickering, M.J., Pearson, J., McLean, J.F.: Linguistic alignment between people and computers. Journal of Pragmatics (in Press)
5. Brennan, S.E.: Conversation with and through computers. User Modeling and User-Adaptive Interaction 1(1), 67–86 (1991)
6. Bromme, R., Jucks, R., Wagner, T.: How to refer to diabetes? Language in online health advice. Applied Cognitive Psychology 19, 569–586 (2005)
7. Buschmeier, H., Bergmann, K., Kopp, S.: Modelling and evaluation of lexical and syntactic alignment with a priming-based microplanner. In: Krahmer, E., Theune, M. (eds.) Empirical Methods in NLG. LNCS (LNAI), vol. 5790, pp. 85–104. Springer, Heidelberg (2010)
8. Clark, H.H.: Using Language. Cambridge University Press, Cambridge (1996)
9. Clark, H.H., Murphy, G.L.: Audience design in meaning and reference. In: Le Ny, J.F., Kintsch, W. (eds.) Language and comprehension. North-Holland Publishing Company, Amsterdam (1982)
10. Dale, R.: Cooking up referring expressions. In: Proc. ACL 1989(1989)
11. van Deemter, K.: Generating referring expressions: Boolean extensions of the Incremental Algorithm. Computational Linguistics 28(1), 37–52 (2002)
12. van Deemter, K.: What game theory can do for NLG: the case of vague language. In: Proc. ENLG 2009 (2009)
13. Gatt, A., Belz, A.: Attribute selection for referring expression generation: New algorithms and evaluation methods. In: Proc. INLG 2008 (2008)
14. Georgila, K., Henderson, J., Lemon, O.: Learning user simulations for information state update dialogue systems. In: Proc. Eurospeech/Interspeech (2005)
15. Heller, D., Skovbroten, K., Tanenhaus, M.K.: Experimental evidence for speakers sensitivity to common vs. privileged ground in the production of names. In: Proc. PRE-CogSci 2009 (2009)
16. Hinds, P.: The curse of expertise: The effects of expertise and debiasing methods on predictions of novice performance. Experimental Psychology: Applied 5(2), 205–221 (1999)
17. Issacs, E.A., Clark, H.H.: References in conversations between experts and novices. Journal of Experimental Psychology: General 116, 26–37 (1987)
18. Janarthanam, S., Lemon, O.: User simulations for online adaptation and knowledge-alignment in troubleshooting dialogue systems. In: Proc. SEMdial 2008 (2008)
19. Janarthanam, S., Lemon, O.: A two-tier user simulation model for reinforcement learning of adaptive referring expression generation policies. In: Proc. SIGDial 2009 (2009)
20. Janarthanam, S., Lemon, O.: A wizard-of-oz environment to study referring expression generation in a situated spoken dialogue task. In: Proc. ENLG 2009 (2009)
21. Komatani, K., Ueno, S., Kawahara, T., Okuno, H.G.: Flexible guidance generation using user model in spoken dialogue systems. In: Proc. ACL 2003 (2003)

22. Komatani, K., Ueno, S., Kawahara, T., Okuno, H.G.: User modeling in spoken dialogue systems to generate flexible guidance. User Modeling and User-Adapted Interaction 15(1), 169–183 (2005)
23. Krahmer, E., van Erk, S., Verleg, A.: Graph-based generation of referring expressions. Computational Linguistics 29(1), 53–72 (2003)
24. Lemon, O.: Adaptive natural language generation in dialogue using reinforcement learning. In: Proc. SEMdial 2008 (2008)
25. Levin, E., Pieraccini, R., Eckert, W.: Learning dialogue strategies within the markov decision process framework. In: Proc. ASRU 1997 (1997)
26. McKeown, K., Robin, J., Tanenblatt, M.: Tailoring lexical choice to the user's vocabulary in multimedia explanation generation. In: Proc. ACL 1993 (1993)
27. Molich, R., Nielsen, J.: Improving a human-computer dialogue. Communications of the ACM 33(3), 338–348 (1990)
28. Pickering, M.J., Garrod, S.: Toward a mechanistic psychology of dialogue. Behavioral and Brain Sciences 27, 169–225 (2004)
29. Porzel, R., Scheffler, A., Malaka, R.: How entrainment increases dialogical efficiency. In: Proc. Workshop on Effective Multimodal Dialogue Interfaces, Sydney (2006)
30. Reiter, E., Dale, R.: Computational interpretations of the Gricean maxims in the generation of referring expressions. Cognitive Science 18, 233–263 (1995)
31. Rieser, V., Lemon, O.: Natural language generation as planning under uncertainty for spoken dialogue systems. In: Proc. EACL 2009 (2009)
32. Rieser, V., Lemon, O.: Learning effective multimodal dialogue strategies from wizard-of-oz data: Bootstrapping and evaluation. In: Proc. ACL 2008 (2008)
33. Rieser, V., Lemon, O.: Natural Language Generation as Planning Under Uncertainty for Spoken Dialogue Systems. In: Krahmer, E., Theune, M. (eds.) Empirical Methods in NLG. LNCS (LNAI), vol. 5790, pp. 105–120. Springer, Heidelberg (2010)
34. Schatzmann, J., Thomson, B., Weilhammer, K., Ye, H., Young, S.J.: Agenda-based user simulation for bootstrapping a POMDP dialogue system. In: Proc. HLT/NAACL 2007 (2007)
35. Schatzmann, J., Weilhammer, K., Stuttle, M.N., Young, S.J.: A survey of statistical user simulation techniques for reinforcement learning of dialogue management strategies. Knowledge Engineering Review, 97–126 (2006)
36. Schlangen, D.: Causes and strategies for requesting clarification in dialogue. In: Proc. SIGDial 2004 (2004)
37. Shapiro, D., Langley, P.: Separating skills from preference: Using learning to program by reward. In: Proc. ICML 2002 (2002)
38. Siddharthan, A., Copestake, A.: Generating referring expressions in open domains. In: Proc. ACL 2004 (2004)
39. Stoia, L., Shockley, D.M., Byron, D.K., Fosler-Lussier, E.: Noun phrase generation for situated dialogs. In: Proc. INLG 2006, pp. 81–88 (July 2006)
40. Sutton, R., Barto, A.: Reinforcement Learning. MIT Press, Cambridge (1998)
41. Williams, J.: Applying POMDPs to dialog systems in the troubleshooting domain. In: Proc. HLT/NAACL Workshop on Bridging the Gap: Academic and Industrial Research in Dialog Technology (2007)
42. Wittwer, J., Nckles, M., Renkl, A.: What happens when experts over- or underestimate a laypersons knowledge in communication? Effects on learning and question asking. In: Proc. CogSci 2005 (2005)

Modelling and Evaluation of Lexical and Syntactic Alignment with a Priming-Based Microplanner

Hendrik Buschmeier, Kirsten Bergmann, and Stefan Kopp

Sociable Agents Group, CITEC, Bielefeld University
PO-Box 10 01 31, 33501 Bielefeld, Germany
{hbuschme,kbergman,skopp}@TechFak.Uni-Bielefeld.DE

Abstract. Alignment of interlocutors is a well known psycholinguistic phenomenon of great relevance for dialogue systems in general and natural language generation in particular. In this chapter, we present the alignment-capable microplanner SPUD *prime*. Using a priming-based model of interactive alignment, it is flexible enough to model the alignment behaviour of human speakers to a high degree. We demonstrate that SPUD *prime* can account for lexical as well as syntactic alignment and present an evaluation on corpora of task-oriented dialogue that were collected in two experiments designed to investigate the alignment behaviour of humans in a controlled fashion. This will allow for further investigation of which parameters are important to model alignment and how the human–computer interaction changes when the computer aligns to its users.

Keywords: interactive alignment model, lexical and syntactic alignment, adaptation, microplanning.

1 Introduction

A well known phenomenon in dialogue situations is *alignment* of the interlocutors. An illustrative example is given by Levelt and Kelter [17], who telephoned shops and either asked the question "What time does your shop close?" or the question *"At* what time does your shop close?". The answers were likely to mirror the form of the question. When asked "At what . . . ?", answers tended to begin with the preposition 'at' (e.g., "At five o'clock."). Conversely, when asked "What . . . ?", answers tended to begin without the preposition (e.g., "Five o'clock."). Similar alignment phenomena can be observed in many aspects of speech production *inter alia* in syntactic and lexical choice.

Pickering and Garrod [19] present the *interactive alignment model* bringing together all alignment phenomena of speech processing in dialogue. According to this model, human language comprehension and production are greatly facilitated by alignment of the interlocutors during conversation. The process of alignment is explained through mutual priming of the interlocutors' linguistic representations. Thus, it is automatic, efficient, and non-conscious. A stronger

E. Krahmer, M. Theune (Eds.): Empirical Methods in NLG, LNAI 5790, pp. 85–104, 2010.

claim of the authors is that alignment — in combination with routines and a dialogue lexicon — is a prerequisite for fluent speech production in humans.

Alignment effects also occur in human–computer interaction. Brennan [7] and Branigan et al. [6] present evidence that syntactic structures and lexical items used by a computer are subsequently adopted by users. For this reason, alignment is an important concept for natural language human–computer interaction in general, and for dialogue systems with natural language generation in particular. Integrating ideas from the interactive alignment model into the microplanning component of natural language generation systems should be beneficial for several reasons. First, microplanning may become more efficient since the subsets of rules or lexical items in the dialogue lexicon that have been used before can be preferentially searched. Second, due to *self*-alignment, the output of the system can become more consistent and thus easier to understand for the user. Finally, mutual alignment of user and dialogue system might make the conversation itself more natural and, presumably, cognitively more lightweight for the user.

In this chapter we present a computational model for parts of the interactive alignment model that are particularly important in the context of natural language generation. We describe how this model has been incorporated into the existing SPUD *lite* system [22,23] to yield the *alignment-capable* microplanner SPUD *prime*. In Sect. 2 we describe previous approaches to integrate alignment into natural language generation. In Sects. 3 and 4, we present our priming-based model of alignment and its implementation in SPUD *prime*. In Sect. 5, we demonstrate that SPUD *prime* works as specified and describe and discuss the results of an empirical evaluation study on two corpora of task-oriented dialogue. In Sect. 6 we discuss our work and in Sect. 7 we conclude and describe possible future directions.

2   Related Work

Computational modelling is an important methodology for evaluating and testing psycholinguistic theories. Thus, it is certainly not a new idea to implement the interactive alignment model computationally. Indeed, a call for 'explicit computational models' is made as early as in the open peer commentary on Pickering and Garrod's paper [19].

Brockmann et al. [9] and Isard et al. [13] present a 'massive over-generation' approach to modelling alignment and individuality in natural language generation. Their system generates a huge number of alternative sentences — up to 3000 — and evaluates each of these sentences with a trigram model consisting of two parts: a default language model computed from a large corpus and a cache model which is calculated from the user's last utterance. The default language model is linearly interpolated with the cache model, whose influence on the resulting combined language model is determined by a weighting factor $\lambda \in [0,1]$ that controls the amount of alignment the system exhibits.

Purver et al. [20] take a more formal approach. They use an implementation of the Dynamic Syntax formalism, which uses the same representations and mechanisms for parsing as well as for generation of natural language, and extend it with a model of context. In their model, context consists of two distinct representations: a record of the semantic trees generated and parsed so far and a record of the transformation actions used for the construction of these semantic trees. *Re-use* of semantic trees and actions is used to model many dialogue phenomena in Dynamic Syntax and can also explain alignment. Thus, the authors declare alignment to be a corollary of context re-use. In particular, re-use of actions is assumed to have a considerable influence on alignment in natural language generation. Instead of looking through the complete lexicon each time a lexical item is chosen, this kind of lexical search is only necessary if no action — which constructed the same meaning in the given context before — exists in the record. If such an action exists, it can simply be re-used, which obviously leads to alignment.

A completely different approach to alignment in natural language generation is presented by de Jong et al. [15], whose goal is to make a virtual guide more believable by aligning to the user's level of politeness and formality. In order to achieve this, the virtual guide analyses several features of the user's utterance and generates a reply with the same level of politeness and formality. According to the authors, lexical and syntactic alignment occur automatically because the lexical items and syntactic constructions to choose from are constrained by the linguistic style adopted.

Finally, Bateman [1] advocates another proposal according to which alignment in dialogue is predictable because communication is an inherently social activity. Following the social-semiotic view of language, Bateman suggests to model alignment as arising from register and micro-register. More specifically, in his opinion priming of a linguistic representation is comparable with pre-selecting a micro-register that must be considered when generating an utterance in a particular social context.

The approaches presented above primarily focus on the linguistic and social aspects of alignment in natural language generation. The work of Brockmann et al. [9] and Isard et al. [13] concentrates on the surface form of language, Bateman [1] sees alignment arising from social-semiotic aspects, and Purver et al. [20] are primarily interested in fitting alignment into a formal linguistic framework. In this paper we adopt a more psycholinguistic and cognitive stance on alignment. Pickering and Garrod [19] suggest that low-level priming is the basic mechanism underlying interactive alignment. Here, we propose that computational modelling of these priming mechanisms also opens up an interesting and new perspective for alignment in natural language generation.

3   A Priming-Based Model of Alignment

We are interested here in those parts of the interactive alignment model that are most relevant for microplanning in natural language generation and it is out of our scope to model all the facets and details of direct/repetition priming in the alignment of linguistic representations. Exact timing effects, for instance, are likely to be not very relevant as, in an actual system, it does not matter how

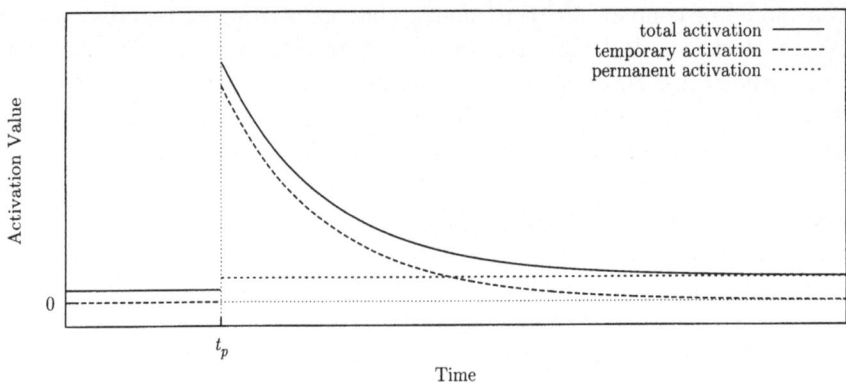

Fig. 1. Change of activation values of a linguistic structure primed at the point of time t_p. In this example, the total activation value is simply the sum of the temporary and the permanent activation values.

many milliseconds faster the retrieval of a primed lexical item is in contrast to the retrieval of an item that is not primed. For this reason we adopt an idealised view, in which priming of linguistic structures results from two basic activation mechanisms:

Temporary activation. This kind of activation should increase abruptly and then decrease slowly over time until it reaches zero again.

Permanent activation. This kind of activation should increase by a certain quantity and then maintain the new level.

Figure 1 shows how the different activation values should change over time when primed at the point of time t_p.

The two mechanisms of priming are in accordance with empirical findings. Branigan et al. [5] present evidence for rapid decay of activation of primed syntactic structures, whereas Bock and Griffin [4] report evidence for their long(er) term activation. In any case, Reitter [21] found both types of priming in his analysis of several corpora, with temporary activation being the more important one. The assumption that both mechanisms play a role in dialogue is also supported by Brennan and Clark [8] whose terminology will be followed in this paper: temporary priming will be called "recency of use effects" and permanent priming will be called "frequency of use effects".

Reitter [21] assumes the repetition probability of primed syntactic structures to depend logarithmically on the distance between priming and usage. Here, we model recency of use effects by a more general *exponential decay* function, modified to meet the needs for modelling activation decay of primed structures:

$$ta(\Delta r) = \exp\left(-\frac{\Delta r - 1}{\alpha}\right), \tag{1}$$

$$\Delta r \in \mathbb{N}^+; \ \alpha > 0; \quad ta \in [0, 1]$$

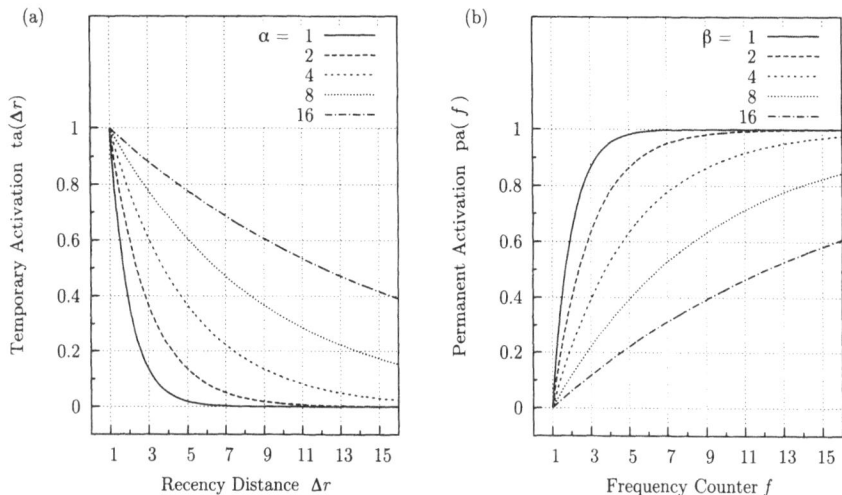

Fig. 2. Plots of the mathematical functions that model recency and frequency effects. Plot (a) displays temporary activation depending on the recency of priming. Plot (b) shows permanent activation depending on the frequency count. Both are shown for different values of the slope parameter α respectively β.

$ta(\Delta r)$ is the temporary activation value of a linguistic structure depending on the distance Δr between the current time T and the time r at which the structure was primed. The slope of the function is determined by the parameter α. Additionally, the function is shifted right in order to yield an activation value of 1 for $\Delta r = 1$. This shift is due to the assumption of discrete time steps with a minimal distance of 1. A plot of $ta(\Delta r)$ with different values for α is given in Fig. 2a.

Using exponential decay to model temporary activation appears to be a sensible choice that is often used to model natural processes. The advantage of this model of temporary activation lies in its flexibility. By changing the slope parameter α, different empirical findings as well as variation among humans can be modelled easily.

Next, a mathematical model for frequency of use effects is needed. To prevent that frequency effects lead to an ever increasing activation value, a maximum activation level exists. This is also found in Reitter's corpus studies [21], which indicate that the frequency effect is inversely connected to the recency effect. Here, we model frequency effects with a general *exponential saturation* function, modified to meet the requirements for modelling permanent activation of linguistic structures:

$$pa(f) = 1 - \exp\left(-\frac{f-1}{\beta}\right),$$ (2)

$$f \in \mathbb{N}^+; \quad \beta > 0; \quad pa \in [0, 1]$$

The most important point to note here is that the permanent activation value $pa(f)$ is not a function of time but a function of the frequency-counter f

attached to each linguistic structure. Whenever a structure is primed, its counter is increased by the value of 1. Again, the slope of the function is determined by the parameter β and the function is shifted right in order to get an activation value of 0 for $f = 1$. A plot of $pa(f)$ with different slope parameters is given in Fig. 2b. Similar to the advantages of the model of temporary activation, this model for frequency effects is very flexible so that different empirical findings and human individuality can be expressed easily.

Now, both priming models need to be combined for a model of alignment. We opted for a weighted linear combination of temporary and permanent activation:

$$ca(\Delta r, f) = \nu \cdot ta(\Delta r) + (1 - \nu) \cdot pa(f), \tag{3}$$

$$0 \leq \nu \leq 1; \quad ca \in [0, 1]$$

Different values of ν allow different forms of alignment. With a value of $\nu = 0.5$ recency and frequency effects are equally important, with a value of $\nu = 1$ alignment depends on recency only, and with a value of $\nu = 0$ alignment is governed solely by frequency. Being able to adjust the influence of the different sorts of priming on alignment is crucial as it has not yet been empirically determined to what extent recency and frequency of use affect alignment (in Sects. 5.3 and 5.4 we will exploit this flexibility for matching empirical data).

In contrast to the models of alignment presented in Sect. 2, the computational alignment model presented here will not only consider alignment between the interlocutors (interpersonal- or *other-alignment*), but also alignment to oneself (intrapersonal- or *self-alignment*). Pickering et al. [18] present results from three experiments which suggest self-alignment to be even more important than other-alignment. In our model, self-alignment is accounted for with the same priming-based mechanisms. To this end, four counters are attached to each linguistic structure:

- Δr_s: recency of use by the system itself
- Δr_o: recency of use by the interlocutor
- f_s: frequency of use by the system itself
- f_o: frequency of use by the interlocutor

The overall activation value of the structure is a linear combination of the combined activation value $ca(\Delta r_s, f_s)$ and the combined activation value $ca(\Delta r_o, f_o)$ from equation (3):

$$act(\Delta r_s, f_s, \Delta r_o, f_o) = \lambda \cdot \big(\mu \cdot ca(\Delta r_s, f_s) + (1 - \mu) \cdot ca(\Delta r_o, f_o)\big), \tag{4}$$

$$0 \leq \lambda, \mu \leq 1; \quad act \in [0, 1]$$

Again, by changing the factor μ, smooth interpolation between pure self-alignment ($\mu = 1$) and pure other-alignment ($\mu = 0$) is possible, which can account for different empirical findings or human individual differences. Furthermore, the strength of alignment is modelled with a scaling factor λ, which determines whether alignment is considered during generation ($\lambda > 0$) or not ($\lambda = 0$).

4   The Alignment-Capable Microplanner SPUD *Prime*

The previously described priming-based model of alignment has been implemented by extending the integrated microplanning system SPUD *lite* [22]. SPUD *lite* is a lightweight Prolog re-implementation of the SPUD microplanning system [23] based on the context-free tree rewriting grammar formalism TAGLET. Not only the microplanner itself, but also the linguistic structures (the initial TAGLET trees) are represented as Prolog clauses.

SPUD *lite* carries out the different microplanning tasks (lexical choice, syntactic choice, referring expression generation and aggregation) at once by treating microplanning as a search problem. During generation it tries to find an utterance that is in accordance with the constraints set by its input (a grammar, a knowledge base and a query). This is done by searching the search space spanned by the linguistic grammar rules and the knowledge base until a goal state is found. Non-goal search states are preliminary utterances that are extended by one linguistic structure in each step until a syntactically complete utterance is found which conveys all the specified communicative goals. Since this search space is large even for relatively small grammars, a heuristic greedy search strategy is utilised.

Our alignment-capable microplanner SPUD *prime* extends SPUD *lite* in several ways. First, we altered the predicate for the initial TAGLET trees by adding a unique identifier ID as well as counters for self/other-recency/frequency values (r_s, f_s, r_o and f_o; see Sect. 3). The activation value of an initial tree is then calculated with equation (4).

Furthermore, we have created a mechanism that enables SPUD *lite* to change the recency and frequency information attached to the initial trees on-line during generation. This is done in three steps with the help of Prolog's meta-programming capabilities: First, the clause of a tree is retrieved from the knowledge base. Second, it is retracted from the knowledge base. Finally, the clause is (re-)asserted in the knowledge base with updated recency and frequency information. As a welcome side effect of this procedure, primed initial trees are moved to the top of the knowledge base and — since Prolog evaluates clauses and facts in the order of their appearance in the knowledge base — they can be accessed earlier than unprimed initial trees or initial trees that were primed longer ago. Thus, in SPUD *prime* recency of priming directly influences the access of linguistic structures.

Most important, the activation values of the initial trees are considered during generation. Thus, in addition to the evaluation measures used by SPUD *lite*'s heuristic state evaluation function, the mean activation value

$$\overline{act}(S) = \frac{\sum_{i=1}^{N} act_{t_i}(\Delta r_{s_{t_i}}, f_{s_{t_i}}, \Delta r_{o_{t_i}}, f_{o_{t_i}})}{N}$$

of the N initial trees $\{t_1, \ldots, t_N\}$ of a given search state S is taken into account as a further evaluation measure. Hence, when SPUD *prime* evaluates (otherwise equal) successor search states, the one with the highest mean activation value is chosen as the next current state.

5   Evaluation

In order to show that our priming-based alignment model and its implementation work as intended, we first demonstrate that SPUD *prime* is in principle capable of lexical and syntactic alignment as well as that it can display recency and frequency of use effects by simulating some — admittedly rather artificial — interactions (Sect. 5.1). Having established these abilities, we then evaluate SPUD *prime* empirically on two corpora collected in two psycholinguistic experiments designed to investigate the alignment behaviour of humans in a controlled fashion (Sects. 5.2–5.4).

5.1   Demonstrating Lexical and Syntactic Alignment

A simple demonstration that SPUD *prime* displays alignment phenomena is to do tests that resemble the course of psychological experiments (e.g., Bock [3]) where subjects are primed and the influence of the prime is observed in their verbal behaviour. This can be done in four steps for, e.g., lexical alignment:

1. Querying SPUD *prime* to generate an utterance u_1 from the communicative goals CG: Utterance u_1 uses lexical item l_1.
2. Priming a lexical item l_2 that is synonymous to lexical item l_1.
3. Querying SPUD *prime* to generate an utterance u_2 from the same communicative goals CG.
4. Analysing utterance u_2: if it uses the primed lexical item s_2, then SPUD *prime* displays lexical alignment, otherwise it does not.

In the following we use this and similar tests in order to demonstrate that SPUD *prime* displays lexical alignment, syntactic alignment as well as recency and frequency of use effects. The steps above are translated into commands for SPUD *prime* (used together with a small German TAGLET grammar in a landmark description domain). The model is set up with the parameters $\alpha = 2, \beta = 16, \lambda = 1$, $\mu = 0.6$ and $\nu = 0.8$ — weighting self-alignment stronger than other-alignment and recency effects stronger than frequency effects. Anyway, the parameter setting is not too important in this demonstration as we just want to show that SPUD *prime* is in principle capable of displaying alignment phenomena.

Lexical Alignment. SPUD *prime*'s ability to display lexical alignment can be demonstrated by following steps 1–4 directly.

(1a)   The following knowledge base is loaded:

```
shared(entity(church-1, single)).
shared(instance_of(church-1, church)).
private(entity(window-7, single)).
private(instance_of(window-7, window)).
private(property(window-7, round)).
private(part_of(church-1, window-7)).
```

(it states that there exists a church church-1 that has a round window window-7).

(1b) SPUD *prime* is requested to generate an utterance that communicates the structure of `church-1`:

```
spudprime(initial_state(structure(church-1), _,
    [part_of(church-1, window-7),
     entity(window-7, single),
     instance_of(window-7, window),
     property(window-7, round)]), W).
```

For this request, SPUD *prime* generates the output *'Die Kirche hat ein rundes Fenster.'* ('The church has got a round window.').

(1c) Now it is pretended that an interlocutor uses the (more or less) synonymous lexical item *'kreisförmig'* ('circular') instead of *'rund'* ('round'). The initial tree of the lemma *'kreisförmig'* (`dynlex-502`) is therefore primed with the following SPUD *prime* query:

```
sp_fake_interlocutor_rule_usage(dynlex-502).
```

(1d) SPUD *prime* is requested to regenerate the utterance with the same query used in (1b). This time the output *'Die Kirche hat ein kreisförmiges Fenster.'* ('The church has got a circular window.') is generated.

To conclude, after priming rule `dynlex-502` the corresponding lexical item in the utterance changes as the model predicts. In (1b) SPUD *prime* uses the word *'rund'* (because it happens to be easier to access in the knowledge base), in (1d) it uses the synonymous word *'kreisförmig'* (because it has a higher activation). Hence, SPUD *prime* displays lexical alignment.

Syntactic Alignment. Since lexicon and syntax are represented uniformly in TAGLET, the steps to demonstrate syntactic alignment are the same as for lexical alignment.

(2a) The following knowledge base is loaded:

```
shared(entity(church-1, single)).
shared(instance_of(church-1, church)).
private(relpos(church-1, left)).
```

(it states that there exists a church `church-1` that is on the left side).

(2b) SPUD *prime* is requested to generate an utterance that communicates the position of `kirche-1`:

```
spudprime(initial_state(position([church-1]), _,
    [relpos(church-1, left)]), W).
```

For this request, SPUD *prime* generates the output *'Die Kirche ist auf der linken Seite.'* ('The church is on the left side.')

(2c) Now it is pretended that an interlocutor uses a different syntactic construction. The initial tree of that construction (`rule-522`) is therefore primed with the following SPUD *prime* query:

`sp_fake_interlocutor_rule_usage(rule-522).`

(2d) Lastly, SPUD *prime* is requested to regenerate the utterance with the same query used in (2b). This time the output *'Auf der linken Seite ist die Kirche.'* ('On the left side is the church.') is generated.

To conclude, after priming `rule-522`, the syntactic structure of the utterance changes as the model predicts. In (2b) SPUD *prime* uses initial tree `rule-526` to generate the utterance (again, because it happens to be easier to access in the knowledge base), in (2d) it uses the primed initial tree `rule-522` (again, because it has a higher activation). Hence, SPUD *prime* displays syntactic alignment.

Recency and Frequency Effects. In the two previous tests, priming with `sp_fake_interlocutor_rule_usage/1` changes both the recency and the frequency information that is attached to initial trees (cf. Sects. 3 and 4). However, the interesting aspect of recency and frequency of use is its behaviour over time, which we test here. To simplify matters, this test is based on the first one.

(3a) See (1a).

(3b) See (1b). Additionally, it is assumed that the self-frequency f_s of `dynlex-9` (*'rund'*) has the value 10 instead of 1. This can be set with the SPUD *prime* query

`sp_set_frecency(dynlex-9, 1, 10, 1,1).`

which sets a given rule's four counters to new values (*frecency* is short for 'frequency and recency'). In this case Δr_s is set to 1, f_s to 10 and Δr_o and f_o to 1. As all counters defaulted to 1, only f_s changed.

(3c) See (1c).

(3d) See (1d).

(3e) Now it is pretended that some time goes by. This can be set with the query

`sp_increase_recency_counter(10).`

which increases the current point of time T by a value of 10.

(3f) Finally, SPUD *prime* is requested to regenerate the utterance with the same query used in (3d). This time the output *'Die Kirche hat ein rundes Fenster.'* ('The church has got a round window.') is generated again.

To conclude, similar to test 1, the lexical item *'kreisförmig'* primed in (3c) is used in the utterance generated in (3d) — although the lexical item *'rund'* has a frequency value five times as high. This demonstrates that recency is more important than frequency of use (given the chosen parameters we use here). After a short period of time (3e) SPUD *prime* uses the word *'rund'* again in (3f): the temporary activation based on the recency value has decayed and permanent activation based on the frequency value is higher again. Hence, SPUD *prime* displays both, recency and frequency of use effects.

The three tests show that SPUD *prime* displays the alignment phenomena predicted by our priming based model of alignment: syntactic and lexical alignment as well as recency and frequency of use effects.

5.2 Empirical Evaluation Method

For the empirical evaluation of our priming-based model of alignment and its implementation in SPUD *prime* we use two small corpora of recorded and transcribed spoken dialogues between human interlocutors. These were collected in two psycholinguistic experiments designed to investigate the alignment behaviour of humans in a controlled fashion. The participants' task was to play the 'Jigsaw Map Game', in which different objects have to be placed correctly on a table. Each participant has a unique set of cards and a box of objects and they take turns in giving each other instructions of how to place the next object in relation to the objects that are already on the table (cf. Weiß et al. [24,25]).

In our evaluation, we concentrate on the generation of the object names (i.e., nouns), by simulating their usage in the dialogues. In each simulation run, SPUD *prime* plays the role of one of the two speakers interacting with a simulated interlocutor who behaves exactly as in the real experiments. With this test setup we examined, first, how well SPUD *prime* can model the alignment behaviour of a real speaker in a real dialogue context and, second, whether our model is flexible enough to consistently emulate different speakers with different alignment behaviour.

In order to find the best model, i.e., the best point $(\alpha, \beta, \mu, \nu)$ in parameter space, for each speaker, we simulated all tests with all parameter combinations and counted the number of mismatches between our model's choice and the real speaker's choice. To make this exhaustive search possible, we limit the set of values for the parameters α and β to $\{1, 2, 4, 6, 8, 10, 14, 18, 24, 30\}$ and the set of values for the parameters μ and ν to $\{0, 0.1, 0.2, \ldots, 1\}$, resulting in a total of $11^2 \times 10^2 = 12100$ points in parameter space. Since we want to investigate alignment, λ is constantly set to 1.

In the next section (5.3) we describe the evaluation on the first corpus and give an example of how it is done. Thereafter we describe the evaluation on the second corpus (Sect. 5.4).

5.3 Corpus 1: Learning of Referring Nouns

The first corpus that we used consists of eight recorded and transcribed dialogues between pairs of two interlocutors — named (A) and (B) — that play the 'Jigsaw Map Game'. Each speaker learned[1] a set of object names before playing the

---

[1] The participants had to learn the object names in the following way: First, a list of the objects and their names was presented to them. Second, after reading the task instructions, the same list was shown to them again. Finally, they had to demonstrate that they memorised the names by naming the objects twice, in a written 'test', and verbally to the experimenter.

Table 1. Sequence of referring target nouns used by participants (A) and (B) in our example dialogue 7

	B: der Klotz		B: der Ball		der Klotz
1	A: die Spielfigur	11	A: der Ball	18	A: das Männchen
2	der Klotz	12	der Ball	19	der Klotz
	B: das Männchen		B: die Kugel		B: das ännchen
	der Klotz		das Männchen	20	A: der Ball
3	A: die Spielfigur	13	A: der Ball	21	A: das Männchen
	B: das Männchen		B: die Kugel		B: der Ball
4	A: das Männchen	14	A: der Klotz		das Männchen
5	das Männchen	**15**	**A: die Kugel**	**22**	**A: die Kugel**
6	das Männchen	16	der Klotz	23	A: der Ball
7	das Männchen		B: der Klotz		B: der Klotz
8	das Männchen		die Kugel	24	A: der Ball
	B: das Männchen		der Klotz		B: der Klotz
9	A: das Männchen	17	A: der Klotz	25	A: der Klotz
10	der Ball		B: das Männchen		———

game, such that both use the same names for all but three objects.[2] Due to this precondition, both speakers use the same lexical referring expressions for most objects and the speaker's lexical alignment behaviour for the differently named objects can be observed easily. The experiment is described in further detail in Weiß et al. [24].

To illustrate our evaluation method, we first present and discuss the simulation of one particular dialogue (number 7) from the corpus from the perspective of participant (A). Both interlocutors learned the object names *'Raute'* ('rhombus'), *'Ring'* ('ring'), *'Schraube'* ('bolt') and *'Würfel'* ('dice'), additionally participant (A) learned *'Spielfigur'* ('token'), *'Ball'* ('sphere') and *'Block'* ('cuboid') and participant (B) learned *'Männchen'* ('token'), *'Kugel'* ('sphere') and *'Klotz'* ('cuboid'). In our simulation, we focus on the use of the differently learned names (the *targets*) and not on the other names (the *non-targets*). Table 1 shows the sequence of target nouns as they occurred in one of the real dialogues (non-targets omitted).

For each point in parameter space $(\alpha, \beta, \mu, \nu)$ the dialogue is simulated in the following way:

– When participant (A) referred to a *target* object in the dialogue, SPUD *prime* is queried to generate a noun for the target object and the corresponding rule(s) are primed automatically. Then it is recorded whether the noun actually generated is the noun used in the actual dialogue (match) or not (mismatch).

---

[2] Note, however, that the participants were not explicitly instructed to use the learned object names during the experiment.

- When participant (A) used a *non-target* object name in the dialogue, self-priming of the corresponding rule(s) in SPUD *prime*'s knowledge base is simulated (i.e., the recency and frequency counters are increased).
- When participant (B) used an object name (target or non-target), priming of the corresponding rule(s) in SPUD *prime*'s knowledge base is simulated.

The evaluation measure for a specific point in parameter space is the number of mismatches it produces when simulating a dialogue. Thus the point (or rather points) in parameter space that produce the least number of mismatches are the ones that best model the particular speaker under consideration. For participant (A) of our example dialogue the distribution of points in parameter space p producing m mismatches is shown in Table 2. Four points in parameter space produce only two mismatches (in phrase 15 and 22; cf. Table 1) and thus our priming-based alignment model can account for 92% of the target nouns produced by speaker (A). However, it must be noted that these two mismatches occur at points in the dialogue where the alignment behaviour of (A) is not straightforward. At target noun 15, both interlocutors have already used the name *'Ball'* and then both switch to *'Kugel'*. The mismatch at target 22 is a special case: (A) used *'Kugel'* and immediately corrected himself to *'Ball'*, the name he learned prior to the experiment. In this case it seems as if (A) suddenly remembers the learning phase before and after the task instructions.

We simulated the noun production for each of the interlocutors from the first corpus. One dialogue has been excluded from the data analysis as the dialogue partners used nouns that none of them had learned in the priming phase. For each of the remaining 14 interlocutors we varied the parameters α, β, μ and ν as described above to identify those point(s) in parameter space that result in the least number of mismatches.

Each interlocutor produced between 18 and 32 target nouns ($N = 14$, $M = 23.1$, SD $= 3.9$). Our simulation runs contain between 0 and 19 mismatches overall ($N = 169400$, $M = 6.4$, SD $= 3.4$). The minimal number of mismatches for each speaker simulation ranges between 0 and 6 ($N = 14$, $M = 2.3$, SD $= 1.7$). That is, our model can simulate a mean of 89.8% of all target nouns ($N = 14$, Min $= 66.7\%$, Max $= 100.0\%$, SD $= 8.2\%$), which is an improvement of 24.6% on the baseline condition (alignment switched off), where 65.3% of the target nouns are generated correctly ($N = 14$, Min $= 36.0\%$, Max $= 100.0\%$, SD $= 7.1\%$). As already illustrated in the example simulation, mismatches typically occur at points in the dialogue where the alignment behaviour of the human interlocutor is not straightforward.

Table 2. Number of points in parameter space p leading to m mismatches for participant (A) in dialogue 7

No. of Mismatches (m)	0	1	2	3	4	5	6	7	8	9	10	...
Points in par. space (p)	0	0	4	833	3777	2248	3204	1105	478	148	294	0

Table 3. Mean parameter values for those simulation runs that result in a minimal number of mismatches for each speaker of the first corpus (T = number of targets, m = least number of mismatches, % = percentage of targets that could be simulated, # p = number of points in parameter space that lead to m mismatches)

	T	m	%	# p	α M	α SD	β M	β SD	μ M	μ SD	ν M	ν SD
VP13	25	2	92.0	4	3.00	1.16	19.50	9.15	0.300	0.000	0.100	0.000
VP14	19	1	94.7	72	5.53	1.52	14.32	9.61	0.819	0.040	0.901	0.108
VP17	25	1	96.0	200	1.66	0.82	12.94	9.53	0.353	0.169	0.955	0.069
VP18	22	3	86.4	2445	15.37	8.76	10.98	9.76	0.597	0.211	0.706	0.236
VP19	22	0	100.0	4321	11.81	9.49	11.01	8.93	0.824	0.148	0.388	0.291
VP20	18	2	88.9	8	1.00	0.00	15.75	9.29	0.738	0.052	0.388	0.146
VP23	18	6	66.7	987	6.85	6.68	12.08	9.35	0.331	0.374	0.400	0.330
VP24	29	3	89.7	256	12.95	9.70	13.63	8.94	0.538	0.201	0.468	0.298
VP39	32	5	84.4	1	1.00	0.00	2.00	0.00	0.900	0.000	0.800	0.000
VP40	26	0	100.0	3504	12.08	9.33	10.30	8.75	0.843	0.147	0.343	0.282
VP41	21	2	90.5	609	11.37	8.48	15.34	8.92	0.770	0.106	0.655	0.213
VP42	22	3	86.4	30	6.00	1.49	17.53	9.02	0.783	0.059	0.760	0.122
VP47	20	2	90.0	326	13.75	7.79	13.53	9.51	0.772	0.095	0.816	0.166
VP48	24	2	91.7	2478	12.87	9.55	10.74	8.54	0.764	0.175	0.166	0.148
M	23.1	2.3	89.8	1089	8.23	5.75	12.83	9.18	0.666	0.137	0.560	0.185
SD	4.1	1.7	8.2	1468	5.20	4.01	4.12	0.37	0.206	0.097	0.274	0.099

As displayed in Table 3 the parameter assignments resulting in least mismatches differ considerably from speaker to speaker. However, there are some remarkable trends to be observed in the data. As concerns the parameter μ, which determines the combination of self- and other-alignment, the majority of values are in the upper range of the interval $[0,1]$. For 8 of 14 speakers the mean is above 0.7 with relatively low standard deviations. Only for one speaker (P13) the mean μ is below 0.3. The overall mean value of μ is 0.666 ($N = 14$, SD = 0.206). Thus, the parameter values indicate a considerable tendency toward self-alignment in contrast to other-alignment.

For the parameter ν, which interpolates between recency and frequency effects of priming, the results are less revealing. For two speaker simulations (P13 and P48) the mean ν is 0.166 or lower, for another four speaker simulations the mean ν is above 0.7. That is, our model produces good matching behaviour in adopting different alignment strategies, depending either primarily on frequency or recency, respectively. All other simulations, however, are characterised by a mean ν in the medium range along with a relatively high standard deviation. The mean ν of all speakers is 0.560 ($N = 14$, SD = 0.274).

One shortcoming of the first experiment and corpus is that participants explicitly learned the object names prior to the game, which is a clear difference from the alignment effects (acquisition of object names by lexical priming) that occur during the game itself. The sudden remembrance of the learning phase (this object is called *'Ball'*) mentioned in the example above might be one consequence of this. Furthermore, it could not be controlled how eager the participants were in learning the names, so it is not clear how the rules in SPUD *prime*'s knowledge base should be initialised. Our — somewhat arbitrary — decision was to prime the rules (i.e., increase their recency and frequency counters prior to the simulation) for the learned object names three times.

5.4   Corpus 2: Implicit Acquisition of Referring Nouns

In order to overcome the shortcomings just mentioned, a second and slightly different study was conducted. The corpus collected in this experiment consists of 12 interactions, each divided into two parts [25]. As before, participants played the 'Jigsaw Map Game', but this time they did not learn the object names explicitly. In the first part of each interaction (the *priming phase*), a naive participant (A) played the game with a confederate (C) that was instructed to use specific object names so that (A) could acquire them implicitly through lexical priming. In the second part of each interaction (the *usage phase*), participant (A) then played the game with a second naive participant (B).

For the evaluation on this corpus, we, again, simulated the dialogues for each point in parameter space $(\alpha, \beta, \mu, \nu)$, first simulating the priming phase followed by a simulation of the usage phase. In the priming phase the dialogue was simulated in the following way:

- When the confederate (C) used an object name (target or non-target) in the dialogue, priming of the corresponding rule(s) in SPUD *prime*'s knowledge base is simulated (i.e., the recency and frequency counters are increased).
- When participant (A) used an object name (target or non-target), self-priming of the corresponding rule(s) in SPUD *prime*'s knowledge base is simulated.

Grounded in the knowledge base from the priming phase, the usage phase dialogue between participants (A) and (B) was simulated — from the perspective of (A) only — in the same way described in Sect. 5.3.

In the usage phase, participants (A) produced between 9 and 22 target nouns ($N = 12$, $M = 14.2$, SD $= 3.5$). Our simulation runs contain between 0 and 15 mismatches overall ($N = 145200$, $M = 5.9$, SD $= 3.5$). The minimal number of mismatches for each speaker simulation ranges between 0 and 7 ($N = 12$, $M = 2.6$, SD $= 1.8$). That is, our model can simulate a mean of 81.9% of all target nouns ($N = 12$, Min $= 56.3\%$, Max $= 100.0\%$, SD $= 12.2\%$), which is an improvement of 17.3% on the baseline condition (alignment switched off), where 64.3% of the target nouns are generated correctly ($N = 12$, Min $= 22.2\%$, Max $= 92.3\%$, SD $= 21.9\%$).

As in the first corpus evaluation, the parameter assignments resulting in least mismatches differ considerably from speaker to speaker (Table 4). Furthermore, comparing the results of this evaluation to the results of the first (Table 3) reveals further similarities. There is no significant difference in the mean least number

Table 4. Mean parameter values for those simulation runs that result in a minimal number of mismatches for each speaker (A) in the usage phase of the second corpus (T = number of targets, m = least number of mismatches, % = percentage of targets that could be simulated, # p = number of points in parameter space that lead to m mismatches)

					α		β		μ		ν	
	T	m	%	# p	M	SD	M	SD	M	SD	M	SD
V1	16	3	81.3	1697	18.06	8.11	13.75	9.80	0.221	0.163	0.676	0.252
V4	9	3	66.7	477	13.20	9.69	12.58	8.38	0.096	0.101	0.100	0.093
V6	13	1	92.3	2967	9.66	7.70	14.10	9.51	0.498	0.179	0.772	0.193
V7	16	7	56.3	6221	12.54	9.35	10.62	8.80	0.678	0.271	0.335	0.282
V8	16	3	81.3	417	18.06	6.88	14.25	9.87	0.114	0.102	0.851	0.120
V19	11	1	90.9	119	5.33	1.67	11.77	9.51	0.761	0.063	0.707	0.136
V33	13	2	84.6	151	24.04	5.58	14.72	9.09	0.313	0.110	0.637	0.195
V34	14	3	78.6	39	24.41	4.50	16.41	0.06	0.621	0.047	0.933	0.081
V35	17	1	94.1	1582	14.38	7.96	12.91	9.51	0.286	0.234	0.663	0.206
V36	11	0	100.0	928	7.63	3.24	11.90	9.40	0.319	0.235	0.619	0.184
V37	12	3	75.0	2774	12.98	8.28	13.69	9.57	0.276	0.215	0.781	0.200
V38	22	4	81.8	2118	7.12	7.32	11.97	9.28	0.193	0.158	0.547	0.211
M	14.2	2.6	81.9	1624	13.95	6.69	13.22	9.40	0.365	0.156	0.635	0.179
SD	3.5	1.8	12.2	1777	6.24	2.46	1.58	0.47	0.222	0.073	0.227	0.061

of mismatches and the mean coverage between the two evaluation studies (2.3 in the first versus 2.6 in the second and 89.8% versus 81.9%). There is also no significant difference in the number of points in parameter space that lead to the least number of mismatches (1089 versus 1624) in the values β (means are 12.83 versus 13.22) and ν (means are 0.560 versus 0.635).

One remarkable difference, however, can be observed between the two μ values which control the relation of self- and other-alignment. Their means for the simulation runs with least number of mismatches for the first corpus evaluation are significantly higher than those for the second corpus evaluation (t-Test: $t = -3.574$, $df = 22.733$, $p < 0.001$), i.e., while participants in the first experiment aligned more to themselves, participants in the second experiment aligned more to their interlocutors. This noteworthy result indicates that the model actually reflects the participants alignment behaviour: Participants in the first experiment were focussed on *their* object names since they activated the corresponding lexical representation through explicit learning prior to the game. Participants in the second experiment on the contrary did not have such highly activated lexical representations and thus they aligned to their interlocutor more easily.

6   Discussion

The evaluation shows that SPUD *prime* and its underlying priming-based model of alignment are capable of simulating alignment phenomena found in psycholinguistic studies. Modelling lexical and syntactic alignment, as well as the effects of recency and frequency of use, it can account for a high degree of the lexical choices participants made in the two 'Jigsaw Map Game' tasks. A few points merit closer inspection.

First, it must be noted that the participants' behaviour (i.e., the behaviour producing the least number of mismatches) could in general be simulated not only by a single point or a compact cluster of points in parameter space, suggesting that the parameters are either not completely independent or that the evaluation method is too simple. Future empirical evaluations should have a wider scope and go beyond the generation of lexical items. SPUD *prime* could for instance be evaluated generating more sophisticated referring expressions. Yet, having several points in parameter space that achieve a certain behaviour is not problematic in an application context where the parameters can simply be set according to the theoretical model and the desired behaviour.

Second and interestingly, the parameters that lead to minimal numbers of mismatches in simulations can differ considerably between participants. Individual differences exist in verbal and non-verbal behaviour: Dale and Viethen [11] (this volume) report individual variation between referring expressions produced by human subjects describing a scene of geometrical objects and Bergmann and Kopp observe in their analysis of an extensive corpus of speech and gesture data that speakers differ significantly in the way they produce iconic gesture [2]. It can be expected that individual differences exist in speakers' alignment behaviour, too (cf. [12] for evidence). Here, data about participants' personalities should be collected so that a correlation with personality traits is possible.

Third, our model could generate a high number of the target nouns correctly, but failed on 10–20%. It should be noted, however, that it tries to give a purely mechanistic explanation of lexical and syntactic choice (in the spirit of Pickering and Garrod's interactive alignment model [19]) and that it, therefore, cannot explain alignment phenomena that are due to social factors (e.g., politeness, relationship, etc.), audience design or cases in which a speaker consciously decides whether to align or not (e.g., whether to use a word or its synonym). This is the main difference between our priming-based alignment model and the model of Janarthanam and Lemon [14] (this volume), which treats alignment from an audience design perspective (cf. Clark [10]). We think that a comprehensive model of alignment that accounts for all phenomena must unify both perspectives: low-level mechanistic alignment that is both rapid and broad in scope as well as more high-level strategic alignment that can account for audience design and social practices. How these two types of alignment could interact and influence each other is an open question.

7   Conclusion

In this paper, we introduced a priming-based model of alignment that focusses more on the psycholinguistic aspects of interactive alignment, and models recency and frequency of use effects — as proposed by Reitter [21] and Brennan and Clark [8] — as well as the difference between intrapersonal and interpersonal alignment [18,19]. The presented model is fully parameterisable and can account for different empirical findings and 'personalities'. It has been implemented in the SPUD *prime* microplanner which activates linguistic rules by changing its knowledge base on-line and considers the activation values of those rules used in constructing the current utterance by using their mean activation value as an additional feature in its state evaluation function.

We evaluated our alignment model and its implementation in SPUD *prime* on two corpora of task-oriented dialogue collected in experimental setups especially designed for alignment research. The results of this evaluation show that our priming-based model of alignment is flexible enough to simulate the alignment behaviour of different human speakers (generating target nouns) in the experimental settings. Our model can reproduce human alignment behaviour to a high degree, but it remains to be investigated which influence each parameter exerts and how exactly the parameters vary across individual speakers.

Nevertheless, the development of the alignment-capable microplanner is only one step in the direction of an intuitive natural language human–computer interaction system. In order to reach this goal, the next step is to combine SPUD *prime* with a natural language understanding system, which should ideally work with the same linguistic representations so that the linguistic structures used by the interlocutor could be primed automatically. This work is underway.

Furthermore, user studies should be carried out in order to evaluate SPUD *prime* in interactive scenarios. Branigan et al. [6] found that human–computer alignment was even stronger than human–human alignment. But how would the alignment behaviour of human interlocutors change if the computer they are speaking to also aligns to them? Further, would integration of an alignment-capable dialogue system into a computer interface make the interaction more natural? And would an embodied conversational agent appear more resonant and more sociable [16], if it aligned to users during conversation? The work presented here provides a starting point for the investigation of these questions.

Acknowledgements. This research is supported by the Deutsche Forschungs-gemeinschaft (DFG) in the Center of Excellence in 'Cognitive Interaction Technology' (CITEC) as well as in the Collaborative Research Center 673 'Alignment in Communication'. We thank Petra Weiß for making the 'Jigsaw Map Game' corpora available and Mariët Theune for some very helpful comments on the draft of this chapter.

References

1. Bateman, J.A.: A social-semiotic view of interactive alignment and its computational instantiation: A brief position statement and proposal. In: Fischer, K. (ed.) How People Talk to Computers, Robots and Other Artificial Communication Partners, Bremen, Germany, pp. 157–170 (2006), SFB/TR 8 Report No. 010-09/2006
2. Bergmann, K., Kopp, S.: GNetIc – Using bayesian decision networks for iconic gesture generation. In: Proceedings of the 9th International Conference on Intelligent Virtual Agents, Amsterdam, The Netherlands, pp. 76–89 (2009)

3. Bock, J.K.: Syntactic persistence in language production. Cognitive Psychology 18, 355–387 (1986)
4. Bock, J.K., Griffin, Z.M.: The persistence of structural priming: Transient activation or implicit learning? Journal of Experimental Psychology: General 129, 177–192 (2000)
5. Branigan, H.P., Pickering, M.J., Cleland, A.A.: Syntactic priming in written production: Evidence for rapid decay. Psychonomic Bulletin & Review 6, 635–640 (1999)
6. Branigan, H.P., Pickering, M.J., Pearson, J., McLean, J.F.: Linguistic alignment between people and computers. Journal of Pragmatics (in Press)
7. Brennan, S.E.: Conversation with and through computers. User Modeling and User-Adapted Interaction 1, 67–86 (1991)
8. Brennan, S.E., Clark, H.H.: Conceptual pacts and lexical choice in conversation. Journal of Experimental Psychology: Learning, Memory, and Cognition 22, 1482–1493 (1996)
9. Brockmann, C., Isard, A., Oberlander, J., White, M.: Modelling alignment for affective dialogue. In: Proceedings of the Workshop on Adapting the Interaction Style to Affective Factors at the 10th International Conference on User Modeling, Edinburgh, UK (2005)
10. Clark, H.H.: Using Language. Cambridge University Press, Cambridge (1996)
11. Dale, R., Viethen, J.: Attribute-centric referring expression generation. In: Krahmer, E., Theune, M. (eds.) Empirical Methods in NLG. LNCS (LNAI), vol. 5790, pp. 163–179. Springer, Heidelberg (2010)
12. Gill, A.J., Harrison, A.J., Oberlander, J.: Interpersonality: Individual differences and interpersonal priming. In: Proceedings of the 26th Annual Conference of the Cognitive Science Society, Chicago, IL, pp. 464–469 (2004)
13. Isard, A., Brockmann, C., Oberlander, J.: Individuality and alignment in generated dialogues. In: Proceedings of the 4th International Natural Language Generation Conference, Sydney, Australia, pp. 25–32 (2006)
14. Janarthanam, S., Lemon, O.: Learning adaptive referring expression generation policies for spoken dialogue systems. In: Krahmer, E., Theune, M. (eds.) Empirical Methods in NLG. LNCS (LNAI), vol. 5790, pp. 67–84. Springer, Heidelberg (2010)
15. de Jong, M., Theune, M., Hofs, D.: Politeness and alignment in dialogues with a virtual guide. In: Proceedings of the 7th International Conference on Autonomous Agents and Multiagent Systems, Estoril, Portugal, pp. 207–214 (2008)
16. Kopp, S.: Social resonance and embodied coordination in face-to-face conversation with artificial interlocutors. Speech Communication (accepted manuscript)
17. Levelt, W.J.M., Kelter, S.: Surface form and memory in question answering. Cognitive Psychology 14(1), 78–106 (1982)
18. Pickering, M.J., Branigan, H.P., McLean, J.F.: Dialogue structure and the activation of syntactic information. In: Proceedings of the 9th Annual Conference on Architectures and Mechanisms for Language Processing, Glasgow, UK, p. 126 (2003)
19. Pickering, M.J., Garrod, S.: Toward a mechanistic psychology of dialogue. Behavioral and Brain Sciences 27(2), 169–226 (2004)
20. Purver, M., Cann, R., Kempson, R.: Grammars as parsers: Meeting the dialogue challenge. Research on Language and Computation 4, 289–326 (2006)
21. Reiter, D.: Context Effects in Language Production: Models of Syntactic Priming in Dialogue Corpora. Ph.D. thesis, University of Edinburgh (2008)

22. Stone, M.: Lexicalized grammar 101. In: Proceedings of the ACL 2002 Workshop on Effective Tools and Methodologies for Teaching Natural Language Processing and Computational Linguistics, Philadelphia, PA, pp. 77–84 (2002)
23. Stone, M., Doran, C., Webber, B., Bleam, T., Palmer, M.: Microplanning with communicative intentions: The SPUD system. Computational Intelligence 19, 311–381 (2003)
24. Weiß, P., Pfeiffer, T., Schaffranietz, G., Rickheit, G.: Coordination in dialog: Alignment of object naming in the Jigsaw Map Game. In: Proceedings of the 8th Annual Conference of the Cognitive Science Society of Germany, Saarbrücken, Germany, pp. 4–20 (2008)
25. Weiß, P., Pustylnikov, O., Mehler, A., Hellmann, S.M.: Patterns of alignment in dialogue: Conversational partners do not always stay aligned on common object names. In: Proceedings of the Conference on Embodied and Situated Language Processing, Rotterdam, The Netherlands, p. 16 (2009)

Natural Language Generation as Planning under Uncertainty for Spoken Dialogue Systems

Verena Rieser[1] and Oliver Lemon[2]

[1] School of Informatics, University of Edinburgh
vrieser@inf.ed.ac.uk
[2] School of Mathematical and Computer Sciences, Heriot-Watt University
o.lemon@hw.ac.uk
http://www.macs.hw.ac.uk/InteractionLab

Abstract. We present and evaluate a new model for Natural Language Generation (NLG) in Spoken Dialogue Systems, based on statistical planning, given noisy feedback from the current generation context (e.g. a user and a surface realiser). The model is adaptive and incremental at the turn level, and optimises NLG actions with respect to a data-driven objective function. We study its use in a standard NLG problem: how to present information (in this case a set of search results) to users, given the complex trade-offs between utterance length, amount of information conveyed, and cognitive load. We set these trade-offs in an objective function by analysing existing MATCH data. We then train a NLG policy using Reinforcement Learning (RL), which adapts its behaviour to noisy feedback from the current generation context. This policy is compared to several baselines derived from previous work in this area. The learned policy significantly outperforms all the prior approaches.

Keywords: Reinforcement Learning, Adaptivity, Spoken Dialogue Systems, Information Presentation, Incremental NLG, Optimisation, data-driven methods.

1 Introduction

Natural language allows us to achieve the same communicative goal ("what to say") using many different expressions ("how to say it"). In a Spoken Dialogue System (SDS), an abstract communicative goal (CG) can be generated in many different ways. For example, the CG to present a number of database results to the user can be realized as a summary [5,20], or by comparing items [30], or by picking one item and recommending it to the user [35].

Previous work on NLG for SDS has shown that it is useful to adapt the generated output to certain features of the dialogue context, for example user preferences, e.g. [5,30], user knowledge, e.g. [10], or predicted TTS quality, e.g. [2,17]. In extending this previous work we treat NLG as a statistical sequential planning problem, analogously to current statistical approaches to Dialogue Management (DM), e.g. [8,21,24] and models of "conversation as action under

E. Krahmer, M. Theune (Eds.): Empirical Methods in NLG, LNAI 5790, pp. 105–120, 2010.

uncertainty" [19]. In NLG we have similar trade-offs and unpredictability as in DM, and in some cases the content planning and DM tasks are interdependent [14,21]. On the one hand, very long system utterances with many actions in them are to be avoided, because users may become confused or impatient. On the other hand, each individual NLG action will convey some (potentially) useful information to the user. Thus, there is an optimization problem to be solved.

Moreover, the user judgements or next (most likely) action after each NLG action are unpredictable, and the behaviour of the surface realizer may also be variable (see Section 6.2). Thus, the decision process has to cope with uncertainty in the environment. NLG could therefore fruitfully be approached as a sequential statistical planning task, where there are trade-offs and decisions to make, in the presence of uncertainty. We investigate NLG choices such as whether to choose another NLG action (and which one to choose) or to instead stop generating. Reinforcement Learning (RL) allows us to optimize such trade-offs in the presence of uncertainty, i.e. the chances of achieving a better state, while engaging in the risk of choosing another action.

In this paper we present and evaluate a new data-driven model for NLG in Spoken Dialogue Systems as planning under uncertainty. This model is incremental at the turn level, and adaptive to a changing generation environment. The paper proceeds as follows. In Section 2 we argue for applying RL to NLG problems and explain the overall framework. In Section 3 we discuss challenges for NLG for Information Presentation. In Section 4 we present results from our analysis of the MATCH corpora [30]. In Section 5 we present a detailed example of our proposed NLG method. In Section 6 we report on experimental results using this framework for exploring Information Presentation policies. In Section 8 we conclude and discuss future directions.

Note that this model is also developed elsewhere in this volume, for the NLG task of adaptive Referring Expression Generation [9].

2   NLG as Planning under Uncertainty

In recent years statistical planning techniques have been recognised as an important approach to the problems of Dialogue Management (DM), e.g. [8,24,35]. In DM there is a standard trade-off between dialogue length and task success. Longer dialogues may increase the chance of task success (for example getting the required information from a user), but longer dialogues are usually rated more poorly and may result in hang-ups. Success therefore depends on the user's actions, which are unpredictable. Similar uncertainty and trade-offs occur in NLG tasks, which leads us to adopt the general framework of NLG as planning under uncertainty (see [13,14] for the initial version of this approach). Some aspects of NLG have been treated as planning, e.g. [11,12], but never before as statistical planning.

We follow a standard NLG architecture similar to [5,16,28,30], where higher-level NLG decisions, e.g. NLG strategy and content choices, are made by a presentation planner. The low level surface form is then realized by a separate module, which we will call the "realizer" in the following.

NLG actions take place in a stochastic (non-deterministic, uncertain) environment, for example consisting of a user, a realizer, and a Text-to-Speech (TTS) system, where the individual NLG actions have uncertain effects on the environment. For example, presenting differing numbers of attributes to the user may make the user more or less likely to choose an item, as shown by [22] for multimodal interaction.

Most SDS employ fixed template-based generation. However, more advanced systems employ stochastic realizers for SDS, with variable behaviour, see for example [25]. One main advantage of using stochastic realization is that prompts are automatically adapted to the current dialogue context, e.g. sentence plans can be adapted to the individual user and the domain [29]. However, this introduces additional uncertainty into the generation environment, which higher level NLG decisions will need to react to. In our framework, the NLG component must achieve high-level Communicative Goals determined by the Dialogue Manager (e.g. to present a number of items) through planning a sequence of lower-level generation steps or actions; for example first summarizing all the items and then recommending the highest ranking one. Each such action has unpredictable effects due to the stochastic realizer and user reactions. For example the realizer might employ all original six attributes when recommending item i_4, but it might use only a subset of two (e.g. price and cuisine for restaurants), depending on its own processing constraints (see e.g. the realizer used to collect the MATCH project data). Likewise, the user may be likely to choose an item after hearing a recommendation, or they may wish to hear more information. In sum, generating appropriate language in context (e.g. attributes presented so far) thus has the following important features in general:

- NLG is *goal driven* behaviour
- NLG must plan a *sequence* of actions
- each action *changes* the environment state or context
- the effect of each action is *uncertain*.

These features make it clear that the problem of planning how to generate an utterance for SDSs (and interactive systems in general) falls naturally into the class of statistical planning problems, rather than rule-based approaches such as [16,30], or supervised learning as explored in previous work, such as classifier learning and re-ranking, e.g. [18,25]. Supervised approaches involve the ranking of a set of completed plans/utterances and as such cannot adapt online to the context or the user. Reinforcement Learning (RL) provides a principled, data-driven optimisation framework for our type of planning problem [27].

In contrast to previous planning approaches to NLG, e.g. [11,12], RL explicitly models uncertainty in the dialogue environment. The RL approach to NLG has recently been followed by Branavan et al. [3] for mapping written instructions into actions, where actions also have uncertain effects in the environment. Furthermore, the idea that NLG actions have context dependent costs attached to them is essential to RL, which is also discussed by van Deemter [4].

3   The Information Presentation Problem

We will tackle the well-studied problem of Information Presentation in NLG to explore the benefits of this approach. The task here is to find the best way to present a set of search results to a user (e.g. some restaurants meeting a certain set of constraints). This is a task common to much prior work in NLG, e.g. [5,20,30,34].

However, we believe that the presented framework can be applied to many other domains which require complex information to be conveyed to the user, for example instruction giving and tutorial dialogue, or other types of Information Presentation tasks, such as sales agents, and tourist or health information systems.

Our suggested framework is well suited to tackle this problem, since there are many Information Presentation decisions available for exploration. For instance, which presentation strategy to apply (*NLG strategy selection*), how many attributes of each item to present (*attribute selection*), how to rank the items and attributes according to different models of user preferences (*attribute ordering*), how many (specific) items to tell them about (*conciseness*), how many sentences to use when doing so (*syntactic planning*), and which words to use (*lexical choice*) etc. All these parameters (and potentially many more) can be varied, and ideally, jointly optimised in a data-driven manner. In this paper we focus on NLG strategy selection given variations in lower level decisions.

We were provided with two corpora from the MATCH project[1][30], with which to study some of the regions of this decision space. See Table 1 for examples of this data.

Note that for a corpus to be useful for RL approaches, it needs to contain alternative actions, which can be explored by the learner, as well as a measure of goodness, e.g. task success or user ratings.

We first use the MATCH corpora to extract an evaluation function (also known as "reward function" or "objective function") for RL. Both corpora contain data from "overhearer" experiments targeted to Information Presentation in dialogues in the restaurant domain. While we are ultimately interested in how hearers *engaged* in dialogues judge and react to different Information Presentations, results from overhearers are still directly relevant to the task.

4   MATCH Corpus Analysis

The MATCH project made two data sets available, see [26] and [32], which we combine to derive an evaluation function for different Information Presentation strategies.

The first data set, see [26], comprises 1024 ratings by 16 subjects (where we only use the speech-based half, $n = 512$) on the following presentation strategies: RECOMMEND, COMPARE, SUMMARY. These strategies are realized using templates as in Table 1, and varying numbers of attributes. In this study

---

[1] Thanks to Prof. Marilyn Walker.

Table 1. NLG strategies present in the MATCH corpus with average no. attributes and sentences as found in the data

Strategy	Example	Av.#attr	Av.#sentence
SUMMARY	"The 4 restaurants differ in food quality, and cost." (#attr = 2, #sentence = 1)	2.07±.63	1.56±.5
COMPARE	"Among the selected restaurants, the following offer exceptional overall value. Aureole's price is 71 dollars. It has superb food quality, superb service and superb decor. Daniel's price is 82 dollars. It has superb food quality, superb service and superb decor." (#attr = 4, #sentence = 5)	3.2±1.5	5.5±3.11
RECOMMEND	"Le Madeleine has the best overall value among the selected restaurants. Le Madeleine's price is 40 dollars and It has very good food quality. It's in Midtown West." (#attr = 3, #sentence = 3)	2.4±.7	3.5±.53

the users rate the individual presentation strategies as significantly different $(F(2) = 1361, p < .001)$. We find that SUMMARY is rated significantly worse $(p = .05$ with Bonferroni correction) than RECOMMEND and COMPARE, which are rated as equally good.

This suggests that one should never generate a SUMMARY. However, SUMMARY has different qualities from COMPARE and RECOMMEND, as it gives users a general overview of the domain, and probably helps the user to feel more confident when choosing an item, especially when they are unfamiliar with the domain, as shown by [20]. SUMMARY can also have the function of implicitly confirming the system's current understanding of the user's goal. However, COMPARE and RECOMMEND explicitly present named restaurants, and thus help the user to choose a specific item. Our hypothesis is that generating SUMMARY is especially useful in combination with any of the other actions.

In order to further describe the strategies, we extracted different surface features as present in the data (e.g. number of attributes realised, number of sentences, number of words, number of database items talked about, etc.) and performed a stepwise linear regression to find the features which were important to the overhearers (following the PARADISE framework [31]).

We discovered a trade-off between the *length* of the utterances (#sentence) and the number of attributes realised (#attr), i.e. their *informativeness*, where overhearers like to hear as many attributes as possible in the most concise way, as indicated by the regression model shown in Equation 1 $(R^2 = .34)$.[2] Note that we will later use this relationship to define a reward function for our Reinforcement Learning model, see Section 6.3.

---

[2] For comparison: [31] report on R^2 between .4 and .5 on a slightly larger data set.

$$score = .775 \times \#attr + (-.301) \times \#sentence; \qquad (1)$$

The second MATCH data set, see [32], comprises 1224 ratings by 17 subjects on the NLG strategies RECOMMEND and COMPARE. The strategies realise varying numbers of attributes according to different "conciseness" values. These conciseness values are determoned by an attribute selection algorithm which takes the relative weights of attributes for this specific user profile into account. A data analysis shows that these conciseness values are equivalent to the following numbers of attributes used, dependent on the user profile: concise (1 or 2 attributes), average (3 or 4), and verbose (4, 5 or 6). Overhearers rate all conciseness levels as significantly different $(F(2) = 198.3, p < .001)$, with verbose rated highest and concise rated lowest, supporting our findings in the first data set. In contrast to the previous findings, we find that not only the number of attributes, but also the number of sentences is positively correlated with the user ratings. However, we find that the relation between number of attributes, number of sentences and user ratings is not strictly linear: ratings drop for $\#attr = 6$ and $\#sentences > 12$. This suggests that there is an upper limit on how many attributes and sentences users like to hear. We expect this to be especially true for real users engaged in actual dialogue interaction, see [34]. We therefore include "cognitive load" as a variable when training the policy (see Section 6).

Note that the number of attributes and the number of sentences are highly positively correlated $(P = .01$, 2-tailed Pearson's correlation) in both corpora. This implies that the cognitive load is most likely to be a function of both sentence length and number of attributes. Furthermore, we have a true optimization task to solve this problem given this positive correlation: a trade-off, by definition, can't be solved by optimising just one aspect (i.e. by only reducing sentence length or increasing attributes), but the number of sentences and number of attributes are interdependent and thus an equilibrium between the two aspects needs to be determined (in a specific context).

In addition to the trade-off between *length* and *informativeness* for single NLG strategies, we are interested in whether this trade-off will also hold for incrementally generating *sequences* of NLG actions. Whittaker et al. [33], for example, generate a *combined strategy* where first a SUMMARY is used to describe the retrieved subset and then they RECOMMEND one specific item/restaurant. For example *"The 4 restaurants are all French, but differ in food quality, and cost. Le Madeleine has the best overall value among the selected restaurants. Le Madeleine's price is 40 dollars and It has very good food quality. It's in Midtown West."*

We therefore extend the set of possible strategies present in the data for exploration: we allow ordered combinations of the strategies, assuming that only COMPARE or RECOMMEND can follow a SUMMARY, and that only RECOMMEND can follow COMPARE, resulting in 7 possible actions, also see Figure 1:

1. RECOMMEND
2. COMPARE
3. SUMMARY
4. COMPARE+RECOMMEND
5. SUMMARY+RECOMMEND
6. SUMMARY+COMPARE
7. SUMMARY+COMPARE+RECOMMEND

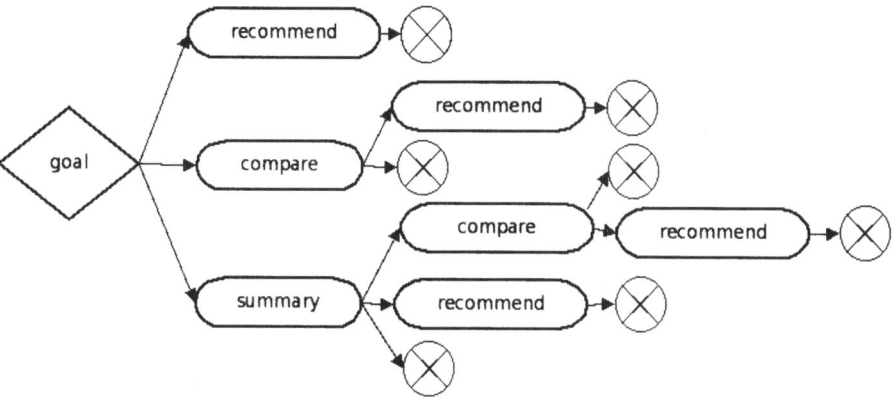

Fig. 1. Possible NLG policies (X=stop generation), for a single system dialogue turn

We then analytically solved the regression model in Equation 1 for the 7 possible strategies using average values from the MATCH data. This is solved by a system of linear inequalities. According to this model, the best ranking strategy is to do all the presentation strategies in one sequence, i.e. SUMMARY+COMPARE+RECOMMEND, see example in Table 2.

However, this analytic solution assumes a "one-shot" generation strategy where there is no intermediate feedback from the environment: users are simply

Table 2. Example utterance presenting information as a sequence: SUMMARY generated as associative clustering with user modeling and 3 attributes; COMPARE 2 restaurants by 2 attributes; and RECOMMEND with 2 attributes

---

115 restaurants meet your query. There are 82 restaurants which are in the cheap price range and have good or excellent food quality. 16 of them are Italian, while 12 of them are Scottish, 9 of them are international and 45 have different types of cuisine. There are also 33 others which have poor food quality. Black Bo's and Gordon's Trattoria are both in the cheap price range. Black Bo's has excellent food quality, while Gordon's Trattoria has good food quality. Black Bo's is a Vegetarian restaurant. This restaurant is located in Old Town. Black Bo's has the best overall quality amongst the selected restaurants.

---

static overhearers (they cannot "barge-in" for example), there is no variation in the behaviour of the surface realizer, i.e. one would use fixed templates as in MATCH, and the user has unlimited cognitive capabilities. These assumptions are not realistic for advanced SDS, and must be relaxed. In the next section we describe a worked through example of the overall framework.

5   Method: The RL-NLG Model

For the reasons discussed above, we treat the NLG module as a statistical planner, operating in a stochastic environment, and optimise it using Reinforcement Learning. The input to the module is a Communicative Goal (CG) supplied by the Dialogue Manager. The CG consists of a Dialogue Act to be generated, for example present_items(i_1, i_2, i_5, i_8), and a System Goal (SysGoal) which is the desired user reaction, e.g. to make the user choose one of the presented items (user_choose_one_of(i_1, i_2, i_5, i_8)). The RL-NLG module must plan a sequence of NLG actions that achieve the goal (at lowest cost) in the current context. The context consists of a user (who may remain silent, supply more constraints, choose an item, or quit), and variation from the sentence realizer as described above.

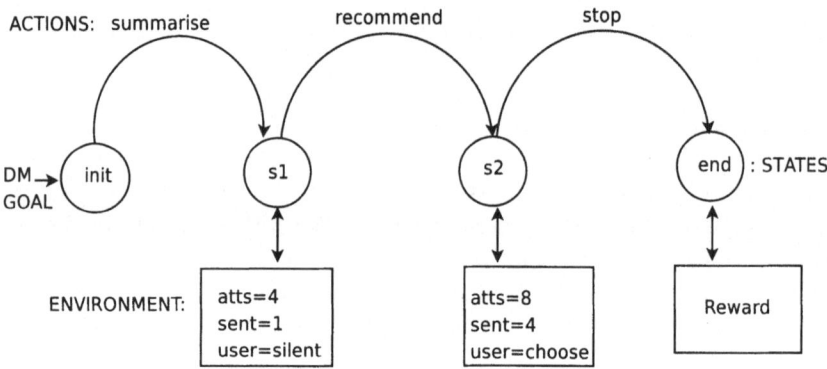

Fig. 2. Example RL-NLG action sequence for Table 3

In order to treat this decision problem using standard statistical planning methods, we now formulate the problem as a Markov Decision Process (MDP). Each state-action pair is associated with a *transition probability*, which is the probability of moving from state s at time t to state s' at time $t + 1$ after having performed action a when in state s. This transition probability is computed by the environment model (i.e. user and realizer), and explicitly captures uncertainty in the environment. This is a major difference to other non-statistical planning approaches. Each transition is also associated with a reinforcement signal (or reward) r_{t+1} describing how good the result of action a was when performed in state s. RL learns via exploring the policy space in the beginning

Table 3. Example utterance planning sequence for Figure 2

State	Action	State change/effect
init	DM set Goal: **present_items**(i_1, i_2, i_5, i_8) & user_choose_one_of(i_1, i_2, i_5, i_8)	initialise state
s1	RL-NLG: SUMMARY(i_1, i_2, i_5, i_8)	att=4, sent=1, user=silent
s2	RL-NLG: RECOMMEND(i_5)	att=8, sent=4, user=choose(i_5)
end	RL-NLG: stop	calculate Reward

of training until it converges to an optimal policy with maximal overall reward. See [27] for details of various RL methods.

Now let us walk-through one simple utterance plan as carried out by this model, as shown in Table 3 and Figure 2. Note that in the following example, the estimated user reactions are determined by a user simulation, as described in Section 6.1. Here, we start with the CG **present_items**(i_1, i_2, i_5, i_8) & **user_choose_one_of**(i_1, i_2, i_5, i_8) from the system's DM. This initialises the NLG state (*init*). The current policy chooses the action SUMMARY and this transitions us to state $s1$, where we observe that 4 attributes and 1 sentence have been generated, and the user is predicted to remain silent. In this state, the current NLG policy is to RECOMMEND the top ranked item (i_5, for this user), which takes us to state $s2$, where 8 attributes have been generated in a total of 4 sentences, and the user is predicted to choose the item. The current policy holds that in states like $s2$ the best thing to do is "stop" and pass the turn to the user. This takes us to the state *end*, where the total reward of this action sequence is computed (see Section 6.3), and used to update the NLG policy in each of the visited state-action pairs via back-propagation.

6   Experiments

We now report on a proof-of-concept study where we train our policy in a simulated learning environment based on the results from the MATCH corpus analysis presented in Section 4. Simulation-based RL allows exploration of action sequences which are not in the original data, and thus less initial data is needed [22]. Note, that we cannot directly learn from the MATCH data, because we would need data from an interactive dialogue. We are currently collecting such data in a Wizard-of-Oz experiment [15].

6.1   User Simulation

User simulations are commonly used to train strategies for Dialogue Management, see for example [35]. A user simulation for NLG is very similar, in that it is a predictive model of the most likely next user act. Note that the user simulation for NLG can be viewed as analogous to the internal user *models* used in POMDP approaches to dialogue management [6,7], in the sense that this user act does not actually change the overall dialogue state (e.g. by filling information slots/search constraints) but it only provides an estimate of the probability of the next user dialogue acts. It changes the generator state, but not the dialogue state. In other words, the NLG user simulation tells us what the user is most likely to do next, *if we were to stop generating now*. It also tells us the probability whether the user chooses to "barge-in" after a system NLG action (by either choosing an item or providing more information).

The user simulation for this study is a simple bi-gram model, which relates the number of attributes presented to the next likely user actions, see Table 4. The user can either follow the goal provided by the DM (SysGoal), for example choosing an item. The user can also do something else (userElse), e.g. providing another search constraint, or the user can quit (userQuit).

For simplicity, we discretise the number of attributes into bins of three concise (1,2,3), average (4,5,6), verbose (7,8,9), in analogy to the conciseness values from the MATCH data, as described in, Section 4.

In addition, we assume that the user's cognitive abilities are limited ("cognitive load"), based on the results from the second MATCH data set in Section 4. Once the number of attributes is more than the "magic number 7" (reflecting psychological results on short-term memory [1]) the user is more likely to become confused and quit. The probabilities in Table 4 are currently manually set heuristics. We are currently analysing data from a Wizard-of-Oz study in order to learn these probabilities (and other user parameters) from real data [15].

Table 4. NLG bi-gram user simulation

	SysGoal	userElse	userQuit
concise	20.0	60.0	20.0
average	60.0	20.0	20.0
verbose	20.0	20.0	60.0

6.2 Realizer Model

The sequential NLG model assumes a realizer, which incrementally updates the context after each generation step (i.e. after each single action). We estimate the realiser's parameters from the mean values we found in the MATCH data (see Table 1). For this study we first (randomly) vary the number of attributes, whereas the number of sentences is fixed (see Table 5). In current work we replace the realizer model with an implemented generator that replicates the variation found in the SPaRKy realizer [25]. A brief description of the current realizer model can be found in [15].

6.3 Reward Function

The reward function defines the final optimization objective of the utterance generation sequence. In this experiment the reward is a function of the various

Table 5. Realizer parameters

	#attr	#sentences
SUMMARY	1 or 2	2
COMPARE	3 or 4	6
RECOMMEND	2 or 3	3

data-driven trade-offs as identified in the data analysis in Section 4: utterance length and number of provided attributes, as weighted by the regression model in Equation 1, as well as the next predicted user action, as shown by the updated reward model in Equation 2. Since we currently only have overhearer data, we manually estimate the reward for the next most likely user act, to supplement the data-driven model. If in the *end* state the next most likely user act is userQuit, the learner gets a penalty of -100, userElse receives 0 reward, and SysGoal gains $+100$ reward. Again, these hand coded scores need to be refined by a more targeted data collection, but the other components of the reward function are data-driven.

$$reward = .775 \times \#attr + (-.301) \times \#sentence + value\_UserAction; \quad (2)$$

Note that RL learns to "make compromises" with respect to the different trade-offs. For example, the user is less likely to choose an item if there are more than 7 attributes, but the realizer can generate 9 attributes. However, in some contexts it might be desirable to generate all 9 attributes, e.g. if the generated utterance is short. Threshold-based approaches, in contrast, cannot (easily) reason with respect to the current context.

6.4 State and Action Space

The state space comprises 9 binary features representing the number of attributes, 2 binary features representing the predicted user's next action to follow the system goal or quit, as well as a discrete feature reflecting the number of sentences generated so far, as shown in Figure 3. This results in $2^{11} \times 6 = 12,288$ distinct generation states. We trained the policy using the well known SARSA algorithm, using linear function approximation [27]. The policy was trained for 3600 simulated NLG sequences.

$$\begin{bmatrix} action: \begin{bmatrix} \text{SUMMARY} \\ \text{COMPARE} \\ \text{RECOMMEND} \\ \text{end} \end{bmatrix} & state: \begin{bmatrix} \text{attributes } |1|\text{-}|9| : \{0,1\} \\ \text{sentence: } \{1\text{-}11\} \\ \text{userGoal: } \{0,1\} \\ \text{userQuit: } \{0,1\} \end{bmatrix} \end{bmatrix}$$

Fig. 3. State-Action space for RL-NLG

Note that binary features are used to present values for some state variables (e.g. number of attributes), rather than using one discrete feature, due to the interactions between the linear function approximation method and the calculation of expected reward, see [23]. Also, representing attributes using binary features allows us to retain information on which attributes have been mentioned. We will use this information for future work.

6.5   Baseline Information Presentation Policies

We derive the baseline policies from Information Presentation strategies as deployed by current dialogue systems. In total we tested 7 different baselines (B1-B7), which correspond to single branches in our policy space (see Figure 1):

B1: RECOMMEND only, e.g. [35]
B2: COMPARE only, e.g. [8]
B3: SUMMARY only, e.g. [20]
B4: SUMMARY followed by RECOMMEND, e.g. [33]
B5: Randomly choosing between COMPARE and RECOMMEND, e.g. [29]
B6: Randomly choosing between all 7 outputs
B7: Always generating whole sequence, i.e. SUMMARY+COMPARE+RECOMMEND, as suggested by the analytic solution (see Section 4).

7   Results

We analyse the test runs (n=200) using an ANOVA with a PostHoc T-Test (with Bonferroni correction). RL significantly ($p < .001$) outperforms all baselines in terms of final reward, see Table 6 and Figure 4. RL is the only policy which significantly improves the next most likely user action by adapting to features in the current context. In contrast to conventional approaches, RL learns to adapt to its environment according to the estimated transition probabilities and the associated rewards.

Table 6. Evaluation Results: Average reward significantly improves for RL-based strategy over baselines B1 – B6 ($p < .001$) and B7 ($p < .01$), also see Figure 4

Policy	Reward ($\pm std$)
B1	99.1 ($\pm$129.6)
B2	90.9 ($\pm$142.2)
B3	65.5 ($\pm$137.3)
B4	176.0 ($\pm$154.1)
B5	95.9 ($\pm$144.9)
B6	168.8 ($\pm$165.3)
B7	242.2 ($\pm$158.9)
RL	283.6 ($\pm$157.2)

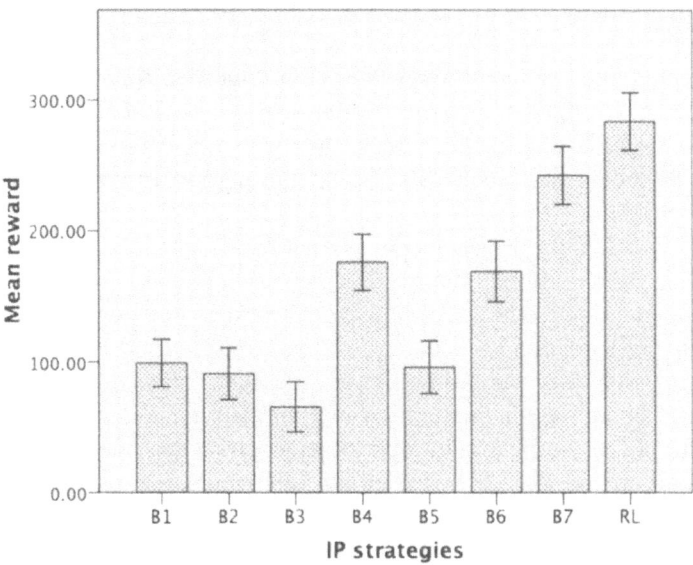

Fig. 4. Bar graphs (with error bars at 95% confidence interval) for mean rewards per Information Presentation (IP) strategy in Table 6

The learnt policy can be described as follows: It either starts with SUMMARY or COMPARE after the *init* state, i.e. it learnt to never start with a RECOMMEND. It stops generating after COMPARE if the `userGoal` is (probably) reached (e.g. the user is most likely to choose an item in the next turn, which depends on the number of attributes generated), otherwise it goes on and generates a RECOMMEND. If it starts with SUMMARY, it always generates a COMPARE afterwards. Again, it stops if the `userGoal` is (probably) reached, otherwise it generates the full sequence (which corresponds to the analytic solution B7).

The analytic solution B7 performs second best, and significantly outperforms all the other baselines ($p < .01$). Still, it is significantly worse ($p < .01$) than the learnt policy as this 'one-shot-strategy' cannot robustly and dynamically adapt to noise or changes in the environment.

In general, generating sequences of NLG actions rates higher than generating single actions only: B4 and B6 rate directly after RL and B7, while B1, B2, B3, B5 are all equally bad given our data-driven definition of reward and environment. Furthermore, the simulated environment allows us to replicate the results in the MATCH corpus (see Section 4) when only comparing single strategies: SUMMARY performs significantly worse, while RECOMMEND and COMPARE perform equally well.

Note that two of the four most successful strategies start by generating a summary; the other two do so regularly. This confirms our hypothesis that generating a SUMMARY in combination with COMPARE or RECOMMEND is useful to the user, even though SUMMARY on its own was rated badly, see Section 4. In future work we will test this hypothesis with real users.

8 Conclusion

We presented and evaluated a new model for adaptive Natural Language Generation (NLG) in Spoken Dialogue Systems, based on statistical planning, that is context-sensitive and incremental at the turn level. After motivating and presenting the model, we studied its use in Information Presentation for Spoken Dialogue Systems.

We derived a data-driven model predicting users' judgements on different information presentation actions, via a regression analysis on MATCH data. We used this regression model to set weights in a reward function for Reinforcement Learning, and so optimize a context-adaptive Information Presentation policy. The learnt policy was compared to several baselines derived from previous work in this area, as well as a non-adaptive analytic solution computed from the data. The learnt policy significantly outperforms all these baselines.

There are many possible extensions to this model, e.g. using the same techniques to jointly optimise choosing the number of attributes, aggregation, word choice, referring expressions, and so on, in a hierarchical manner.

We are currently collecting data in targeted Wizard-of-Oz experiments [15], to derive a fully data-driven training environment and test the learnt policy with real users, following [22]. The trained NLG strategy will also be integrated in an end-to-end statistical system within the CLASSiC project (www.classic-project.org).

In future work we also plan to learn lower level NLG decisions, such as lexical adaptation based on the vocabulary used by the user, following a hierarchical optimisation approach, also see [9] (this volume) and [22].

Acknowledgments. We are grateful to Marilyn Walker for making the MATCH corpora available. The research leading to these results has received funding from the European Community's Seventh Framework Programme (FP7/2007-2013) under grant agreement no. 216594 (CLASSiC project: www.classic-project.org) and from the EPSRC (project nos. EP/E019501/1 and EP/G069840/1).

References

1. Baddeley, A.: Working memory and language: an overview. Journal of Communication Disorder 36(3), 189–208 (2001)
2. Boidin, C., Rieser, V., van der Plas, L., Lemon, O., Chevelu, J.: Predicting how it sounds: Re-ranking dialogue prompts based on TTS quality for adaptive spoken dialogue systems. In: Proceedings of the Interspeech Special Session: Machine Learning for Adaptivity in Spoken Dialogue (2009)
3. Branavan, S., Chen, H., Zettlemoyer, L., Barzilay, R.: Reinforcement learning for mapping instructions to actions. In: Proceedings of ACL, pp. 82–90 (2009)
4. van Deemter, K.: What game theory can do for NLG: the case of vague language (keynote paper). In: 12th European Workshop on Natural Language Generation (ENLG), pp. 154–161 (2009)
5. Demberg, V., Moore, J.D.: Information presentation in spoken dialogue systems. In: Proceedings of EACL, pp. 65–72 (2006)

6. Gasic, M., Keizer, S., Mairesse, F., Schatzmann, J., Thomson, B., Young, S.: Training and Evaluation of the HIS POMDP Dialogue System in Noise. In: Proceedings of SIGdial Workshop on Discourse and Dialogue, pp. 112–119 (2008)
7. Henderson, J., Lemon, O.: Mixture Model POMDPs for Efficient Handling of Uncertainty in Dialogue Management. In: Proceedings of ACL, pp. 73–76 (2008)
8. Henderson, J., Lemon, O., Georgila, K.: Hybrid reinforcement / supervised learning of dialogue policies from fixed datasets. Computational Linguistics 34(4), 487–512 (2008)
9. Janarthanam, S., Lemon, O.: Learning Adaptive Referring Expression Generation Policies for Spoken Dialogue Systems. In: Krahmer, E., Theune, M. (eds.) Empirical Methods in NLG. LNCS (LNAI), vol. 5790, pp. 67–84. Springer, Heidelberg (2010)
10. Janarthanam, S., Lemon, O.: User simulations for online adaptation and knowledge-alignment in Troubleshooting dialogue systems. In: Proceedings of SEMdial, pp. 133–134 (2008)
11. Koller, A., Petrick, R.: Experiences with planning for natural language generation. In: ICAPS (2008)
12. Koller, A., Stone, M.: Sentence generation as planning. In: Proceedings of ACL, pp. 336–343 (2007)
13. Lemon, O.: Adaptive Natural Language Generation in Dialogue using Reinforcement Learning. In: Proceedings of SEMdial (2008)
14. Lemon, O.: Learning what to say and how to say it: joint optimization of spoken dialogue management and Natural Language Generation. Computer Speech and Language (to appear)
15. Liu, X., Rieser, V., Lemon, O.: A Wizard-of-Oz interface to study Information Presentation strategies for Spoken Dialogue Systems. In: Proceedings of the 1st International Workshop on Spoken Dialogue Systems Technology (2009)
16. Moore, J., Foster, M.E., Lemon, O., White, M.: Generating tailored, comparative descriptions in spoken dialogue. In: Proceedings of FLAIRS (2004)
17. Nakatsu, C., White, M.: Learning to say it well: Reranking realizations by predicted synthesis quality. In: Proceedings of ACL (2006)
18. Oh, A., Rudnicky, A.: Stochastic natural language generation for spoken dialog systems. Computer, Speech & Language 16(3/4), 387–407 (2002)
19. Paek, T., Horvitz, E.: Conversation as action under uncertainty. In: Proceedings of the 16th Conference on Uncertainty in Artificial Intelligence, pp. 455–464 (2000)
20. Polifroni, J., Walker, M.: Intensional Summaries as Cooperative Responses in Dialogue Automation and Evaluation. In: Proceedings of ACL, pp. 479–487 (2008)
21. Rieser, V., Lemon, O.: Does this list contain what you were searching for? Learning adaptive dialogue strategies for Interactive Question Answering. J. Natural Language Engineering 15(1), 55–72 (2008)
22. Rieser, V., Lemon, O.: Learning Effective Multimodal Dialogue Strategies from Wizard-of-Oz data: Bootstrapping and Evaluation. In: Proceedings of ACL, pp. 638–646 (2008)
23. Rieser, V., Lemon, O.: Learning and evaluation of dialogue strategies for new applications: Empirical methods for optimization from small data sets. Computational Linguistics (subm)
24. Singh, S., Litman, D., Kearns, M., Walker, M.: Optimizing dialogue management with Reinforcement Learning: Experiments with the NJFun system. Journal of Artificial Intelligence Research (JAIR) 16, 105–133 (2002)
25. Stent, A., Prasad, R., Walker, M.: Trainable sentence planning for complex information presentation in spoken dialog systems. In: Proceedings of ACL, pp. 79–86 (2004)

26. Stent, A., Walker, M., Whittaker, S., Maloor, P.: User-tailored generation for spoken dialogue: an experiment. In: Proceedings of ICSLP (2002)
27. Sutton, R., Barto, A.: Reinforcement Learning. MIT Press, Cambridge (1998)
28. Wahlster, W., Andre, E., Finkler, W., Profitlich, H.J., Rist, T.: Plan-based integration of natural language and graphics generation. Artificial Intelligence 16(63), 387–427 (1993)
29. Walker, M., Stent, A., Mairesse, F., Prasad, R.: Individual and domain adaptation in sentence planning for dialogue. Journal of Artificial Intelligence Research (JAIR) 30, 413–456 (2007)
30. Walker, M., Whittaker, S., Stent, A., Maloor, P., Moore, J., Johnston, M., Vasireddy, G.: User tailored generation in the match multimodal dialogue system. Cognitive Science 28, 811–840 (2004)
31. Walker, M.A., Kamm, C.A., Litman, D.J.: Towards developing general models of usability with PARADISE. Natural Language Engineering 6(3) (2000)
32. Whittaker, S., Walker, M., Maloor, P.: Should I Tell All? An Experiment on Conciseness in Spoken Dialogue. In: Proceedings of Eurospeech (2003)
33. Whittaker, S., Walker, M., Moore, J.: Fish or Fowl: A Wizard of Oz evaluation of dialogue strategies in the restaurant domain. In: Proceedings of LREC (2002)
34. Winterboer, A., Hu, J., Moore, J.D., Nass, C.: The influence of user tailoring and cognitive load on user performance in spoken dialogue systems. In: Proceedings of Interspeech/ICSLP (2007)
35. Young, S., Schatzmann, J., Weilhammer, K., Ye, H.: The Hidden Information State Approach to Dialog Management. In: ICASSP (2007)

Generating Approximate Geographic Descriptions

Ross Turner[1,*], Somayajulu Sripada[2], and Ehud Reiter[2]

[1] Nokia Gate5 GmbH, Berlin, Germany
ross.turner@nokia.com
[2] Department of Computing Science,
University of Aberdeen, UK
{yaji.sripada,e.reiter}@abdn.ac.uk

Abstract. Georeferenced data sets are often large and complex. Natural language generation (NLG) systems are beginning to emerge that generate texts from such data. One of the challenges these systems face is the generation of geographic descriptions that refer to the location of events or patterns in the data. Based on our studies in the domain of meteorology we present an approach to generating approximate geographic descriptions involving regions, which incorporates domain knowledge and task constraints to model the utility of a description. Our evaluations show that NLG systems, because they can analyse input data exhaustively, can produce more fine-grained geographic descriptions that are potentially more useful to end users than those generated by human experts.

Keywords: data-to-text systems, georeferenced data, geographic descriptions.

1 Introduction

Disciplines such as environmental studies, geography, geology, planning and business marketing make extensive use of Geographical Information Systems (GIS); however, despite an explosion of available mapping software, GIS remains a specialist tool with specialist skills required to analyse and understand the information presented using map displays. Complementing such displays with textual summaries therefore provides an immediate niche for NLG systems.

Recently, research into NLG systems that generate text from georeferenced data has begun to emerge [22,27,29]. These systems (also known as data-to-text) are required to textually describe the geographic distribution of domain variables such as road surface temperature and unemployment rates. For example, descriptions such as 'road surface temperatures will fall below zero in some places in the southwest' and 'unemployment is highest in the rural areas' need to be generated by these systems. One of the main challenges these systems face is the generation of geographic descriptions such as 'in some places in the southwest' and 'in the rural areas'. Such a task is challenging for a number of reasons:

- many geographic concepts are inherently vague (see for example [32] for a discussion on this topic);

---

* The work described was carried out while Turner was at the Department of Computing Science, University of Aberdeen, UK.

E. Krahmer, M. Theune (Eds.): Empirical Methods in NLG, LNAI 5790, pp. 121–140, 2010.

- often the underlying data sets contain little explicit geographic information for a generation system to make use of [29];
- as input to a generation system, georeferenced data is often complex and constraints imposed on the output text (such as length) may make the traditional approach to the Generation of Referring Expressions (GRE) of finding a distinguishing description implausible [29].

This chapter looks at the problem in the context of work the authors have carried out on summarising georeferenced data sets in the meteorology domain. The main feature of our approach is that geographic descriptions perform the dual function of referring to specific geographic locations unambiguously (traditional function of GRE) and also communicate the relationship between the domain information and the geography of the region (novel function of geographic descriptions).

We present a two staged approach to generating geographic descriptions that involve regions. The first stage involves using domain knowledge (meteorological knowledge in our case) to select a frame of reference and the second involves using constraints imposed by the end user to select values within a frame of reference. While generating geographic descriptions it is not always possible to produce a distinguishing description because of the inherent vagueness in geographic concepts. Therefore, we aim to produce a distinguishing description wherever possible, but more often allow non-distinguishing descriptions in the output text, which approximate the location of the event being described as accurately as possible.

After a short overview of the background in Section 2, some empirical observations on geographic descriptions from knowledge acquisition (KA) studies we have carried out are discussed in Section 3. Taking these observations into account, in Section 4 we describe how this problem is approached using examples from RoadSafe [29], which generates spatial references to events in georeferenced data in terms of regions that approximate their location. It pays particular attention to the use of different frames of reference to describe the same situation and how factors that affect what makes a good reference in this domain are taken into account by the system. In Section 5 we present a qualitative discussion of aspects of geographic description from the evaluations of Road-Safe that were carried out, and how this relates to future possible work on this topic.

2 Background

RoadSafe is an NLG system that was operationally deployed at Aerospace and Marine International (AMI) to produce weather forecast texts for winter road maintenance. It generates forecast texts describing various weather conditions on a road network as shown in Figure 1. The input to the system is a data set consisting of numerical weather predictions (NWP) calculated over a large set of point locations across a road network. An example static snapshot of the input to RoadSafe for one parameter is shown in Figure 2. The complete input is a series of such snapshots for a number of parameters (see [29] for details). In applications such as RoadSafe, the same geographical situation can be expressed in a variety of different ways dependent upon the perspective employed, henceforth termed as a frame of reference. Space (geographic or otherwise) is inherently tied to a frame of reference that provides a framework for assigning different values to, and determining relations between, different locations in a space. The notion of a frame of reference has various meanings across a number of fields.

In GIS a reference frame is any framework that realises a coordinate system for defining the location of something in geographic space. The process for establishing the location of something within that coordinate system is known as Georeferencing. One of the most common forms of georeference is latitude and longitude. Georeferenced data

---

Overview: Road surface temperatures will fall below zero <u>on all routes</u> during the late evening until around midnight.

Wind (mph): NE 15-25 gusts 50-55 this afternoon <u>in most places</u>, backing NNW and easing 10-20 tomorrow morning, gusts 30-35 during this evening until tomorrow morning <u>in areas above 200M</u>.

Weather: Snow will affect <u>all routes</u> at first, clearing at times then turning moderate during tonight and the early morning <u>in all areas</u>, and persisting until end of period. Ice will affect <u>all routes</u> from the late evening until early morning. Hoar frost will affect <u>some southwestern and central routes</u> by early morning. Road surface temperatures will fall slowly during the evening and tonight, **reaching zero <u>in some far southern and southwestern places</u>** by 21:00. Fog will affect <u>some northeastern and southwestern</u> routes during tonight and the early morning, turning freezing <u>in some places above 400M</u>.

Fig. 1. RoadSafe forecast text showing geographic descriptions underlined and example weather event description highlighted in bold face

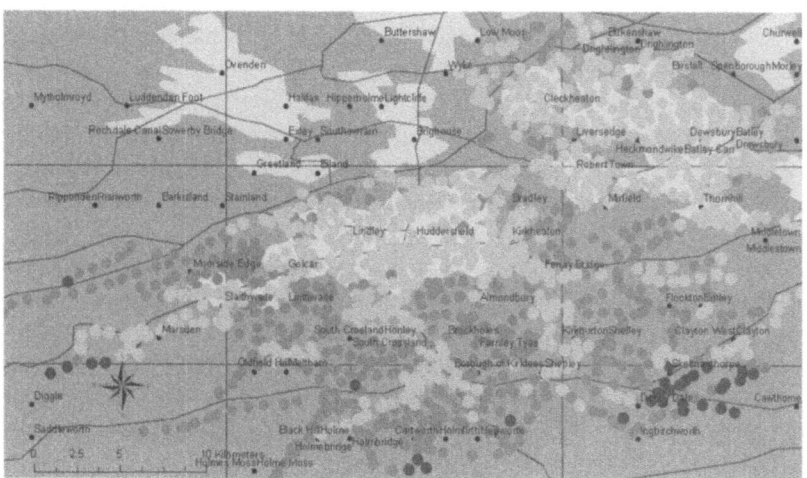

Road Surface Temperatures at 21:00 22/03/08

- marginal
- near critical
- zero

Fig. 2. RoadSafe input data

however, is communicated in terms of geographic features such as altitude and population. These geographic features provide both a framework with which to partition a geographic space and an alternative georeference such as altitude or cardinal directions.

In this context a frame of reference defines how the domain should be partitioned based upon a particular feature and the set of geographic descriptors applied to each element of the partition. Used in this way the frame of reference grounds the meaning of geographic terms that are relevant to describing the data for both the author and user. Psycholinguists typically differentiate between three broad categories of frame of reference according to its origin: intrinsic (object-centred), relative (viewer-centred) and absolute (environment-centred) [15]. Taking into account that geographical data is communicated via maps, imposing a fixed survey perspective upon the scene, the frames of reference we employ are of the absolute variety.

Much work on generation of spatial descriptions has concentrated on smaller scale spaces that are immediately perceivable. For example, spatial descriptions have been studied from the perspective of robot communication [12], 3D animation [24] and basic visual scenes [6,33]. In a more geographical context route description generation systems such as [3] and [17] have had wide appeal to NLG researchers. [31] also generate landmark based spatial descriptions using maps from the map task dialogue corpus. Many georeferenced data sets contain large numbers of observations and their underlying geography may contain many landmarks. In this regard they can be considered as complex scenes. [12] point out that existing GRE algorithms assume the availability of a predefined scene model that encodes all the relations between objects, and the construction of such models can be costly for complex scenes. For example, in the graph model of [14], objects are represented as nodes of a graph while relations between them are encoded as arcs between nodes. In a large geographic environment such as a road network consisting of thousands of points, where the task is to refer to an event occurring at a small subset of those points, it is impractical (generated descriptions may be long and complex) and prohibitively expensive (large numbers of spatial relations between objects may have to be computed) to take this approach. If the classic assumption in GRE that referring expressions are always distinguishing is taken in this context, the logical completeness of an algorithm becomes an issue. [5] define a GRE algorithm as complete if it is successful in every situation in which a distinguishing description exists.

The most interesting aspect of RoadSafe from an NLG perspective is that it generates spatial referring expressions that refer to the target referent (in this case regions approximating a set of points) approximately. Approaches to GRE to date have concentrated on distinguishing descriptions; more specifically, given a domain they look to generate a description of a target object that uniquely distinguishes it from all other objects within that domain. RoadSafe relaxes this common assumption: while it does produce distinguishing descriptions, it more often allows non-distinguishing descriptions in the output text, which approximate the location of the event being described as accurately as possible. This departure from the traditional formulation of the GRE problem raises the question of what makes an adequate reference strategy in this case? We argue that this is reliant, to a large extent, on the communication goal of the system. This chapter investigates this problem in the context of the RoadSafe application, which uses a simple spatial sublanguage to generate the types of descriptions required in this application domain.

3 Observations on Geographic Descriptions from the Weather Domain

In this section we summarise some empirical observations on how meteorologists use geographic descriptions in weather forecasts. It describes work carried out over the course of the RoadSafe project involving knowledge acquisition (KA) studies with experts on summarising georeferenced weather data, observations from data-text corpora (one aimed at the general public and one aimed at experts) and a small study with people from the general public. During RoadSafe we built two prototype georeferenced data-to-text systems that summarised georeferenced weather data: one that produces pollen forecasts based on very simple data [27], and the RoadSafe system, which generates road ice forecasts based on complex data. Small corpora consisting of forecast texts and their underlying NWP data were collected in both application domains. Using techniques described in [19] these corpora have been analysed to understand how experts describe georeferenced weather data.

The major finding from our studies is the fact that experts tailor their geographic descriptions to the task context. Not only does the geographic knowledge of the end user have to be taken into account in their descriptions, but also how the geography of the region causes events and patterns in the data. The latter consideration has a large affect on the frame of reference experts employ to describe particular geographic situations. Section 3.1 looks at these observations from the point of view of end users of weather forecasts, while Section 3.2 looks at the descriptive strategies of experts. In Section 3.3 we discuss some domain specific factors that affect the utility of a description in road ice forecasting.

3.1 End Users' Geographic Knowledge

It is a well known and accepted fact that geographic knowledge varies greatly between individuals. To illustrate this point 24 students of a further education college near Glasgow, Scotland, were asked a geography question without reference to a map. Which of four major place names in Scotland (Ayr, Glasgow, Isle of Arran and Stirling) did they consider to be in the south west of the country? The responses showed a great variation in the subject's geographic knowledge. Half of all subjects considered Glasgow and Ayr to be in the south west, while one third considered this to be true of Stirling. Most surprisingly only four considered this to be true of the Isle of Arran. The fact that more participants thought that Stirling, the least south westerly place, was in the south west as opposed to the Isle of Arran, was surprising, because the question was designed to test their knowledge of local geography. In general the experiment showed that subjects did not have a clear mental representation of their geographic environment. Such errors in memory and judgement of environmental knowledge are common [30].

Contrast this with the detailed knowledge of road engineers who rely upon a large amount of local geographic knowledge and experience when treating roads. Indeed, their spatial mental models are specified at a much finer detail. For example, they get to know where frost hollows tend to form and also come to learn of particular unexpected black spots, such as where garages allow hose water to cover part of a road during winter. More importantly, they are more likely to have a clear idea of the area a term like south west refers to. This is an important point to be taken into account when communicating georeferenced data because spatial descriptions should be sensitive to end users' geographic knowledge, because it dictates how accurately such descriptions will be interpreted.

Both task context and structural features of data (e.g. number of observations, granularity of measurement), as well as functional features of data (how the entities being described function in space) influence how it is described geographically. Analysis of a small pollen forecast corpus [27] revealed that forecast texts contain a rich variety of spatial descriptions for a location despite the data containing only six data points for the whole of Scotland. In general, the same region could be referred to by its proper name, e.g. *'Sutherland and Caithness'*; by its relation to a well known geographical landmark, e.g. *'North of the Great Glen'*; or using fixed bearings, e.g. *'the far North and North-west'*. Proper names for regions and landmarks are not included in the underlying data set and experts make use of their geographic knowledge to characterise the limited geographic information contained within the data. In this case task context plays a role because consumers of these forecasts are the general public and there is a greater onus on the expert to make the texts more interesting, unlike more restricted domains such as marine (see [19]) or road ice forecasts that require consistent terminology.

3.2 Experts' Descriptive Strategy

The most striking observation about the expert strategy is that the geographic descriptions in the corpora are approximations of the input [28]. The input is highly overspecified with 1000s of points for a small forecast region, sampled at sub hourly intervals during a forecast period. Meteorologists use vague descriptions in the texts to refer to weather events such as:

– *'in some places in the south, temperatures will drop to around zero or just above zero.'*

There are a number of reasons they use this descriptive strategy: the forecasts are highly compressed summaries, a few sentences describes megabytes of data; very specific descriptions are avoided unless the pattern in the data is very clear cut; experts try to avoid misinterpretation because road engineers often have detailed local geographic knowledge and experts may not be aware of the more provincial terminology they use to refer to specific areas.

Work in psychology has suggested that meteorologists use a dynamic mental model to arrive at an inference to predict and explain weather conditions [25]. Vital to this process is also their ability to take into account how the geography of a region influences the general weather conditions. Understanding the weather's interaction with the terrain enables them to make reliable meteorological inferences particularly when a certain pattern in the data may appear random. It is often unfeasible for a human forecaster to spend large amounts of time inspecting every data point in a detailed visual display.

Using experience and expertise a forecaster can use her mental model to *'play out different hypothetical situations'* [25, p.2] and thus arrive at a plausible explanation for an apparently random weather pattern. Consider the following example description of a weather event by an expert taken from our road ice corpus:

- *'exposed locations may have gales at times.'*

This is a good example of a forecaster using her meteorological expertise to make an inference about a random weather pattern. Clearly there is no way from inspection of a map one can ascertain with certainty where the exposed locations are in a region. However, an expert's knowledge of how the referent entity (the wind parameter) is affected by geographical features allow her to make such an inference. Extending this concept further, one can easily think of similar examples from other domains. Consider the task of summarising georeferenced crime statistic data and the following description:

- *'crime rates are highest on the low ground.'*

Even to a non expert, a description such as this seems intuitively wrong (even if it is a factually correct), or at the very least rather strange. Again, this assertion may be the most geographically accurate, but what effect does the relative altitude of a region have on the amount of crime that takes place there? The key point to make here is that of the avoidance of unwanted conversational implicatures, i.e. causing the user to infer things that are not true of the situation depicted by the data. Conversational implicatures were proposed by [10], where he hypothesised speakers obey four maxims of conversation: quantity, quality, relevance and manner. This work has been very influential in NLG (e.g. [4,20]) and many other fields. In this description of crime rates scenario the speaker can be considered as violating the maxims of quality and relevance. As discussed at the beginning of this section, communicating georeferenced data not only involves describing where things are happening, but also drawing conclusions about why and how they are happening. In this context causality is an important factor that contributes to the relevance of an utterance. It can be argued therefore that perspective, as imposed by selection of a frame of reference in a geographic description, plays a role in not only helping a speaker to make reliable inferences about the events they are trying to communicate, but help them ensure a description may be interpreted correctly by the hearer.

These pragmatic factors play a large part in determining an expert's descriptive strategy, where certain frames of reference may be considered more appropriate to describe certain weather events [28]. This comes from weather forecasters' explicit knowledge of spatial dependence (the fact that observations in georeferenced data at nearby locations are related, and the values of their non-spatial attributes will be influenced by certain geographical features). This is one of the most important and widely understood facts about spatial data from an analysis point of view, and one of the main reasons that it requires special treatment in comparison to other types of non-spatial data. This fact is most clearly outlined by an observation made in [23, p.3] that *'everything is related to everything else, but near things are more related than distant things'*. This is commonly known as the first law of geography or Tobler's first law (TFL), and still resonates strongly today amongst geographers [16]. The implication of TFL is that samples in spatial data are not independent, and observations located at nearby locations are more likely to be similar. Recasting this into meteorological terms, exposed locations are more likely to be windier and elevated areas colder for example.

In fact, an analogy can be drawn between how meteorologists consider frames of reference in their descriptive strategy and the preferred attribute list in the seminal work on GRE by [4]. In their specification of an algorithm for generating referring expressions content selection is performed through the iteration over a pre-determined and task specific list of attributes. In our context, preferred attributes are replaced by preferred frames of reference. This means describing georeferenced weather data requires situational knowledge of when to apply a particular frame of reference given a particular geographic distribution to describe.

3.3 Domain Specific Constraints on Description

In addition to taking into account the geographic knowledge of individuals into the GRE strategy of an NLG system, domain specific aspects also play a role. The purpose of road ice forecasting is to provide road maintenance personnel with the necessary information on which they can base their decisions on whether or not to go out and treat a road network. Descriptions in forecasts should highlight the important weather events (ice, sub-zero temperatures) that affect road safety occurring at sites across a road network. Analysing descriptions in our corpus such as *'most western parts of the forecast area dropping below freezing by the morning'*, shows that errors are likely to be introduced: false positives constitute points covered by the description that do not drop below zero by the specified time, while false negatives constitute points that are wrongly omitted from the description. For road gritting purposes, costs can be assigned to each type of error: road accidents in the case of false negatives and wasted salt in the case of false positives. Where it is not practical to single out each individual location precisely, based on cost an optimal description would eliminate all false negatives, while attempting to minimise false positives as far as possible. Next we discuss how these observations are applied in the RoadSafe system.

4   Generating Approximate Geographic Descriptions

A standard assumption in GRE is that domain entities are characterised by properties that represent attribute-value pairs, which may be used to generate a reference to a domain entity. For example, [4] assume a knowledge base where every entity in the domain is characterised by a collection of properties, such as $\langle colour, red \rangle$. Georeferenced data sets very rarely contain an explicit model of the underlying geography they reference; therefore, a mechanism is required to describe the domain in terms of a collection of attributes that characterise it. The mechanism we employ to accomplish this task are frames of reference.

For the purposes of GRE, a frame of reference provides a set of properties that partitions a geographic space based upon a common attribute. Where multiple categorisations of an entity exist, each attribute can be considered as a perspective. [8] show that to identify a set of referents coherently, a reference should adopt the same perspective on each referent in the set. More specifically, it should include properties that can be lexicalised in a way that most emphasises the semantic relatedness between elements of the referent set. In other words, make use of a common attribute that is most relevant to the discourse. We make the distinction here between a frame of reference and an attribute because a frame of reference is a more complex construct, which also contains a spatial component that maps descriptions to the spatial framework of the domain. This provides an explicit representation of a partition of a geographic space into a set of non-overlapping sub regions.

The summaries produced by RoadSafe are intended to give a brief synopsis of conditions over a forecast region. The intended users of the system are meteorologists and road engineers with good geographical knowledge, as discussed in Section 3.1. A descriptive strategy similar to that of the meteorologists discussed in Section 3.2 was adopted in RoadSafe because, unlike non-experts, end users were not expected to have problems with resolving more general descriptions. Such a strategy is unconventional in comparison to traditional GRE approaches that aim to rule out all distractors in the domain (properties that are not true of the referent). The notion of underspecified references (those that contain insufficient information to distinguish the target referent precisely) is not new in GRE. [13] propose modeling underspecification as a trade-off between clarity (lack of ambiguity) and brevity, through a general cost function which scores both aspects separately. Amongst others, they sketch out how this might be applied to situations where the target referent may be fuzzy, as well as dialogue where negotiation is necessary. Underspecified descriptions are typically produced by common GRE algorithms as a consequence of failure. RoadSafe takes a similar approach to [13] in that it models the utility of the descriptions it generates.

The constraints we have discussed have been realised in a two stage approach to generating geographic descriptions involving regions. The first stage involves using domain knowledge (meteorological knowledge in our case) to select a frame of reference, while the second accounts for task constraints to select values within that frame of reference. We illustrate our approach through use of a specific example description indicated by boldface font in Figure 1. The weather event this description applies to is the set of sub-zero points highlighted on the map in Figure 3. Before we describe the individual stages, two necessary pre-processing stages for generation are described.

4.1 Geographic Characterisation

The RoadSafe input data consists of a finite set of observations. Geographic characterisation is responsible for assigning a set of qualitative descriptors to each observation in the data set, based upon a set of reference frames, so that observations can be collectively distinguished from each other. This provides both a criterion for partitioning the data, and a set of properties to generate geographic descriptions. Each observation in the data set has a georeference that maps it to a latitude longitude coordinate pair. The set of coordinates defines a forecast region R. A frame of reference in this context is the triple $F = \langle K, \{R\}, m \rangle$, where K is a non-empty set of descriptors based upon a common topic such as coastal proximity e.g. $\{inland, coastal\}$ or population e.g. $\{urban, rural\}$. $\{R\}$ is a partition of the forecast region and m is a function that maps every element of every element of $\{R\}$ to an element of K. In RoadSafe four frames of reference have been implemented, based on the corpus: altitude, coastal proximity, population and direction. Those that make use of human (population) and physical

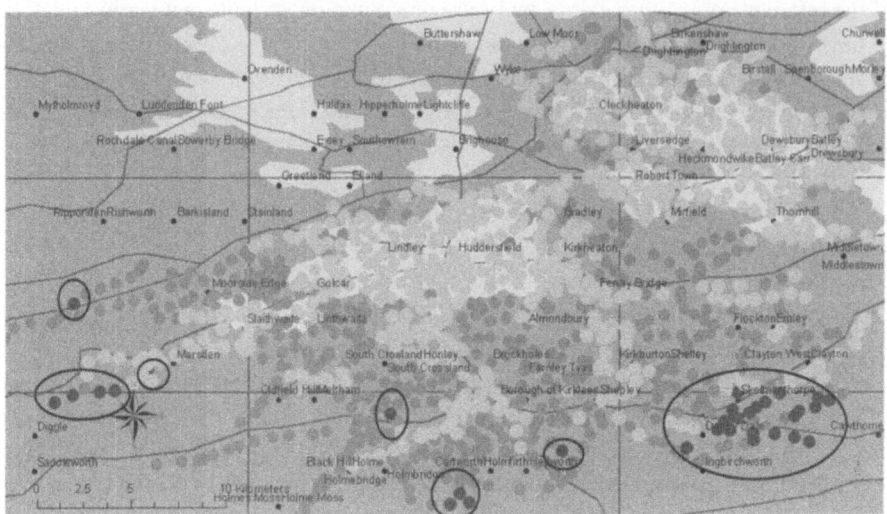

Road Surface Temperatures at 21:00 22/03/08

  * marginal
  * near critical
  * zero

Fig. 3. Sub-zero road surface temperatures

Fig. 4. Directional grid overlayed over road network data

geographical features (altitude, coastal Proximity) can be represented by existing GIS data sets; therefore, in these cases geographic characterisation is responsible for mapping observations to descriptors that describe regions defined by these data sets. In contrast, directions are abstract and require definition. In RoadSafe, geographic characterisation maps each observation to a set of directional regions with crisp boundaries, described next.

The partitioning of the region for the direction frame of reference is accomplished by computing a grid overlaying the forecast region, which splits it into sixteen equally sized directional regions based on their latitude longitude extents, shown in Figure 4. The individual regions are combined in various ways (see subsequent sections) to create larger regions such as North. The set of descriptors applicable to each element of the grid is shown in Table 1. The frame of reference associates each observation in the region in Figure 4 with the descriptor in the corresponding position of Table 1.

Table 1. Direction frame of reference descriptors

TrueNorthWest	NorthNorthWest	NorthNorthEast	TrueNorthEast
WestNorthWest	CentralNorthWest	CentralNorthEast	EastNorthEast
WestSouthWest	CentralSouthWest	CentralSouthEast	EastSouthEast
TrueSouthWest	SouthSouthWest	SouthSouthEast	TrueSouthEast

4.2   Data Interpretation

To generate descriptions, the geographic distribution of the event to be communicated has to be approximated using data analysis techniques such as clustering. While not new to data-to-text systems, the novel aspect here is that the data is partitioned based upon the frames of reference that make up the spatial sublanguage of the system. This process summarises the location of the event by measuring its density within each frame of reference's set of descriptors. An example of the distribution of the subzero points in the map in Figure 3 is shown in Figure 5.

Reference Frame	Description	Density
Altitude		
	100M	0.033
	200M:	0.017
	300M	0.095
	400M	0.042
Direction		
	SSE	0.037
	SSW	0.014
	WSW:	0.048
	TSE	0.489
	TSW	0.444
Population		
	Rural:	0.039

Fig. 5. Density of zero temperatures in Figure 2 (descriptors with zero density removed)

Data interpretation organises descriptors within the direction frame of reference into sets that can be used for pattern matching. These are required by the GRE module in the microplanner to generate referring expressions such as 'in the far west'. The direction descriptors apply to regions defined by the grid described in Section 4.1. A natural way to represent the spatial neighbourhood of these regions is by using an adjacency matrix, which are commonly used to summarise relationships in spatial analysis [18]. Moreover, the distribution of an event can be summarised using this representation. For example, the distribution in Figure 5 can be summarised using the following matrix:

$$\begin{bmatrix} 0 & 0 & 0 & 0 \\ 0 & 0 & 0 & 0 \\ 1 & 0 & 0 & 0 \\ 1 & 1 & 1 & 1 \end{bmatrix}$$

Each element of the matrix represents a descriptor within the direction frame of reference, where a value of 1 indicates that the event has a non-zero density in the region that descriptor applies to. Data interpretation represents subsets of direction descriptors that form templates whose elements define a greater overall description such as 'north' or 'central'. The templates are used by the GRE module to generate references using the direction frame of reference. Combinations of templates are used to find the best match of a spatial distribution of an event, such as the distribution shown above. The templates are further organised into covers that represent descriptions relating to cardinal (N, S, E, W), intercardinal (NE, NW, SE, SW) and gradeable directions (Far North, Far South, Far East, Far West, Central). The term gradeable is used here to refer to directions that are not fixed by points of the compass. A example template from each cover is shown below:

Gradable

- Far South:

$$\{TSW, SSW, SSE, TSE\} = \begin{bmatrix} 0 & 0 & 0 & 0 \\ 0 & 0 & 0 & 0 \\ 0 & 0 & 0 & 0 \\ 1 & 1 & 1 & 1 \end{bmatrix}$$

Intercardinal

- South West:

$$\{TSW, WSW, SSW, CSW\} = \begin{bmatrix} 0 & 0 & 0 & 0 \\ 0 & 0 & 0 & 0 \\ 1 & 1 & 0 & 0 \\ 1 & 1 & 0 & 0 \end{bmatrix}$$

Cardinal

- South:

$$SouthEast \cup SouthWest = \begin{bmatrix} 0 & 0 & 0 & 0 \\ 0 & 0 & 0 & 0 \\ 1 & 1 & 1 & 1 \\ 1 & 1 & 1 & 1 \end{bmatrix}$$

4.3   Frame of Reference Selection

The selection of a frame of reference is based upon both the location of the event as discussed previously, and situational knowledge about which frame of reference to use when describing certain events. Following [4] we model this knowledge as a preference order over a set of frames of reference. Our model implements different preference orders based upon the parameter and geographic region being described. Similar to the GRE framework outlined by [14], who associate costs with each domain attribute, our model associates weights with each frame of reference that determine the order. The weight applied to each frame of reference is the proportion of spatial descriptions in the corpus that used that frame of reference to describe a particular parameter in a geographic region. Weights represent the likelihood of using a frame of reference in a particular context. Frames of reference that may result in a poor inference are therefore avoided by low weightings. An example of the ordering used to describe road surface temperatures in the region provided in the example is shown below:[1]

$$order(RoadSurfaceTemperature, Kirklees) = \begin{cases} 1.\ Altitude & 0.58 \\ 2.\ Direction & 0.29 \\ 3.\ Population & 0.13 \end{cases}$$

The purpose of the algorithm is to select frames of reference to describe the location of an event occurring at a subset of observations. Recall that individual descriptors within a frame of reference apply to subregions of the domain. Therefore, our algorithm is describing a set of points in terms of geographic regions. In GRE terms, the set of descriptors within each frame of reference constitute properties to describe the referent. Descriptors characterise the set distributively rather than collectively; that is, they describe properties that hold true of its individual elements rather than properties that hold true only of the set as a whole. Approaches to GRE involving collective properties have been proposed in [9,21]. [5] extends the incremental algorithm of [4] to handle reference to sets involving distributive properties, as well as negation and disjunctive combination (set union) of properties. [7] presents an alternative constraint-based algorithm that addresses the tendency for incremental approaches to produce descriptions that contain redundant information and are unnecessarily long and ambiguous. [11] also notes that combinations of properties produced by GRE algorithms for referring to sets of objects are not always adequately expressed in natural language. He addresses this by incorporating restrictions which express constraints on preferred surface forms into his algorithm.

    In comparison to the work outlined above, the algorithm presented here generates approximate rather than distinguishing descriptions. It also focuses upon incorporating a measure of the utility of a description based on constraints imposed by a specific task; namely, communicating the extent of a weather event over a geographic region. Furthermore, the input has abstracted away from the geometric details of the domain, and hence the algorithm is concerned only with the density values of the event associated with each descriptor rather than the individual elements of the target set. With the

---

[1] Note that this example is taken from a forecast region that is landlocked and therefore coastal proximity is not considered.

assumption that a referring expression should be distinguishing relaxed, the aim is to select the frame of reference(s) that can maximise the utility of a reference, which is based upon three main constraints:

1. Complete coverage of the event being described (no false negatives)
2. Minimise false positives.
3. Meteorological correctness (inferencing about causal relationships).

The algorithm proceeds as follows: firstly, descriptors with zero densities are removed from each frame of reference. Next, any frames of reference where each descriptor has non-zero density is removed. For example, where a frame of reference has the set of descriptors $K = \{rural, urban\}$, if each descriptor has a non-zero density a subset of the region cannot be distinguished using it. The next step calculates a utility score for each individual remaining frame of reference by first averaging over their densities, and then multiplying the result by the appropriate weight. More formally for an event e: F is the frame of reference; $L \subset K$ is the set of descriptions in F that apply to regions where e has a non zero density; w is the weight from the preference order in this context and $n = |L|$, utility is calculated as:

$$Utility(F, e) = \frac{\sum_{l \in L}^{n} density(e, l)}{n} \cdot w(F, e)$$

The candidate frames of reference are ranked according to their utility. Frames of reference with a score equal to zero are not considered for selection. The frame of reference with the highest score is automatically selected, while each remaining frame of reference is only added if it reduces the number of false positives. The direction frame of reference is selected in this example because it has both the highest utility score and introduces the least number of false positives. Informally, the algorithm can be summarised as follows:

1. Remove descriptors with zero densities from each frame of reference
2. Remove frames of reference where no descriptors have been removed (all have non-zero density)
3. Calculate $Utility(F, e)$ for each remaining frame of reference and rank them in descending order, removing any with scores equal to zero.
4. Select the top ranking frame of reference or nothing if all have been removed.
5. For each remaining frame of reference add it if it reduces the number of false positives.

4.4 Attribute Selection

Once a frame of reference has been selected the microplanner maps the descriptors to abstract syntax templates. This is fairly trivial for most frames of reference in RoadSafe, because they contain a limited number of descriptors. Direction descriptors however, need to be combined by applying pattern matching before they are mapped to syntactic templates. The input to the microplanner is a structure comprised of the density of the event within the containing area plus a subset of descriptors, shown in Figure 6.

Location

$$\{ \begin{array}{ll} Density: & 0.206 \\ Relation: & in \\ Container: & \begin{bmatrix} 0 & 0 & 0 & 0 \\ 0 & 0 & 0 & 0 \\ 1 & 0 & 0 & 0 \\ 1 & 1 & 1 & 1 \end{bmatrix} \end{array}$$

}

Fig. 6. GRE input to describe Figure 2

The attribute selection algorithm is based upon four constraints incorporating the first two principles of the descriptive strategy outlined at the beginning of this section:

1. Minimise false positives: The description describing the distribution should introduce the least number of distractors. For the above example distribution the set {South} ensures coverage but introduces three distractors: CSW, CSE and ESE, while the set {Far South, South West} only introduces one: CSW. In general, a measure of how distinguishing a description x is of a distribution y is given by:

$$distinguishing(x,y) = \frac{|x \cap y|}{|x|}$$

   Thus, for a distribution z and descriptions x and y, x is a more distinguishing description of z than y iff distinguishing(x,z) > distinguishing(y,z).

2. Coverage (no false negatives): The description should completely describe the distribution. The set of directions {Far South,South West} completely describes the above example distribution while {Far South} does not. A measure of the coverage a description x has of a target distribution y is given by:

$$coverage(x,y) = \frac{|x \cap y|}{|y|}$$

   For description x and distribution y, $covers(x,y)$ is true iff $coverage(x,y) = 1$.

3. Brevity: The set of directions should yield the shortest description of the distribution. For the above example distribution there is only one set of directions that ensures complete coverage. But when faced with a choice for example {South} and {South West, South East}, the brevity constraint favours {South}. In general, the set x should be chosen over y because it is a shorter description. For the distribution z and sets of directions x, y with equal coverage of z, x is a shorter description of z than y iff $|x| < |y|$.

4. Ordering: Where ties occur, descriptions are chosen based upon a preference ordering imposed over their properties. Based on the direction descriptions in the RoadSafe corpus, cardinal directions are used more frequently than intercardinal directions, while intercardinal directions are more frequent than gradeable directions. Assigning each type of property a score: Cardinal = 3, Intercardinal = 2 and Gradeable = 1, the score of a description x is given by:

$$score(x) = \sum_{p\in x}^{n} order(p)$$

Therefore, if two descriptions x, y have equal coverage, brevity and are equally distinguishing for a given distribution, x is preferred over y iff $score(x) > score(y)$.

The process to generate a direction reference is composed of three stages: ranking orders the individual properties by how well they describe the target; overgeneration combines individual properties to produce a set of candidate descriptions; and selection chooses the best description from the set of candidates. Each stage makes use of the constraints outlined above. Returning to the input in Figure 6, the ranking stage produces the following list of properties: $ranking = \langle FarSouth,\ SouthWest,\ South,\ SouthEast,\ FarWest,\ West,\ East,\ FarEast \rangle$. Informally, the ranking can be summarised as follows:

1. Remove properties that do not intersect the target set.
2. Rank individual properties based on the following constraints:
   (a) Rank properties higher if they introduce fewer false positives.
   (b) If two properties introduce the same number of false positives add the property with greater coverage of the target first.
   (c) If two properties introduce the same number of false positives and have equal coverage of the target use the specified order.

After properties have been ranked, overgeneration iterates over the list of properties to produce a set of candidate descriptions. The algorithm recursively tries to find combinations of properties using the constraints outlined above, removing one property from the list after each iteration. The algorithm stops when an iteration fails to produce a description that covers the target, or all properties have been removed. For efficiency overgeneration does not produce an exhaustive list of candidates, because there is no backtracking. This means that once a property has been removed it cannot be used again. Based on the example input, overgeneration produces the set: $candidates = \{\{FarSouth,$

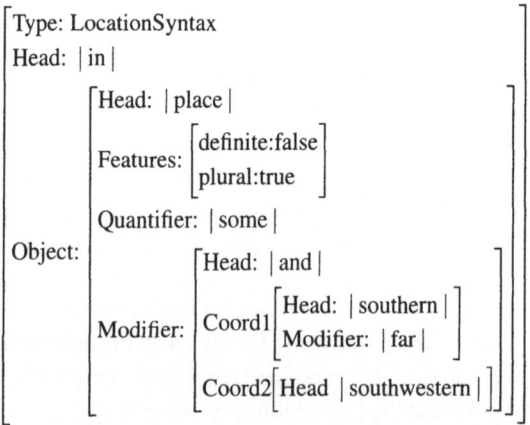

Fig. 7. Phrase syntax for input in Figure 6

$SouthWest\}, \{South\}, \{SouthWest, SouthEast\}, \{SouthEast, West\}\}$. Finally, selection chooses the best description from *candidates* using the constraints. For the current example, the description $\{FarSouth, SouthWest\}$ is selected because it introduces the least number of false positives. This description is mapped to the abstract syntax template shown in Figure 4, which would be realised as 'in some far southern and southwestern places'.

5   Evaluation and Discussion

As complex map displays can often be difficult to interpret for a user, the primary added value of generating textual summaries is providing clarification where required. Variations in the geographic descriptions used in the text may have a profound effect on how the information is interpreted. RoadSafe was evaluated with the two main user groups: meteorologists, who can utilise the system as an authoring aid; and road maintenance personnel, who require forecasts texts to base their operational decisions upon. Meteorologists at AMI were asked to post-edit generated texts as they analysed the input data in a Road Weather Information System during their normal forecasting operations. Potential users were contacted using a questionnaire and asked to compare the quality of generated summaries to corpus texts based on the same data. Consequently, unlike the meteorologists, they did not have access to visualisations of the input data.

Details of the evaluations can be found in [26]. Despite the fact these evaluations were intended to test the overall quality of the texts we have received much feedback on the geographic descriptions the system generates. Meteorologists have commented that generally the geographic descriptions generated by the system are accurate but should be more general. Of 97 post-edited texts generated by the system 20% of the geographic descriptions were edited. Most notable was feedback from twenty one road maintenance personnel who responded to the questionnaire. The respondents pointed out that one of the main reasons they preferred the style of the computer generated texts over the corresponding human ones was because they contained more detailed geographic descriptions. The fact that a data-to-text system can analyse every data point is an advantage. In contrast, it is not possible for a meteorologist to inspect every data point; instead they rely upon knowledge and experience, which reflects in a more general and conservative approach in their geographic descriptions. Perhaps one of their biggest criticisms of the system as a whole was that it doesn't do a good job of generating geographic descriptions that involve motion, such as 'a band of rain works east across the area'. Indeed, this was the most edited type of generated phrase during the post-edit evaluation. There has been little work to our knowledge on describing motion in the NLG literature.

We have also carried out some comparison of the direction descriptions to those in the corpus, by annotating the corpus descriptions with our direction descriptors and running them through the system. Descriptions were compared by calculating the Jaccard coefficient between the two sets. Overall the mean score was 0.53, with a low perfect recall percentage (Jaccard coefficient equal to 1) of 30%. The low precision score is perhaps not surprising because the corpus descriptions are not solely based on the input data we have available (experts had access to additional weather information at the time of writing). However, the majority (67%) of partial alignments were the result of RoadSafe producing a subset of the human description, e.g. northwest versus north, which indicates the system descriptions are more fine grained. In terms of the human descriptions, what was most apparent from this evaluation is the fact that they almost exclusively used the eight major points of the compass.

This chapter has focused upon aspects of perspective choice and describing geographic regions. The approach we have presented here is limited in its aspirations because landmarks are not considered. This is sufficient for the application we describe because the texts are only required as a brief synopsis of weather conditions over a geographic region. If texts were needed for individual route summaries (which would appear to be a natural future requirement) more complex descriptions involving projective relations between landmarks (e.g. areas south of x) would have to be generated. This is non-trivial because it requires complex spatial reasoning and the computation of landmark salience. A natural extension would be to use this approach as a first approximation of a target, restricting the search space for a subsequent call to a subsequent GRE algorithm. Furthermore, we only consider frames of reference that define regions with crisp boundaries. As the boundaries of regions described by concepts such as urban or southern are inherently uncertain, one obvious extension is to extend the model to include frames of reference with indeterminate boundaries.

There are many aspects of the generation of spatial descriptions at the geographical scale that have not been addressed in this work and warrant further exploration. At the content level, there is a need to consider how to account for semantic composition effects caused by overlaying frames of reference. For example, in descriptions such as 'in higher urban areas' and how these may affect interpretation by the hearer. Another question that arises is when is it best to use an intensional rather than extensional description, i.e. 'North Eastern Scotland' versus 'Aberdeen, Aberdeenshire and Moray'. Some research in question answering systems have looked at this problem e.g. [1,2]. There is also the question of when to use descriptions that involve projective relations ('areas west of Aberdeen') or gradeable properties ('the more western areas'). At the lexical choice level there is also the issue how to choose appropriate quantifiers. These are all choices that a data-to-text system can make that will affect how the summary is interpreted.

6   Conclusions

We have described an approach to generating approximate geographic descriptions involving regions which takes into account both general domain knowledge and task specific constraints, based upon empirical work carried out in the weather domain. Our strategy takes into account constraints on what constitutes a good reference in the application domain described in two stages: the first uses a utility function incorporating weights learnt from a domain corpus to select an appropriate frame of reference, while the second ranks candidate descriptions based on task specific measures of quality. What is most apparent from our empirical studies is that geographic descriptions describing georeferenced data are influenced not only by location but also task context. An important observation based on our evaluation studies is that NLG systems, by virtue of their ability to analyse input data exhaustively, can generate descriptions that are potentially more useful to end users than those generated by human experts.

References

1. Benamara, F.: Generating intensional answers in intelligent question answering systems. In: Belz, A., Evans, R., Piwek, P. (eds.) INLG 2004. LNCS (LNAI), vol. 3123, pp. 11–20. Springer, Heidelberg (2004)
2. Cimiano, P., Hartfiel, H., Rudolph, S.: Intensional question answering using ILP: What does an answer mean? In: Kapetanios, E., Sugumaran, V., Spiliopoulou, M. (eds.) NLDB 2008. LNCS, vol. 5039, pp. 151–162. Springer, Heidelberg (2008)
3. Dale, R., Geldof, S., Prost, J.P.: Using natural language generation in automatic route description. Journal of Research and Practice in Information Technology 37(1), 89–105 (2005)
4. Dale, R., Reiter, E.: Computational interpretations of the Gricean maxims in the generation of referring expressions. Cognitive Science 19, 233–263 (1995)
5. van Deemter, K.: Generating referring expressions: Boolean extensions of the incremental algorithm. Computational Linguistics 28(1), 37–52 (2002)
6. Ebert, C., Glatz, D., Jansche, M., Meyer-Klabunde, R., Porzel, R.: From conceptualization to formulation in generating spatial descriptions. In: Schmid, U., Krems, J., Wysotzki, F. (eds.) Proceedings of the 1st European Workshop on Cognitive Modeling, pp. 235–241 (1996)
7. Gardent, C.: Generating minimal definite descriptions. In: Proceedings of the 40th Annual Meeting on Association for Computational Linguistics (ACL 2002), pp. 96–103 (2002)
8. Gatt, A., van Deemter, K.: Lexical choice and conceptual perspective in the generation of plural referring expressions. Journal of Logic, Language and Information 16, 423–443 (2007)
9. Gatt, A.: Generating collective spatial references. In: Proceedings of the 28th Annual Meeting of the Cognitive Science Society, pp. 255–260 (2006)
10. Grice, H.: Logic and conversation. In: Cole, P., Morgan, J. (eds.) Syntax and Semantics, Speech Acts, vol. 3, pp. 43–58. Academic Press, New York (1975)
11. Horacek, H.: On referring to sets of objects naturally. In: Belz, A., Evans, R., Piwek, P. (eds.) INLG 2004. LNCS (LNAI), vol. 3123, pp. 70–79. Springer, Heidelberg (2004)
12. Kelleher, J.D., Kruijff, G.J.M.: Incremental generation of spatial referring expressions in situated dialog. In: Proceedings of the 21st International Conference on Computational Linguistics and 44th Annual Meeting of the Association for Computational Linguistics (COLING-ACL 2006), pp. 1041–1048 (2006)
13. Khan, I.H., Ritchie, G., van Deemter, K.: The clarity-brevity trade-off in generating referring expressions. In: Proceedings of the 4th International Natural Language Generation Conference (INLG 2006), pp. 89–91 (2006)
14. Krahmer, E., van Erk, S., Verleg, A.: Graph-based generation of referring expressions. Computational Linguistics 29(1), 53–72 (2003)
15. Levinson, S.C.: Space in language and cognition: explorations in cognitive diversity. Cambridge University Press, Cambridge (2003)
16. Miller, H.J.: Tobler's first law and spatial analysis. Annals of the Association of American Geographers 93(3), 574–594 (2004)
17. Moulin, B., Kettani, D.: Route generation and description using the notions of object's influence area and spatial conceptual map. Spatial Cognition and Computation 1, 227–259 (1999)
18. O'Sullivan, D., Unwin, D.J.: Geographic Information Analysis. John Wiley & Sons, Chichester (2003)
19. Reiter, E., Sripada, S., Hunter, J., Yu, J., Davy, I.: Choosing words in computer-generated weather forecasts. Artificial Intelligence 67, 137–169 (2005)

20. Sripada, S., Reiter, E., Hunter, J.R.W., Yu, J.: Generating English summaries of time series data using the Gricean maxims. In: Proceedings of the 9th ACM SIGMOD International Conference on Knowledge Discovery and Data Mining (KDD-2003), pp. 187–196 (2003)
21. Stone, M.: On identifying sets. In: Proceedings of the 1st International Conference on Natural Language Generation (INLG 2000), pp. 116–123 (2000)
22. Thomas, K.E., Sripada, S.: What's in a message? Interpreting geo-referenced data for the visually-impaired. In: Proceedings of the 5th International Conference on Natural Language Generation (INLG 2008), pp. 113–120 (2008)
23. Tobler, W.: A computer movie simulating urban growth in the Detroit region. Economic Geography 46(2), 234–240 (1970)
24. Towns, S., Callaway, C., Lester, J.: Generating coordinated natural language and 3D animations for complex spatial explanations. In: Proceedings of the 15th National Conference on Artificial Intelligence, pp. 112–119 (1998)
25. Trafton, J.G.: Dynamic mental models in weather forecasting. In: Proceedings of the Human Factors and Ergonomics Society 51st Annual Meeting, pp. 311–314 (2007)
26. Turner, R.: Georeferenced Data-To-Text: Techniques and Application. Unpublished PhD thesis, University of Aberdeen (2009)
27. Turner, R., Sripada, S., Reiter, E., Davy, I.: Generating spatio-temporal descriptions in pollen forecasts. In: Proceedings of the 11th Conference of the European Chapter of the Association for Computational Linguistics (EACL 2006), pp. 163–166 (2006)
28. Turner, R., Sripada, S., Reiter, E., Davy, I.: Building a parallel spatio-temporal data-text corpus for summary generation. In: Proceedings of the LREC 2008 Workshop on Methodologies and Resources for Processing Spatial Language, pp. 28–35 (2008)
29. Turner, R., Sripada, S., Reiter, E., Davy, I.: Using spatial reference frames to generate grounded textual summaries of georeferenced data. In: Proceedings of the 5th International Conference on Natural Language Generation (INLG 2008), pp. 16–24 (2008)
30. Tversky, B.: Cognitive maps, cognitive collages, and spatial mental models. In: Frank, A., Campari, I. (eds.) Spatial Information Theory, pp. 14–24. Springer, Berlin (1993)
31. Varges, S.: Spatial descriptions as referring expressions in the MapTask domain. In: Proceedings of the 10th European Workshop on Natural Language Generation (ENLG 2005), pp. 207–210 (2005)
32. Varzi, A.C.: Vagueness in geography. Philosophy & Geography 4(1), 49–65 (2001)
33. Viethen, J., Dale, R.: The use of spatial relations in referring expressions. In: Proceedings of the 5th International Conference on Natural Language Generation (INLG 2008), pp. 59–67 (2008)

A Flexible Approach to Class-Based Ordering of Prenominal Modifiers

Margaret Mitchell

Computing Science Department, University of Aberdeen
Aberdeen, Scotland, U.K.
m.mitchell@abdn.ac.uk
http://www.csd.abdn.ac.uk

Abstract. This chapter introduces a class-based approach to ordering prenominal modifiers. Modifiers are grouped into broad classes based on where they tend to occur prenominally, and a framework is developed to order sets of modifiers based on their classes. This system is developed to generate several orderings for sets of modifiers with more flexible positional constraints, and lends itself to bootstrapping for the classification of previously unseen modifiers. The approach to modifier classification outlined here is useful for automated language generation tasks, and the proposed modifier classes may be useful within constraint-based grammars.

Keywords: natural language processing, natural language generation, adjective ordering, modifier ordering.

1 Introduction

Ordering prenominal modifiers is a necessary task in the generation of natural language. For a system to effectively generate natural-sounding utterances, the system must determine the proper order for any given set of modifiers. For example, consider the alternation between "big red ball" and "?red big ball"[1] with "white floppy hat" and "floppy white hat". "Big red ball" sounds fine, while "red big ball" sounds marked. On the other hand, "floppy white hat" and "white floppy hat" both seem fine, and perhaps may be used interchangeably. One can imagine situations where all of the given examples would be produced naturally; however, some prenominal modifier orderings sound equally natural, while others sound more marked.

Determining ways to order modifiers prenominally and what factors underlie modifier ordering has been an area of considerable research, including work in natural language processing [14,27], linguistics [31,32], and psychology [8,24]. This chapter continues this work, developing and evaluating a system to automatically order prenominal modifiers in a class-based framework. Below, we

---

[1] Here, the ? symbol signifies something technically grammatical that fluent speakers judge as sounding somewhat awkward.

E. Krahmer, M. Theune (Eds.): Empirical Methods in NLG, LNAI 5790, pp. 141–162, 2010.
© Springer-Verlag Berlin Heidelberg 2010

1. big beautiful white wooden house
2. big white beautiful wooden house
3. ?white wooden beautiful big house

4. big rectangular green Chinese silk carpet
5. big green rectangular Chinese silk carpet
6. ?Chinese big silk green rectangular carpet

Fig. 1. Grammatical modifier alternations (examples 1, 3, 4, 6 in [31])

establish and evaluate a classification system that may be used to order prenominal modifiers automatically. Each modifier is mapped to one of a small group of classes, and we outline ordering constraints between the modifier classes. Using the proposed system, we are able to correctly order the vast majority of modifiers in an unseen corpus.

The modifier classification model built by this system may be utilized in a surface realization component of a natural language generation (NLG) system, or may be implemented in a microplanning component of an NLG system to help specify the ordering of properties that feed into a referring expression generation (REG) algorithm. Predictions of prenominal modifier ordering based on these classes are shown to be robust and accurate.

To date, there has been no one modifier classification system that is able to order modifiers much higher than chance levels when trained and tested on different corpora.[2] In light of this, no modifier classification exists that may be used to reliably order modifiers for a wide variety of topics. This chapter addresses this issue, and creates an approach to modifier ordering that is accurate across different domains.

We build from the work outlined in [21], presenting a newer version of the system that makes ordering decisions for all unordered sets of modifiers in the test data. We also further refine our evaluation approach, judging accuracy on three different levels.

The work here diverges from the approaches commonly employed in modifier classification by assuming no underlying relationship between semantics and prenominal order. We do not use information from a tagger or a parser within our classification, and no external resources are used to modify the testing data. The classification method instead relies on generalizing empirical evidence from a corpus, without precluding any underlying linguistic constraints that may affect order.

We approach the task in this way for several reasons. First, the development of linguistic resources in any language benefits from linguistic classification approaches that lend themselves to automatic lexical acquisition. Lexical acquisition tasks generally require a seed lexicon to extract further lexical information from a new source, and these two sources are used together to extract further

---

[2] This may be an effect of the available corpora; if larger datasets with more modifiers were available, perhaps the ability to automatically order modifiers in proposed systems would improve.

lexical information, and so forth. The type of classes developed for this project are constructed to be useful for such an application, so that bootstrapping based on a set of known modifiers is a straightforward task.

Second, we construct broad classes for modifiers instead of developing ordering constraints between individual modifiers in part to examine the feasibility of ordering modifiers within a constraint-based grammar. Grammars of this sort, such as HPSG and LFG, provide rules for combining lexical items to generate natural utterances. These rules operate on lexical and syntactic classes, and a lexicon specifies which words belong to which classes. It is therefore optimal to have a usable set of modifier classes that can operate within such a grammar. In hopes of addressing this issue, each modifier in the training data is mapped to a class, and ordering constraints operate on the classes themselves.

Third, the extent to which a given NLP technique is usable in languages other than the language of origin is an issue in the development of any NLP system. The majority of the world's languages have a very limited amount of linguistic resources (such as stemmers or lemmatizers), and these so-called resource-poor languages may benefit from NLP approaches that do not rely heavily on the data available in resource-rich languages. Previous approaches to automatic modifier ordering have made good use of the rich linguistic tools available, such as a morphology module to transform plural nouns and comparative and superlative adjectives into their base forms [27]. However, approaching the problem of modifier ordering in this way limits the ability to extend to languages without such tools. To this end, although the work here is far from resource-independent (NPs are extracted from a corpus), the classification system itself determines modifier classes without reliance on any external resource.

The classification system discussed in this chapter requires as input only lists of noun phrases containing modifiers. The resulting model may then be used to order new modifier instances in a variety of domains.

In the next section, we discuss prior work on this topic, and address the differences in our approach. Section 3 discusses the relationship between modifier ordering and referring expression generation. Section 4 describes the ideas behind the modifier classification system. Sections 5 and 6 present the materials and methodology of the current study, with a discussion of the corpus involved and the basic modules used in the process. In Section 7 we discuss the results of our study. Finally, in Section 8, we outline areas for improvement and possible future work.

2   Related Work

Discerning the rules governing the ordering of adjectives has been an area of research for quite some time, with some work dating back thousands of years (cf. [23]). However, there is no consensus on the exact qualities that affect position. It is clear that some combinations of modifiers before a noun sound more natural than others, although all are strictly speaking grammatical (see Fig. 1).

Most approaches assume an underlying relationship between semantics and prenominal position, and most related research has looked specifically at adjectives. We now turn to a discussion of these early approaches. This is followed by a discussion of computational approaches to prenominal modifier ordering, and how the current approach builds off this prior work.

2.1 Early Approaches

Whorf [32], in a discussion of grammatical categories, defines adjectives as *cryptotypes*, a category for word groups where each member follows similar rules, but there is no overt linguistic marking for which words fall into which group. Whorf suggests that there are two main cryptotypes of adjectives, each with subclasses. These correspond to *inherent* and *non-inherent* qualities of the noun they modify. Whorf proposes an ordering based on these categories, where non-inherent adjectives precede inherent adjectives. Thus the modifiers in the phrase "large red house" exhibit a standard ordering, and the modifiers in the phrase "red large house" do not. Whorf points out that the order may be reversed to make a balanced contrast, but only by changing the normal stress pattern, and the form is "at once sensed as being reversed and peculiar" (5).

A similar approach to predicting the ordering of adjectives based on the semantic relationship between adjective and head noun is discussed by Ziff [35]. Ziff explains that adjectives that are semantically close to the head nouns they modify are tied to these nouns and so are selected with them when language is produced (e.g., "bright light", "playful child"). Adjectives that are not tied to head nouns in this way will appear with a greater range of nouns (e.g., "good", "pretty"), and so appear farther from the head noun. Adjectives that appear farther from the head noun are thus those that are applied to a wider class of nouns and have less semantic similarity to the head noun, comparable to Whorf's analysis. In complementary work, Posner [25] argues in favor of the idea that more noun-like modifiers tend to occur closer to the head noun than less no unlike modifiers, a phenomenon coined the *nouniness principle*.

Martin [19] introduces the *definiteness of denotation principle*, which proposes that adjectives that are more definite in their meaning (e.g., less reliant on surrounding context) come closer to the noun than adjectives that are not – as such, size generally precedes color. This mirrors the earlier work done by Whorf [32] and Ziff [35], and further psychological work on the topic builds off this study.

For example, in a psychological study by Danks and Glucksberg [8], the authors examine human reference to objects and conclude that prenominal adjective ordering is "determined by the pragmatic demands of the communication situation". This is called the *pragmatic communication rule*, and specifies that adjectives that are more discriminating precede less discriminating ones, but the specific ordering of adjectives is dependent on the context.

The field of NLG has been influenced in particular by the psychological work of Pechmann [24], who distinguishes two kinds of modifiers: those that are derived from properties that are easily cognizable and those that are derived from comparison of the intended referent with surrounding objects. In this proposal, modifiers derived from easily cognizable properties appear closer to the head noun than modifiers derived from comparison processes (again explaining why size modifiers would tend to appear farther from the head noun than color modifiers).

Not all approaches to prenominal modifier ordering are purely semantic; for example, Wulff [33] examines using word length to determine order. Torrent and Alemany [28] examine syntactic and orthographic cues. Boleda et al. [3] look at syntactic and morphological patterns, in particular using derivational markings. And Vendler [31] views the natural order of adjectives as a function of transformational operations applying to different classes, where adjectives are transformed from a predicate structure into a nominalization, and the task of prenominal modifier ordering is a task of transformational rule ordering.

However, almost all work exploring prenominal adjective order focuses in part on adjective semantics, proposing that adjective ordering is affected by some sort of semantic scale. The exact nature of the scale differs, but research tends to favor an analysis where adjectives closer to the head noun have a stronger association with the noun than adjectives that are farther away. These approaches can be characterized as predicting modifier order based on degrees of semantic closeness to the noun. This follows what is known as **Behaghel's First Law** [1]:

> *Word groups:* What belongs together mentally is placed close together syntactically. [5]

Semantically-based approaches are satisfying when the goal is to explain word order generally, but are less satisfying when the goal is to order any given set of modifiers in an open domain. A semantically-based approach to modifier ordering requires determining the semantic properties of each modifier used, relative to the context in which it occurs, with ordering decisions based on these constraints.

Such requirements are difficult to implement in a generation system, and do not lend themselves to an automatic approach that aims to order modifiers in novel circumstances. They provide little guidance on how to order modifiers that have not been seen before and do not tell us when and how more than one modifier ordering is natural. Further, such approaches do not fit in well with constraint-based frameworks, where syntactic rules must operate on syntactic types, not individual words. If a modifier classification scheme is to be implemented in a system, it should be able to create a variety of natural, unmarked orders; be robust enough to handle a wide variety of modifiers; and be flexible enough to allow different natural orderings.

In this chapter, the approach to determining the ordering of prenominal modifiers therefore backs away from proposing a relationship between modifier position and any underlying factors. However, it does not preclude such interpretations; modifier ordering is examined from a perspective of form rather than function. Future work may examine mapping the classes derived here to semantic (or other linguistic) categories in order to determine further principles of language use.

2.2 Computational Approaches

In one of the earliest attempts at automatically ordering prenominal modifiers, Shaw and Hatzivassiloglou [27] develop ways to order several modifier types that appear before nouns. These include adjectives as well as nouns, such as "baseball" in "baseball field"; gerunds, such as "running" in "running man"; and past participles, such as "heated" in "heated debate". The authors present three different methods for prenominal modifier ordering that may be implemented in a generation system. These are the *direct evidence* method, the *transitivity* method, and the *clustering* method.

Briefly, given prenominal modifiers a and b in a training corpus, the direct evidence method utilizes probabilistic reasoning to determine whether the frequency count of the ordered sequence $<a, b>$ or $<b, a>$ is stronger. The transitivity method makes inferences about unseen orderings among prenominal modifiers; given a third prenominal modifier c, where a precedes b and b precedes c, the authors can conclude that a precedes c. In the clustering method, an order similarity metric is used to group modifiers together that share a similar relative order to other modifiers.

Shaw and Hatzivassiloglou achieve their highest prediction accuracy of 90.67% for prenominal modifiers (including nouns) using their transitivity technique on prenominal modifiers from a medical corpus. They test how well their system does when trained and tested on different domains, and achieve an overall prediction accuracy of only 56%, not much higher than random guessing. The authors conclude that prenominal modifiers are used differently in the two domains, and so domain-specific probabilities are necessary in such an automated approach.

Malouf [14] continues this work, determining the order for sequences of prenominal adjectives by examining several different statistical and machine learning techniques. These achieve good results, ranging from 78.28% to 89.73% accuracy. Malouf achieves the best results by combining direct evidence, memory-based learning, and positional probability, which reaches 91.85% accuracy at predicting the prenominal adjective orderings in the first 100 million tokens of the BNC.

In Malouf's memory-based learning approach, each unordered modifier pair $\{a, b\}$ seen in the training data is stored in a vector as some canonical ordering ab. If a precedes b more than b precedes a, ab belongs to class X. If b precedes a more, ab belongs to class Y. The feature set for each instance corresponds to the last 8 letters of a and the last 8 letters of b. An unknown modifier pair seen in the testing data can therefore be classified as class X or Y by determining its 16 character feature set and examining the similarity to known instances.

In the positional probability approach, which is similar to our own, he posits that the probability of an adjective appearing in a particular position depends only on that adjective. With this independence assumption, the probability of the ordered set $<a, b>$ given the unordered set $\{a, b\}$ is the product of $P(<a, x>|\{a, x\})$ and $P(<x, b>|\{b, x\})$, where x is any other adjective in the data. This is used for unordered pairs that have not been seen, and achieves 89.73% prediction accuracy when used along with direct evidence.

However, the proposed ordering approaches do not extend to other kinds of prenominal modifiers, such as gerund verbs and nouns. They are also difficult to extend to sets of modifiers larger than a pair. The model is also not tested on a different domain.

As noted in the introduction, the work here builds off of [21], who showed that a system where modifiers belong to independent classes can be used to correctly order most modifiers across several corpora. In this approach, each modifier is mapped to a class based on the frequency with which it occurs in different prenominal positions. The resultant model may then be used to order new sets of modifiers. Modifiers with strong positional preferences are in a class separate from modifiers with weaker positional preferences, and different ordering constraints apply to different classes. Ordering is therefore based on the learned modifier classes, not the individual modifiers. This approach does not rely on detailed linguistic information to construct classes or to order modifiers, and appears relatively stable across different corpora.

Grouping modifiers into classes based on prenominal positions mitigates the problems noted by Shaw and Hatzivassiloglou that ordering predictions cannot be made (1) when both a and b belong to the same class, (2) when either a or b are not associated to a class that can be ordered with respect to the other, and (3) when the evidence for one class preceding the other is equally strong for both classes: When two modifiers belong to the same class, we may propose both orders, and all classes can be ordered with respect to one another.

This approach has several other benefits over previous automated modifier ordering methods. Any prenominal modifiers a and b seen in the training corpus can be ordered, regardless of which particular modifiers they appear with and whether they occur together in the training data at all. Only the model developed from the training corpus is necessary to order new sets of modifiers, which, as pointed out in [14], is considerably resource-saving over approaches that require a minimum of all training data available during classification. The modifier classes are usable across many different domains, which permits sets of modifiers from new corpora to be ordered naturally in most cases. This method is also conceptually simple and easy to implement.

3 The Problem of Ordering Prenominal Modifiers in NLG

Generating referring expressions in part requires generating the adjectives, verbs, and nouns that modify head nouns. In order for these expressions to clearly convey the intended referent, the modifiers must appear in an order that sounds natural and mimics human language use. As discussed above, it is not clear exactly how to order any given set of modifiers, particularly within an NLG system.

Referring expression generation algorithms generally produce structures with properties to be mapped to modifiers, and these modifiers must be ordered to produce a final referring expression [7,9,13]. Developments in REG research have provided ways to select which modifiers will be used to refer to specific referents, but the method used to determine the ordering of these modifiers is an open issue.

The difficulty with capturing the ordering of modifiers stems from the problem of data sparsity. In the example in Fig. 1, the modifier "silk" may be rare enough in any corpus that finding it in combination with another modifier is nearly impossible. This makes creating generalizations about ordering constraints difficult. Malouf [14] examined the first million sentences of the British National Corpus and found only one sequence of adjectives for every twenty sentences. With sequences of adjectives occurring so rarely, the chances of finding information on any particular sequence is small. The data is just too sparse.

4   Towards a Solution

To create a flexible system capable of predicting a variety of natural orderings, we use several corpora to build broad modifier classes. Modifiers are classified by where they tend to appear prenominally, and ordering constraints between the classes determine the order for any set of modifiers. This system incorporates three main ideas:

1. Not all modifiers have equally stringent ordering preferences.
2. Modifier ordering preferences can be learned empirically.
3. Modifiers can be grouped into classes indicative of their ordering preferences.

The classification scheme therefore allows rigid as well as more flexible orders. In the case where several orderings are natural, the system generates several orderings. Classification is not based on any mapping between position and semantics, morphology, or phonology, but does not exclude any such relationship: This classification scheme builds on what there is clear evidence for, independently of why each modifier appears where it does.

To create our model, all simplex noun phrases (NPs) are extracted from parsed corpora. A *simplex NP* is defined as a maximal noun phrase that includes premodifiers such as determiners and possessives but no post-nominal constituents such as prepositional phrases or relative clauses [27]. From these simplex NPs, we extract all those headed by a noun and preceded by only prenominal modifiers. This includes modifiers tagged as adjectives (JJ, JJS), nouns (NN, NNS), gerunds (VBG), base verbs (VB), past tense verbs (VBD), and past participles (VBN). We also filter out the quantity adjectives "few", "many", "much", and "several". These largely pattern with the words "most" and "some", which are usually tagged as determiners. This keeps our accuracy from becoming inflated due to the presence of modifiers that occur in very fixed positions. The counts and relative positions of each modifier are stored, and these are converted into position probabilities in vector file format. Modifiers are classified based on the positions in which they have the highest probabilities of occurring.

5   Materials

To create the training and test data, we utilize the Penn Treebank-3 [17] releases of the parsed Wall Street Journal corpus, the parsed Brown corpus, and the parsed Switchboard corpus.[3] The parsed Wall Street Journal corpus is a selection of over one million words collected from the Wall Street Journal over a three-year period. From this, we extract 13,866 noun phrases. The parsed Brown corpus is half a million words of prose written in various genres, including mystery, humor, and popular lore, collected from newspapers and periodicals in 1961. This provides us with 2,429 noun phrases. The Switchboard corpus is over one million words of spontaneous speech collected from thousands of five-minute telephone conversations, from which we extract 2,474 noun phrases. Several programs were constructed to then analyze the information provided by these data. The details of each module are outlined below.

5.1   Code Modules

The following five components were developed (in Python) for this project.

Modifier Extractor – This function takes as input a parsed corpus and outputs a list of all occurrences of each noun phrase. First, all simplex noun phrases are extracted. These phrases are then chopped from their determiners and the quantity adjectives discussed above. Single letters and words that appear prenominally and are occasionally mistagged as modifiers are removed as well. A complete list of these filtered words is available in Table 1.
  input: Parsed corpus
  output: List of NPs

Modifier Organizer – This function takes as input a list of NPs preceded by only prenominal modifiers are returns a vector with frequency counts for all positions in which each observed modifier occurs.
  input: List of NPs
  output: Vector with distributional information for each modifier position

Modifier Classifier – This function takes as input a vector with distributional information for each modifier's position, and from this builds a classification model by determining the class for each modifier.
  input: Vector with distributional information for each modifier position
  output: Ordering model: File with each modifier associated to a class

Prenominal Modifier Ordering Predictor – This program takes as input two files: an ordering model and a test set of NPs with prenominal modifiers. The program then uses the model to assign a class to each modifier seen in the testing data, and predicts the ordering for all the modifiers that appear prenominally. A discussion of the ordering decisions is given below. This program then compares its predicted ordering of modifiers prenominally to

---

[3] Clearly, such data would not be available in resource-poor languages, but we use these corpora to extract noun phrases; our modifier ordering system works the same regardless of how the list of noun phrases is created.

Table 1. Filtered words

a	about	above	after
at	down	every	few
in	inside	just	like
many	most	much	near
no	off	only	outside
past	several	such	the
through	under	up	very

the observed ordering of modifiers prenominally. It returns precision and recall values for its predictions.

input: Ordering model, list of NPs

output: Precision and recall for modifier ordering predictions

6   Method

6.1   Classification Scheme

It is rare to find more than three modifiers before a noun in most corpora (see, for example, the distribution of modifiers shown in Table 5), and we therefore assume a maximum of four primary modifier positions. This assumption covers 99.69% of our data. The four broad positions may be labelled as *one*, *two*, *three* and *four*. Position one is the modifier slot closest to the head noun. Position two is the modifier slot one slot removed from the head noun; position three is the modifier slot two slots removed from the head noun, etc. The longest noun phrases for this experiment are therefore those with five words: Four modifiers followed by a noun (see Fig. 2).

small	smiling	white	fuzzy	bunny
four	three	two	one	

Fig. 2. Example NP with prenominal positions

A modifier's class is determined by counting the frequency of the modifier in each position. This is turned into a probability over all four positions. All position probabilities $\geq 25\%$ determine the modifier class. Since there are four positions in which a modifier can occur, using only those with a 25% probability or higher provides a baseline cutoff: A single modifier preceding a noun has approximately equal probability of being in each of the four positions. So, a modifier that appears in position one 25% or more of the time and position two 25% or more of the time, but appears in positions three and four less than 25% of the time, will be classified as a modifier that tends to be in positions one and two (a "one-two" modifier).

To calculate modifier position for each phrase, counts are incremented for all feasible positions. This is a way of sharing evidence among several positions. For example, in the phrase "clean wooden spoon", the adjective "clean" was counted as occurring in positions two, three, and four, while the adjective "wooden" was counted as occurring in positions one, two, and three.

This method is preferred over assigning probabilities based on single positions in order to distribute the probability mass over the four possible positions. If we assign "clean" a 50% probability of being in position two and "wooden" a 50% probability of being in position one, then we skew probability towards positions closer to the head nouns – all prenominal modifier orderings will evidence a position one, but not all prenominal modifier orderings will evidence a position four.

In light of this approach, positions two and three are slightly more frequent than positions one and four. It is clear that the 25% positional probability cutoff is a simplifying assumption, and an alternative approach to determining a modifier's class could use all positions in which the modifier has a higher than average probability of occurring.[4] The average positional probability ends up being around 21% for positions one and four, and around 29% for positions two and three. In practice, using these cutoffs slightly lowers our accuracy, particularly when training and testing on different domains, and so 25% is a useful simple probability cutoff point.

The classification that emerges after applying this technique to a large body of data gives rise to the broad positional preferences of each modifier. In this way, a modifier with a strict positional preference can emerge as occurring in just one position; a modifier with a less strict preference can emerge as occurring in three. There are nine derivable modifier classes in this approach, listed in Table 2.

Table 2. Modifier classes

Class 1: one	**Class 4:** four	**Class 7:** three-four
Class 2: two	**Class 5:** one-two	**Class 8:** one-two-three
Class 3: three	**Class 6:** two-three	**Class 9:** two-three-four

Examples of the intermediary data structures in this process are given in Tables 3 and 4. Table 3 shows a vector of modifiers with frequency counts in each prenominal position, which serves as input to the Modifier Classifier function. Each modifier is then associated to a class based on its positional probabilities. An example of this is shown in Table 4. The first column lists the modifier,

---

[4] E.g., if M is the set of all modifiers, and mod is a single modifier, then for a given $position$, if

$$P(position|mod) > \frac{\sum_{mod_i \in M} P(position|mod_i)}{|M|} \quad (1)$$

then that $position$ may be used to determine the modifier class.

Table 3. Example modifier classification intermediate structure

corrupt	four 2	three 2	two 1	one 0
historical	four 0	three 7	two 7	one 7
large	four 111	three 106	two 91	one 6
red	four 13	three 42	two 43	one 28
uninsured	four 4	three 4	two 2	one 0

Table 4. Example modifier classification intermediate structure

corrupt	three-four	four 0.40	three 0.40	two 0.20	one 0.00
historical	one-two-three	four 0.00	three 0.33	two 0.33	one 0.33
large	two-three-four	four 0.35	three 0.34	two 0.29	one 0.02
red	two-three	four 0.10	three 0.33	two 0.34	one 0.22
uninsured	three-four	four 0.40	three 0.40	two 0.20	one 0.00

the second the class, and the remaining columns list the probabilities of each position used to determine the modifier class.

A diagram of how a modifier is associated to a class is shown in Fig. 3. In this example, "red" appears in several NPs. In each sequence, we associate "red" to its possible positions within the four prenominal slots. We see that "red" occurs in positions one, two and three; two, three, and four; and two and three. With only this data, "red" has a 12.5% probability of being in position one; a 37.5% probability of being in position two; a 37.5% probability of being in position three; and a 12.5% probability of being in position four. It can therefore be classified as belonging to Class 6, the class for modifiers that tend to occur in positions two and three.

This kind of classification allows the system to be flexible to the idea that some modifiers exhibit stringent ordering constraints, while others have more

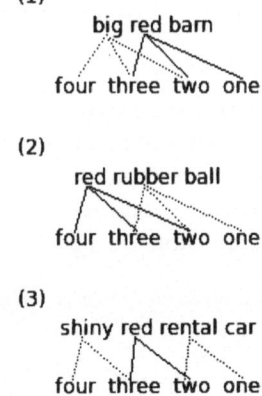

(1)

big red barn

four three two one

(2)

red rubber ball

four three two one

(3)

shiny red rental car

four three two one

Fig. 3. Constructing the class of the modifier "red"

loose constraints. Some modifiers may always appear immediately before the noun, while others may generally appear close to or far from the noun. This gives us a platform to develop different ordering constraints for different kinds of modifiers.

The frequencies of all extracted groupings of prenominal modifiers used to build our model are shown in Table 5. The frequencies of the extracted classes are shown in Table 6.

Table 5. Frequency of prenominal modifier amounts

Mods	Count	Percentage
2	16578	88.33
3	1919	10.22
4	214	1.14
5	53	0.28
6	5	0.03

Table 6. Modifier class distribution

Class	Position	Count	Percentage
1	one	12	0.16
2	two	41	0.55
3	three	41	0.55
4	four	16	0.21
5	one-two	317	4.24
6	two-three	1168	15.63
7	three-four	276	3.69
8	one-two-three	3010	40.27
9	two-three-four	2589	34.64

Modifiers of Class 8, the class for modifiers that show a general preference to be closer to the head noun but do not have a strict positional preference, make up the largest portion of the data. An example of a modifier from Class 8 is "historical". The next largest portion of the data are modifiers of Class 9, the class for modifiers that show a general preference to be farther from the head noun. An example of a modifier from Class 9 is "large". With these defined, "large historical house" is predicted to sound grammatical and unmarked, but "?historical large house" is not.

Some expected patterns also emerge in these groupings. For example, "red", "orange", "yellow", and "green", are determined to be Class 6. "56-year-old", "41-year-old" and "four-year-old", are determined to be Class 7, and "big" and "small" are both determined to be Class 9. These sorts of suggest that the system is clustering modifiers with similar semantic and morphological features together, without explicit knowledge of these features.

Once classified, modifiers may be ordered according to their classes. The proposed ordering constraints for these classes are listed in Table 7. Modifiers belonging to Class 3 are placed closer to the head noun than modifiers belonging to Classes 4 and 7; modifiers belonging to Class 6 are placed closer to the head noun than modifiers belonging to Classes 3, 4, 5, and 9; etc.

Table 7. Proposed modifier ordering

Class	Position	Generated Before Class							
1	one	2	3	4	5	6	7	8	9
2	two	3	4	6	7	9			
3	three	4	7						
4	four								
5	one-two	2	3	4	6	7	8	9	
6	two-three	3	4	7	9				
7	three-four	4							
8	one-two-three	4	6	7	9				
9	two-three-four	4	7						

Note that using this classification scheme, the ordering of modifiers that belong to the same class is not predicted. This seems to be reflective of natural language use. For example, both "uninsured" and "corrupt" are predicted to be Class 7. This seems reasonable; whether "corrupt uninsured man" or "uninsured corrupt man" is a more natural ordering of prenominal modifiers is at least debatable. Indeed, it may be the case that both are equally natural, but mean slightly different things. We leave a discussion of such phenomena for future research. 22.75% of the pairs ordered within our three combined corpora (Table 8) had known modifiers that belong to the same class.

When the class of a modifier is not known (e.g., when the testing corpus contains a modifier that was not seen in the training corpus), the modifiers may be ordered based on the class of the known modifiers: For example, when ordering a pair of modifiers with one unknown and one known modifier, if the known modifier belongs to Classes 1, 2, 5, or 8, it may be placed closer to the head noun than the unknown modifier, while if it belongs to Classes 3, 4, 7, and 9, it may be placed farther from the head noun than the unknown modifier. This reflects the idea that Classes 1, 2, 5, and 8 are all classes for modifiers that broadly prefer to be closer to the head noun, while Classes 3, 4, 7, and 9 are all classes for modifiers that broadly prefer to be farther from the head noun.

When the class of none of the modifiers is known, we rely on a back-off strategy that utilizes word similarity to make ordering decisions. The last characters of each modifier in the training data are compared against each unknown modifier. The unknown modifiers are then mapped to the classes of the modifiers they are most similar to, with the highest number of character matches in each position, and ordered based on these classes. If no match is found for one of the modifiers, the modifiers are ordered based on the classes of the known modifiers, as discussed in the last paragraph. If no match is found for any of the modifiers, all orderings are returned. This is similar to the form-based similarity metric proposed in [14], but may be used for sets of modifiers larger than a pair.

In a handful of cases, the distribution of the positions for a given modifier did not combine to form a class (for example, if positions "one", "two", "three", and "four" had equal probability). In these cases, we classified the modifier as "unknown", and treated it as unseen in our analysis.

6.2   Evaluation

In order to test how well the proposed system works, 10-fold cross-validation was used on the three combined corpora. The held-out data was selected as random lines from the corpus, with a list storing the index of each selected line to ensure no line was selected more than once. In each trial, modifier classification was learned from 90% of the data and the resulting model was used to predict the prenominal ordering of modifiers in the remaining 10%.

In each trial, the modifiers from all noun phrases in the test data were stored in unordered groups. The ordering for each unordered prenominal modifier pair $\{a, b\}$ was then predicted based on the model built from the training data, and the predictions were compared to the actual orderings seen in the test data. The ordering predictions followed the constraints listed in Table 7.

The system accuracy may be measured on three different dimensions:

1. Proportion of unordered modifier pairs $\{a, b\}$ that can be predicted.
2. Proportion of ordered modifier pairs $<a, b>$ seen in the testing data that are correctly predicted by the system.
3. Proportion of ordered modifier pairs $<a, b>$ predicted by the system that are seen in the testing data.

The first two measurements listed are both measures of recall, while the last is a measure of precision. Since this system produces an ordering for each given set of modifiers, we achieve 100% recall for all unordered modifier pairs. Below, we report the values of precision and recall for ordered pairs.

Precision is the number of true positives divided by the number of true positives plus false positives. This is calculated as $tp/(tp + fp)$, where tp is the number of predicted orderings found in the test data, and fp is the number of predicted orderings not found in the test data. This measure provides information about how accurate the modifier classification is. Recall is the number of true positives divided by the number of true positives plus false negatives. This is calculated as $tp/(tp + fn)$, where tp again is the number of predicted orderings found in the test data, and fn is the number of orderings in the test data that are not predicted by our system. This measure provides information about the proportion of modifiers in the training data that can be correctly ordered.

We evaluated precision and recall in terms of type and token. Evaluating the accuracy of our system by token tells us the overall proportion of orderings that we correctly predict, while evaluating by type tells us the proportion of unique orderings that we correctly predict. However, token precision can necessarily not be calculated (we do not predict how many times a pair appears in the dataset, but rather if it appears at all), and we found that type and token recall were quite close (with less than 1% higher for token than type across the different evaluations). The values listed below are therefore the precision and recall values for type.

Our precision and recall values are similar to the values reported by [27] and [14] for accuracy. In these earlier approaches, one order is generated for each pair in the test set, and so one incorrect ordering is also one ordering in the test set that is not predicted. This means that the previously reported values for accuracy correspond to both token precision and token recall. Since the approach outlined here sometimes generates more than one ordering for a set of modifiers, and looks for matches throughout the test data, our results are best measured using type precision and type/token recall measurements. It has been shown that precision of a lexicon has less impact on grammar accuracy than recall [34], and so we hope our approach achieves a high recall for ordered pairs.

In our evaluation, we consider an unordered set of modifiers to be one modifier set for which there may be one or more prenominal orders. For example, given modifiers a and b, which appear in the test set as the ordered sequences $<a, b>$ and $<b, a>$, we distinguish a single set of unordered modifiers that manifest in two prenominal modifier ordering types. Storing the test data in this way allows us to better calculate precision and recall for all orderings found in the test data.

7   Results

First, in order to compare our approach with previous approaches, we examined training our system on a randomly selected 90% of the Wall Street Journal (WSJ) corpus and testing on the remaining 10%. This gives us quite good results, with a token recall of 96.70%, a type recall of 96.57% and a type precision of 82.75%. Shaw and Hatzivassiloglou achieve their highest prediction accuracy of 71.04% on the same corpus, and so the method outlined here appears to offer an improvement, with our token recall value being most directly comparable to Shaw and Hatzivassiloglou's prediction accuracy. We now examine how well our system does when applied across different corpora.

Results are shown in Table 8. The precision values illustrate that 70.40% of all the orderings generated by the model have corresponding types in the test data. The recall values show that the model produces ordered modifier pairs for 84.81% of the modifier pair types seen in the test data. The values given are averages over each trial. The standard deviation for each average is given in parentheses. On average, 227 unordered modifier pairs were ordered in each trial, based on the assigned classes of 1,676 individual modifiers.

Table 8. Type precision & recall percentages for prenominal modifier ordering

Recall	Precision
84.81 (2.69)	70.40 (1.55)

The results above show the overall precision and recall when predicting orders for all modifier pairs in the test set. However, 33.47% of the modifier pairs in the test data had at least one unknown modifier, and so some sort of back-off strategy was used. Generating precision and recall values for only those pairs in the test data with known modifiers yields a higher type precision of 73.85% and a type recall of 91.16%.

In light of the fact that nouns preceded by just two modifiers make up more than 88% of the data, and Class 8 ("one-two-three") and Class 9 ("two-three-four") compose the majority of the assigned classes, it is interesting to see how well the system does when only these two classes are assigned. By changing Classes 1, 2, 5, and 6 to Class 8 in our models, and Classes 3, 4, and 7 to Class 9 in our models, the system precision and recall goes down, but still maintains relatively good values – 82.48% recall and 66.57% precision.

For comparison, a simple baseline was constructed that predicted both possible orders for each unordered set of modifiers. This achieves a recall of 100%, but a type precision of 50.16%. We also examined the upper bound of accuracy for this task, by training and testing on the same sets of data. This gave us a type precision of 83.49% and a type recall of 96.84%. It is clear from this that even a perfect method for ordering adjectives within this framework will be well below 100% precision and recall. This was also noted by Malouf [14].

As discussed in the introduction, previous attempts have achieved very poor results when training and testing their models on different domains, resulting in values that were about equal to random guessing (56% as reported in [27]). We conclude our analysis by testing the accuracy of our models on this task. To do this, we combine two corpora to build our model and then test this model on the third. Results of this task are shown in Table 9. We find that our system does relatively well on this task, with the best results reaching 87.21% recall and 70.53% precision.

Combining the Brown and WSJ corpus to build our modifier classes and then testing on the Switchboard (Swbd) corpus yields a high recall and a precision with more than 20% improvement over the baseline. Other training and testing combinations produce similar results, although slightly lower scores: A model built from the Switchboard corpus and the WSJ corpus achieves 83.50% type recall and a 64.81% type precision, while a model built from the Switchboard corpus and the Brown corpus achieves 80.80% type recall and 64.60% type precision.

As noted in Section 2.2, the system generates both modifier orderings for modifiers belonging to classes without any ordering constraints (e.g., when both

Table 9. Type precision & recall for prenominal modifier ordering of a new domain

Training Corpus	Testing Corpus	Recall	Precision
Brown+WSJ	Swbd	87.21%	70.53%
Swbd+WSJ	Brown	83.50%	64.81%
Swbd+Brown	WSJ	80.80%	64.60%

Table 10. Orderings not predicted and predicted orderings not found

Did Not Predict	Did Not Find
high-risk high-yield	same exact
blatant unsubstantiated	brokerage commercial
abortion clinic	industrial primitive
ship spare	young sweet
indicated new	holding testing
permanent titanium	monthly second

modifiers belong to the same class). But with individual modifier pairs occurring so infrequently, it is unlikely to find a given pair within the test set in both
orders, even if both orders sound natural. Lower precision values for each individual domain may therefore be a trade-off for system stability across different
domains.

Examples of decisions affecting the system accuracy are listed in Table 10. This
helps to give a sense of how "natural" the proposed orderings are. Sequences
the system does not correctly predict include (1) modifying compounds (e.g.,
"abortion clinic"); (2) modifiers that modify a phrase (e.g., "second" in "second
monthly payment" is not just modifying "payment", but "monthly payment");
(3) modifiers that may be equally "natural" in either order, but appear in the
corpus in just one order ("same exact").

8   Discussion

This chapter describes a new approach to ordering prenominal modifiers. The
outlined method achieves a type recall of 96.57% and a type precision of 82.75%
when training and testing on the same domain, and an average type recall of
84.81% and average type precision of 70.40% when training and testing on different domains. Evaluated across different corpora, the system precision and recall
values remain relatively stable, suggesting that this modifier ordering approach
may be used to order modifiers in a variety of domains.

These results show that groups of modifiers can successfully be ordered by
modifier class, when these classes are built from empirical evidence drawn from
a corpus. Such a system could be used in natural language generation, particularly within referring expression generation, and the classes utilized within an
automatic lexical acquisition task. Having a class-based modifier ordering approach may also be useful in constraint-based grammars. We will now briefly
discuss each of these points. More work would need to be done to adapt such a
system to a resource-poor language, but this provides a clear starting point.

One promising effect of these simple modifier classes is that the same classes
could be used to classify unseen modifiers based on the known classes of
surrounding modifiers. This could in turn be used for further bootstrapping,
expanding a lexical database automatically. For example, given the following

noun phrase, we could develop a proposed class for the unknown modifier "shining":

grey   shining metallic   chain
three-four unknown one-two head-noun

Given its position and the classes of the surrounding modifiers, unknown could be two-three.

Grouping modifiers into classes that determine their order offers the possibility of automatically ordering modifiers within a constraint-based grammar. Ordering rules that operate on the modifier classes outlined in this system, perhaps by using the ordering constraints presented in Table 7, could operate on modifiers whose classes are specified in the lexicon. The classification developed here may be a starting point for further detailed modifier classification to be incorporated into such grammars.

Modifier ordering in this way also helps further the goals of referring expression generation. Given a final referring expression with an unordered set of modifiers, this system may be used to produce a final ordered expression. Alternately, in a framework where the ordering or weighting of properties determines the ordering of output modifiers, a given set of properties could be ordered automatically before referring expression generation begins. REG algorithms generally require a prespecified weighting or listing or properties, which determines the order in which each property is considered by the algorithm. If these properties could first be mapped to their corresponding modifiers, the ordering among modifiers could determine the ordering or weighting among properties that serve as input to the algorithm.

It bears mentioning that this same system was attempted on the Google Web 1T 5-Gram corpus [4], where we used WordNet [20] to extract sequences of nouns preceded by modifiers. The precision and recall were similar to the values reported here, however, the proportions of prenominal modifiers betrayed a problem in using such a corpus for this approach: 82.56% of our data had two prenominal modifiers, 16.79% had four, but only 0.65% had three. This pattern was due to the many extracted sequences of modifiers preceding a noun that were not actually simplex NPs. That is, the 5-Grams include many sequences of words in which the final one has a use as a noun and the earlier ones have uses as adjectives, but the 5-Gram itself may not be a noun phrase. We found that many of our extracted 5-Grams were actually lists of words (for example, "Chinese Polish Portuguese Romanian Russian" was observed 115 times). In the future, we would like to examine ways to use the 5-Gram corpus to supplement our system.

We also examined ordering using a few other strategies. We implemented a version of Behaghel's Law of Increasing Terms, which states that, given two phrases, when possible, the shorter precedes the longer. Applying this to our study, longer modifiers would therefore be predicted to be closer to the head noun than shorter modifiers. We examined determining length both by number of vowels present, which is closely related to the number of syllables in a word, and by number of letters. However, this approach does not appear to be better than guessing. Applying this rule alone to our dataset, we achieve a type recall of 65.36% and a type precision of 52.60% when length is determined by vowels, and a type recall of 59.30% and type precision of 53.53% when length is determined by number of letters.

We also implemented a version of the Syllable Contact Law [22], which predicts that two adjacent syllables will generally be structured such that the phone ending the first syllable is higher in sonority than the phone beginning the second syllable, and the ordering between these two syllables will favor the greatest sonority drop. Applying this to word boundaries across modifier pairs, given an unordered modifier pair {a, b}, modifier a can be predicted to precede modifier b if the sonority of the last phone in a is higher in sonority than the first phone in b, and, further, the ordering between a and b will be the one with the greatest sonority drop between the two modifiers. Using a rough mapping of characters to phones, we achieve relatively good results, with a type recall of 71.64% and a type precision of 51.69%. Further work could examine a more detailed application of this theory.

In future work, we hope to measure how acceptable the prenominal modifier sequences that display the predicted orders are. We would like to pay particular attention to larger groups of modifiers and develop ways to order word compounds that modify another head noun. Prenominal modifier ordering provided by this system also serves as a starting point to add rules about how prosody, length, morphology, and semantics may affect order.

Acknowledgements. I am grateful to Emily Bender, Brian Roark, Aaron Dunlop, Advaith Siddharthan, Kees van Deemter, Ehud Reiter, and Emiel Krahmer for useful discussions and guidance throughout this project.

References

1. Behaghel, O.: Von Deutscher Wortstellung, vol. 44. Zeitschrift Für Deutschen, Unterricht (1930)
2. Bever, T.G.: The cognitive basis for linguistic structures. In: Hayes, J.R. (ed.) Cognition and the Development of Language. Wiley, New York (1970)
3. Boleda, G., Badia, T., Schulte im Walde, S.: Morphology vs. syntax in adjective class acquisition. In: Proceedings of the ACL-SIGLEX Workshop on Deep Lexical Acquisition, Ann Arbor, MI, pp. 77–86 (2005)
4. Brants, T., Franz, A.: Web 1t 5-gram version 1 (2006), http://www.ldc.upenn.edu/Catalog/CatalogEntry.jsp?catalogId
5. Clark, H.H., Clark, E.V.: Psychology and Language: An Introduction to Psycholinguistics. Harcourt Brace Jovanovich, New York (1976)
6. Daelemans, W., van den Bosch, A.: Generalization performance of backpropagation learning on a syllabification task. In: Proceedings of TWLT3: Connectionism and Natural Language Processing. University of Twente, Enschede (1992)

7. Dale, R., Reiter, E.: Computational interpretations of the gricean maxims in the generation of referring expressions. Cognitive Science 18, 233–263 (1995)
8. Danks, J.H., Glucksberg, S.: Psychological scaling of adjective order. Journal of Verbal Learning and Verbal Behavior 10, 63–67 (1971)
9. van Deemter, K.: Generating referring expressions: Boolean extensions of the incremental algorithm. Computational Linguistics 28(1), 37–52 (2002)
10. Greenberg, J.H.: Some Universals of Grammar with Particular Reference to the Order of Meaningful Elements, pp. 73–113. MIT Press, London (1963)
11. Group, T.L.S.: Specialist lexicon (2009)
12. Halliday, M., Matthiessen, C.: Construing Experience As Meaning: A Language-Based Approach to Cognition. Cassell, London (1999)
13. Krahmer, E., van Erk, S., Verleg, A.: Graph-based generation of referring expressions. Computational Linguistics 29(1), 53–72 (2003)
14. Malouf, R.: The order of prenominal adjectives in natural language generation. In: Proceedings of the 38th ACL, Hong Kong, pp. 85–92 (2000)
15. Manning, C.D.: Automatic acquisition of a large subcategorization dictionary from corpora. In: Meeting of the Association for Computational Linguistics, pp. 235–242 (1993)
16. Marcus, M.P., Santorini, B., Marcinkiewicz, M.A.: Building a large annotated corpus of English: The Penn Treebank. Computational Linguistics 19, 313–330 (1993)
17. Marcus, M.P., Santorini, B., Marcinkiewicz, M.A., Taylor, A.: Treebank-3 (1999)
18. Martin, J.E.: Semantic determinants of preferred adjective order. Journal of Verbal Learning and Verbal Behavior 8, 697–704 (1969)
19. Martin, J.E.: Some competence-process relationships in noun phrases with prenominal and postnominal adjectives. Journal of Verbal Learning and Verbal Behavior 8, 471–480 (1969)
20. Miller, G.A., Fellbaum, C., Tengi, R., Wakefield, P., Langone, H., Haskell, B.R.: WordNet: A lexical database for the English language (2006)
21. Mitchell, M.: Class-based ordering of prenominal modifiers. In: Proceedings of ENLG 2009 (2009)
22. Murray, R., Vennemann, T.: Sounds change and syllable structure in Germanic phonology. Language 59, 514–528 (1983)
23. Panini: Astadhyayi. India (ca 350 BCE)
24. Pechmann, T.: Incremental speech production and referential overspecification. Linguistics 27, 89–110 (1989)
25. Posner, R.: Iconicity in syntax: The natural order of attributes. In: Bouissac, P., Herzfeld, M., Posner, R. (eds.) Iconicity. Essays on the Nature of Culture, pp. 305–337. Stauffenburg Verlag, Tübingen (1986)
26. Sag, I., Wasow, T., Bender, E.: Syntactic Theory: A Formal Introduction. CSLI Publications, Stanford University (2003)
27. Shaw, J., Hatzivassiloglou, V.: Ordering among premodifiers. In: Proceedings of the 37th ACL, pp. 135–143. Association for Computational Linguistics, Morristown (1999)
28. Torrent, G.B., i Alemany, L.A.: Clustering adjectives for class acquisition. In: Proceedings of the EACL 2003 Student Session, Budapest, pp. 9–16 (2003)
29. Toutanova, K., Klein, D., Manning, C., Singer, Y.: Feature-rich part-of-speech tagging with a cyclic dependency network. In: Proceedings of HLT-NAACL 2003, pp. 252–259 (2003)
30. Toutanova, K., Manning, C.D.: Enriching the knowledge sources used in a maximum entropy part-of-speech tagger. In: Joint SIGDAT Conference on EMNLP/VLC-2000, Hong Kong (2000)

31. Vendler, Z.: Adjectives and Nominalizations, Mouton, The Netherlands (1968)
32. Whorf, B.L.: Grammatical categories. Language 21(1), 1–11 (1945)
33. Wulff, S.: A multifactorial corpus analysis of adjective order in English. International Journal of Corpus Linguistics 8, 245–282 (2003)
34. Zhang, Y., Baldwin, T., Kordoni, V.: The corpus and the lexicon: Standardizing deep lexical acquisition evaluation. In: Proceedings of the 45th ACL, Ann Arbor, pp. 152–159 (2005)
35. Ziff, P.: Semantic Analysis. Cornell University Press, Ithaca (1960)

Attribute-Centric Referring Expression Generation

Robert Dale and Jette Viethen

Centre for Language Technology
Macquarie University
Sydney, Australia
rdale@science.mq.edu.au, jviethen@science.mq.edu.au

Abstract. In this chapter, we take the view that much of the existing work on the generation of referring expressions has focused on aspects of the problem that appear to be somewhat artificial when we look more closely at human-produced referring expressions. In particular, we argue that an over-emphasis on the extent to which each property in a description performs a discriminatory function has blinded us to alternative approaches to referring expression generation that might be better-placed to provide an explanation of the variety we find in human-produced referring expressions. On the basis of an analysis of a collection of such data, we propose an alternative view of the process of referring expression generation which we believe is more intuitively plausible, is a better match for the observed data, and opens the door to more sophisticated algorithms that are freed of the constraints adopted in the literature so far.

1 Introduction

Ever since at least the late 1980s, the generation of referring expressions has been a key focus of interest in the natural language generation community (see, for example, [3,4,5,6,7,8,10,11,13,14,15,16]). A glance at the proceedings of workshops in natural language generation over the last ten years demonstrates that the topic attracts significantly more attention than other aspects of the generation process, such as text structuring, sentence planning, and linguistic realisation; and, largely because of this critical mass of interest, the generation of referring expressions has served as the focus of the first major evaluation efforts in natural language generation (see, for example, [1,9]).

This level of attention is due in large part to the consensus view that has arisen as to what is involved in referring expression generation: the task is widely agreed to be one that involves a process of selecting those attributes of an intended referent that distinguish it from other potential distractors in a given context, resulting in what is often referred to as a *distinguishing description*. Based on this agreement on the nature of the task, a large body of work has developed over the last 20 years that has focused on developing algorithms that encompass an ever-wider range of referential phenomena. Key issues that have underpinned

E. Krahmer, M. Theune (Eds.): Empirical Methods in NLG, LNAI 5790, pp. 163–179, 2010.

much of this work are the need to take account of the computational complexity of the algorithms developed; the production of descriptions which are in some sense minimal (in that they do not contain unnecessary information); and, occasionally, a recognition of some of the phenomena that characterise the kinds of referring expressions that humans produce.

Our key point is that the last of these concerns has not occupied the central position that it should, and that the other criteria that have been considered in order to determine what counts as a good algorithm have been given undue weight. Our position is based on two observations. First, it is clear (and this observation is not new) that humans do not always produce what are referred to as *minimal distinguishing descriptions*, i.e. referring expressions whose content walks the line between being both necessary and sufficient, despite this having served as a concern for much algorithmic development in the past. As has long been recognised, human-produced referring expressions are in many cases informationally redundant. The Incremental Algorithm [5], which serves as the basis for many algorithmic developments in the literature, is occasionally given credit because it can lead to referring expressions that contain redundancy; but even its authors were careful not to claim that the redundancy it produces is the same as that produced by humans. The kinds of redundancy evident in human-produced referring expressions have never, in our view, been properly explored, and this has led to algorithms which at best pay lip-service to the need to account for redundancy.

Our second observation (also not particularly new, but surprisingly ignored in the literature) is that different people do different things when faced with the same reference task. This poses serious questions for both the development of algorithms and their evaluation: as has been noted for other tasks that involve natural language output (such as document summarisation), in such circumstances we clearly cannot evaluate an algorithm by comparing its results against a single gold-standard answer. Even with a range of possible candidate answers, it is still possible that an algorithm might produce a perfectly acceptable solution that is not present amongst this set. This forces us to consider more carefully what it is that we are doing when we develop algorithms for the generation of referring expressions (or, for that matter, for any generation task): are we trying to emulate or predict the behavior of a single given speaker in a given situation? Or are we trying to produce a solution which might somehow rate as optimal in a task-based evaluation scenario (such as might be measured by the amount of time it takes a listener to locate a referred-to object), recognizing that human-produced referring expressions are not necessarily optimal in this sense? What counts as a good solution may be quite different in each case.

In this chapter, we examine some human-produced data in order to observe the variety that it exhibits (Section 2). We then posit a different way of thinking about the process of referring expression generation (Section 3), and go on to demonstrate how some machine-learning experiments run on the human-produced data are supportive of this view (Section 4). Although this way of looking at the problem is, we argue, more explanatory than previous approaches, it still leaves a number of important questions unanswered; we discuss these in Section 5, before drawing some conclusions in Section 6.

2 What Do People Do?

For the purposes of the explorations discussed in this chapter, we use a corpus of human-produced referring expression data called the GRE3D3 Corpus. This corpus is introduced and discussed in significant detail elsewhere (see [20,21]); here we summarise its key characteristics.

The data in question was gathered via a web-based experiment where participants were asked to produce referring expressions that would enable a listener to identify one of a number of objects shown on the screen. The purpose of the experiment was to explore how relations were used in referring expressions, and the design of scenes was carefully controlled so that the use of relations was encouraged but not strictly necessary in order to identify the intended referent.

Participants visited a website, where they first saw an introductory page with a set of instructions and a sample stimulus scene. The task was to describe the target referent in the scene (marked by a grey arrow) in a way that would enable a friend looking at the same scene to pick it out from the other objects. Figure 1 shows an example stimulus.

Each participant was assigned one of two trial sets of ten scenes each; the two trial sets are superficially different (involving colour variations and mirror-image orientations), but the elements of the sets are pairwise identical in terms of the factors explored in the research. The complete set of 20 scenes is shown in Figure 2: Trial Set 1 consists of Scenes 1 through 10, and Trial Set 2 consists of Scenes 11 through 20.[1] Each scene contains three objects, which we refer to as the *target* (the intended referent), the potential *landmark* (a nearby object), and the *distractor* (a further-away object).

The experiment was completed by 74 participants from a variety of different backgrounds and ages. One participant asked for their data to be discarded. We also disregarded the data of one other participant who reported to be colour-blind. One participant consistently produced very long and syntactically complex referring expressions including reference to parts of objects and the onlooker, such as *the red cube which rests on the ground and is between you and the yellow cube of equal size*. While these descriptions are very interesting, they are clearly outliers in our data set.

Eight participants consistently only used **type** to describe the target object, for example simply typing *cube* for the target in Scene 5. These descriptions were excluded from the corpus under the assumption that the participants had not understood the instructions correctly or were not willing to spend the time

---

[1] Scene 1 is paired with Scene 11, Scene 2 with Scene 12, and so on; in each pair, the only differences are (a) the colour scheme used and (b) the left–right orientation, with these variations being introduced to make the experiment less monotonous for participants; in [20], we report that these variations appear to have no significant effect on the forms of reference used.

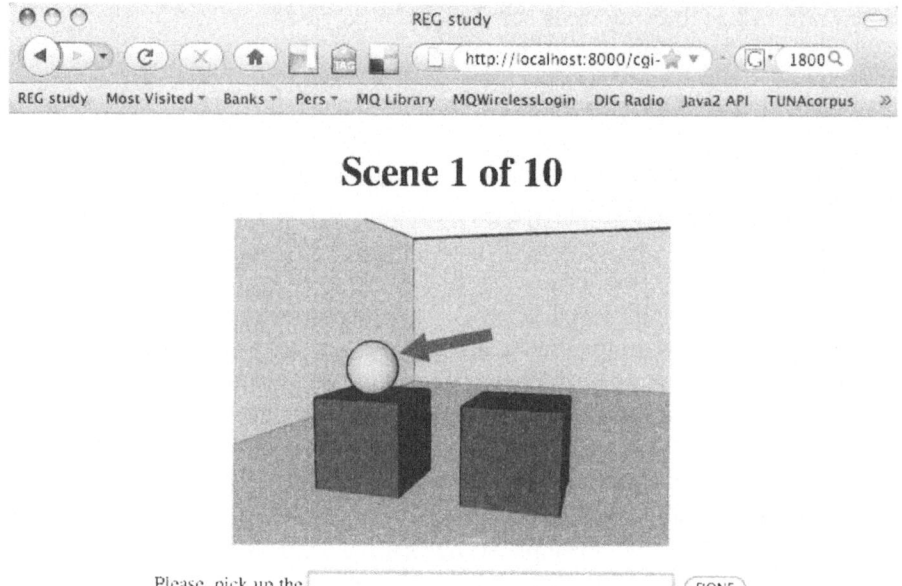

Fig. 1. An example stimulus scene

required to type fully distinguishing referring expressions for each trial. For the research presented here we also removed 7 descriptions which failed to uniquely distinguish the target referent from the other two objects.

The resulting corpus consists of 623 descriptions. As we are at this point only interested in the semantic content of these descriptions, we carried out a rigourous lexical and syntactic normalisation. In particular, we corrected spelling mistakes; normalised names for *colour* values and head nouns (such as *box* instead of *cube*); and replaced complex syntactic structures such as relative clauses with semantically equivalent simpler ones such as adjectives.

This normalisation makes it apparent that every one of the descriptions is an instance of one of the 18 content patterns shown in Table 1; for ease of reference, we label these patterns A through R. Each pattern indicates the set of attributes used in the description, where each attribute is identified by the object it describes (tg for target, lm for landmark) and the attribute used (col, size and type for colour, size and type, respectively).

The description patterns in Table 1 are collected from across all the stimuli scenes; but of course not all patterns are equally common for every scene. Table 2 shows the distribution of the different patterns across the different scenes. We

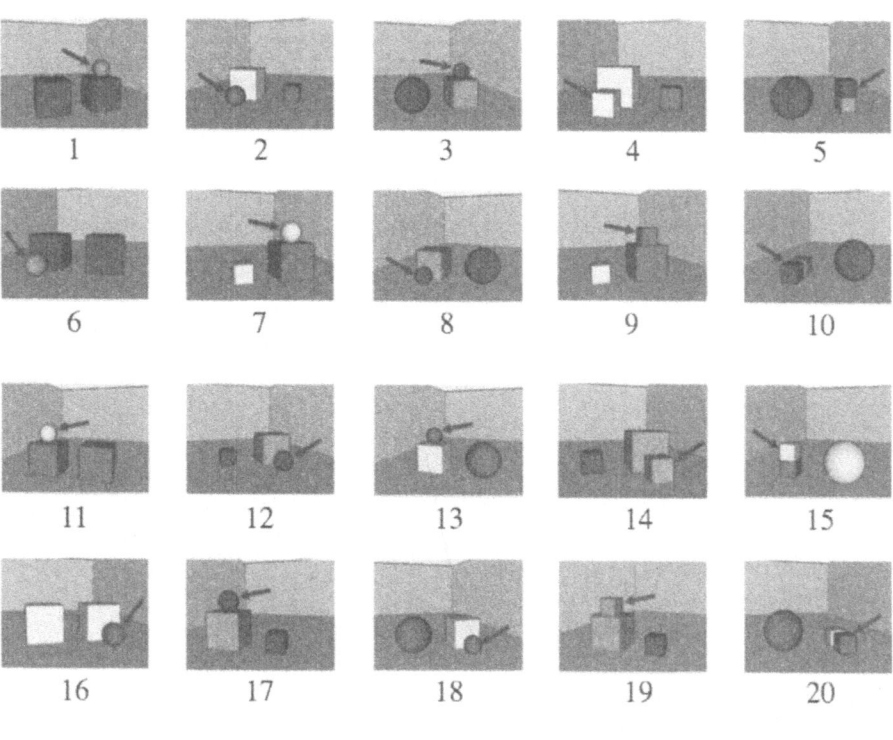

Fig. 2. The complete set of stimulus scenes

collapse the pairwise equivalent scenes from the two stimulus sets into one, so that we are now only dealing with ten scenes. The primary observation of relevance here is that there is no one correct answer for how to refer to the target in any given scene: for example, Scenes 4, 5, 9 and 10 result in five semantically distinct referring expression forms, and Scene 7 results in 12 distinct referring expression forms. All of these are distinguishing descriptions, so all are acceptable forms of reference, although some contain more redundancy than others. Most obvious from the table is that, for many scenes, there is a predominant form of reference used; for example, pattern F ($\langle$tg_size, tg_col, tg_type$\rangle$) accounts for 43 (68%) of the descriptions used in Scene 4, and pattern A ($\langle$tg_col, tg_type$\rangle$) is very frequently used in a number of scenes.

Faced with such variation, the question arises: what exactly should we be trying to model when we develop algorithms for the generation of referring expressions? Clearly we could try to model an individual speaker, although this would require considerably more data than we have available here, and it remains unclear just what the utility of such a modelling exercise would be; we would also need to have some way of accounting for the fact that speakers are not necessarily consistent with themselves. An alternative would be to try to

Table 1. The 18 different patterns corresponding to the different forms of description that occur in the GRE3D3 corpus

Label	Pattern	Example
A	⟨tg_col, tg_type⟩	*the blue cube*
B	⟨tg_col, tg_type, rel, lm_col, lm_type⟩	*the blue cube in front of the red ball*
C	⟨tg_col, tg_type, rel, lm_size, lm_col, lm_type⟩	*the blue cube in front of the large red ball*
D	⟨tg_col, tg_type, rel, lm_size, lm_type⟩	*the blue cube in front of the large ball*
E	⟨tg_col, tg_type, rel, lm_type⟩	*the blue cube in front of the ball*
F	⟨tg_size, tg_col, tg_type⟩	*the large blue cube*
G	⟨tg_size, tg_col, tg_type, rel, lm_col, lm_type⟩	*the large blue cube in front of the red ball*
H	⟨tg_size, tg_col, tg_type, rel, lm_size, lm_col, lm_type⟩	*the large blue cube in front of the large red ball*
I	⟨tg_size, tg_col, tg_type, rel, lm_size, lm_type⟩	*the large blue cube in front of the large ball*
J	⟨tg_size, tg_col, tg_type, rel, lm_type⟩	*the large blue cube in front of the ball*
K	⟨tg_size, tg_type⟩	*the large cube*
L	⟨tg_size, tg_type, rel, lm_size, lm_type⟩	*the large cube in front of the large ball*
M	⟨tg_size, tg_type, rel, lm_type⟩	*the large cube in front of the ball*
N	⟨tg_type⟩	*the cube*
O	⟨tg_type, rel, lm_col, lm_type⟩	*the cube in front of the red ball*
P	⟨tg_type, rel, lm_size, lm_col, lm_type⟩	*the cube in front of the large red ball*
Q	⟨tg_type, rel, lm_size, lm_type⟩	*the cube in front of the large ball*
R	⟨tg_type, rel, lm_type⟩	*the cube in front of the ball*

determine what characteristics of human referential behaviour, whether widely-used or specific to a select few 'good referrers', are particularly effective from the hearer's point of view, and then to try to model such an 'optimal' generation strategy.

Ultimately, practical applications of NLG techniques are more likely to be able to make use of algorithms developed from the latter perspective. To get to such a point, we have to first dissect referential behaviour in order to determine what kinds of characteristics might be relevant; the aim of this chapter is to reframe the referring expression generation problem in such a way as to shed light on the processes involved.

Table 2. The number of occurrences of each description pattern across the ten stimulus scenes. Counts of 5 or more are shown in bold to make it easier to see the patterns, on the basis that low counts might be considered noise; of course, 5 is a fairly arbitrary cut-off point.

Pattern	1	2	3	4	5	6	7	8	9	10
					Scene #					
A	**17**	**23**			**37**	**38**	**25**			**39**
B	**14**	**8**	3		**16**	**7**	**8**	3		**10**
C		4		1			3		1	
D						1	1		1	
E	4		1		2					
F	1	1	**14**	**43**	4	3	2	**24**	**39**	**8**
G	1		**14**		2		1	**14**		1
H		1	1	**13**	2	1	2	1	**17**	2
I				3					1	
L				1						
M	1		**7**					4		
N	**11**	**13**				**7**	**14**			
O		4					1			
P							1			
Q		3					2			
R	**13**	**5**	**9**			2	2	1		

3   An Alternative Paradigm

The Incremental Algorithm and a considerable number of other algorithms proposed in the literature share the structure shown in Figure 3. Starting with an empty set of properties D to be used in the referring expression and a set of distractors C, they add one property at a time to D (Line 3), and check after each step whether all distractors from C are ruled out yet (Lines 4 and 5).

What makes one algorithm different from another, in terms of this schema, is the particular means by which the next attribute to use is selected (Line 3, the first step inside the repeat loop): in the Greedy Algorithm [3], for example, the most useful attribute (i.e., that which rules out most potential distractors) is chosen next, whereas in the Incremental Algorithm the next attribute chosen is selected according to a pre-determined preference order.

In earlier work [19], we observed that the kinds of variation found in a particular set of human data might be modelled by using different preference orderings in the Incremental Algorithm. But this requires not only that potentially different preference orders are used by different speakers, but even that different preference orders are used by the same speaker on different occasions. Constantly switching preference orders does not seem to us to be a very convincing explanation for what humans do. More fundamentally, our view is that the schema shown in Figure 3 does not seem very convincing as a characterisation of the cognitive processes involved in producing referring expressions, requiring as it does a check of the scene (the second step in the repeat loop) after each attribute

---

Given an intended referent R, a set of distractors C, a set of attributes L_R, and the set of attributes to use in a description D:

1   Let $D = \emptyset$
2   repeat
3   add a selected attribute $\in L_R$ to D
4   recompute C given D
5   until $C = \emptyset$

Fig. 3. The structure of referring expression generation algorithms

is added in order to determine whether further work should be done. There may be contexts where such a careful strategy is pursued, but it seems less likely that this is what people do each and every time they construct a referring expression. We do not, therefore, consider the Incremental Algorithm or its derivatives (or any other algorithms that share the schematic structure shown in Figure 3) to be convincing models of human behaviour.

One element of the Incremental Algorithm that we do find appealing, however, is its notion that there might be a 'force of habit' element to referential behaviour, as encoded in the preference ordering: for example, a preference order which proposes that an object's colour should be the first candidate attribute to be tried is just a way of saying that, for this speaker (or the speaker represented by such a preference ordering), 'colour is often useful so we'll try it first'. Clearly such a heuristic will not always be used—if all the objects in a scene are of the same colour, we might think it unlikely that a speaker would even consider the use of colour before ruling it out. However, in very many circumstances, it happens to be the case that colour is useful.

This leads us to hypothesise that there may be straightforwardly-apparent aspects of scenes that a speaker uses to determine what information is likely to be useful in producing a referring expression. For example, at least up to a certain number of objects in a scene, if the target referent is of one colour and all the other objects are of another, then, rather than a careful strategy of considering each attribute on its merits as embodied in conventional algorithms, it seems to us more plausible that some kind of gestalt perceptual strategy might be in play: the speaker knows, just by looking at the scene and without any algorithmic computation, that colour is a useful attribute for picking out the intended referent.

Of course, this is somewhat vague. To try to make the idea more concrete, we can look at the data available at a different grain-size than is usually done in algorithmic studies: rather than considering the overall content of the referring expressions produced by speakers, we can look at the individual attributes that occur in our data set. We can then establish how the gestalt perceptual characteristics of the scenes correspond to the use of each attribute, to see whether there are patterns across speakers that suggest more commonality than is apparent in the variety of data represented in Table 2.

4   What We Can Learn from the Data

4.1   Learning Algorithms for Description Construction

It seems obvious that the visual context of reference must play at least some role in the choice of attributes in a given referring expression. An obvious question is then whether we can learn the description patterns in this data from the contexts in which they were produced. To explore this, we can capture the relevant aspects of context by means of a notion of *characteristics of scenes*. The characteristics of scenes which we hypothesise might have an impact on the choice of referential form in our data are those summarised in Table 3.[2] Each of these captures some aspect of the scene which we consider to be readily apparent without complex computation on the part of the perceiver, a point we will return to later. Each feature compares two of the three objects in the scene to each other with respect to their values for one of the attributes type, colour and size. In this way we capture how common the landmark's and the target's properties are overall, which is a simple way of approximating their visual salience.

We used the implementation provided in the Weka toolkit [22] of the C4.5 decision tree classifier [17] to see what correspondences might be learned between these characteristics of scenes and the forms of referring expression shown in Table 1. The pruned decision tree learned by this method predicted the actual form of reference used in only 48% of cases under 10-fold cross-validation. On the basis of the discussion in the previous section, this is not surprising: Given that there are many 'gold standard' descriptions for each scene, a low score is to be expected. A mechanism which learns only one answer will inevitably be 'wrong'—in the sense of not replicating the human-produced description— in many cases, even if the description produced is still a valid distinguishing description.

More revealing, however, is the rule learned from the data:

if tg_type = dr_type
then use F $(\langle$tg_size, tg_col, tg_type$\rangle)$
else use A $(\langle$tg_col, tg_type$\rangle)$

Patterns A and F are the two most prevalent patterns in the data, and indeed one or the other appears at least once in the human data for each scene. If we analyse the learned rule in more detail, we see that it predicts Pattern F for Scenes 4, 5, 8, 9, 14, 15, 18, and 19, and A for all other scenes. The pattern distribution in Table 2 shows that this means that the rule is able to produce a 'correct' answer (in the sense of emulating what at least one speaker did) for every scene.

As there is clearly another factor at play causing variation in the data, we then re-ran the classifier, this time using the participant ID as well as the scene characteristics in Table 3 as features. This improved pattern prediction to 57.62%.

---

[2] Note that the spatial relations between the distractor and the other two objects are not listed as characteristics of the scenes because they are the same in all scenes. Whether the distractor is to the left or the right of the other two objects has no impact.

Table 3. The 10 characteristics of scenes

Label	Attribute	Values
tg_type = lm_type	Target and Landmark share Type	TRUE, FALSE
tg_type = dr_type	Target and Distractor share Type	TRUE, FALSE
lm_type = dr_type	Landmark and Distractor share Type	TRUE, FALSE
tg_col = lm_col	Target and Landmark share Colour	TRUE, FALSE
tg_col = dr_col	Target and Distractor share Colour	TRUE, FALSE
lm_col = dr_col	Landmark and Distractor share Colour	TRUE, FALSE
tg_size = lm_size	Target and Landmark share Size	TRUE, FALSE
tg_size = dr_size	Target and Distractor share Size	TRUE, FALSE
lm_size = dr_size	Landmark and Distractor share Size	TRUE, FALSE
rel	Relation between Target and Landmark	on_top_of, in_front_of

Table 4. Accuracy of learning attribute inclusion. Statistically significantly increases (p<0.01) are marked in bold.

Attribute to Include	Baseline (0-R)	Using Scene Characteristics	Using Scene Characteristics and Participant
tg_col	78.33%	78.33%	**89.57%**
tg_size	57.46%	**90.85%**	90.85%
rel	64.04%	65.00%	**81.22%**
lm_col	74.80%	**87.31%**	**93.74%**
lm_size	88.92%	**95.02%**	95.02%

This suggests that individual differences may indeed be capturable from the data, although we would need more data than the mere 10 examples we have from each participant to learn a good predictive model for a single speaker.

4.2   Learning Heuristics for Attribute Inclusion

Attribute Inclusion Based on Scene Characteristics. The experiments just described demonstrate that there is indeed considerable variation across speakers, and put into question any attempt to model human referring behaviour that ignores this. On the other hand, it seems implausible that there are no commonalities whatsoever between speakers. The alternative approach we propose here is to look for commonalities in the data in terms of the *constituent elements* of the different forms of reference used for each scene, rather than at the level of complete descriptions: are there characteristics of scenes which are highly likely to result in *specific attributes* being used in descriptions? This way of thinking about the data was foreshadowed in [21], where we observed that our participants could be separated into those who always used relations, those who never used relations, and those who sometimes used relations.

Target Colour:
include tg_col

Target Size:
if tg_type = dr_type then include tg_size

Relation:
if rel = on_top_of and lm_size = dr_size then include rel

Landmark Colour:
if we have used a relation then include lm_col

Landmark Size:
if we have used a relation and tg_col = lm_col then include lm_size

---

Fig. 4. Rules learned on the basis of scene characteristics

As a baseline here, we use the success rate of simply predicting the majority class. We might think of this as a 'context-free' approach, in the sense that the particular context of reference plays no role in the decision as to whether or not an attribute should be used. Table 4 compares the results for this approach with one model that is trained on the characteristics of scenes, and another that takes both the characteristics of scenes and the participant ID into account.[3]

The 'context-free' strategies work surprisingly well for predicting the inclusion of some attributes in the human data. As has been noted in other work, colour is often included in referring expressions irrespective of its discriminatory power, and this is borne out by the data here. Perhaps more surprising is the large degree to which the inclusion of landmark size is captured by a context-free strategy.

Improvement on all attributes other than target colour increases when we take into account the characteristics of the scenes, confirming the widely-held assumption that the context of reference does indeed make a difference. When we add participant ID to the features used in the learner, performance improves further still, indicating that there are speaker-specific consistencies across contexts. The numbers suggest that colour is effectively a participant property, whereas size is a scene property.

It is interesting to look at the rules learned on the basis of the scene characteristics alone; these are shown in Figure 4. Not surprisingly, the rule derived for target colour inclusion is simply to always include the colour (i.e., the same context-free colour inclusion rule that proves most effective in modelling the data without reference to scene characteristics). The target's size is included if target and distractor are of the same type (Scenes 2, 4, 7, 9, 12, 14, 17, 19); the spatial

---

[3] As before, the results reported are for the accuracy of a pruned decision tree, under 10-fold cross-validation.

relation between the target and the landmark is included if the target is on top of the landmark and the landmark is of the same size as the distractor (Scenes 1, 3, 11, 13); the landmark's colour is included in all relational descriptions; and the landmark's size goes into all relational descriptions if the target and landmark cannot be distinguished by colour (Scenes 4, 9, 14 and 19).

Attribute Inclusion on a Speaker-by-Speaker Basis. The rules learned when we include participant ID are more complex, but can be summarised in a way that demonstrates how this approach can reveal something about the variety of ways in which speakers might be approaching the task of referring expression generation.

Focussing on the question of whether or not to use *the target object's colour* in a referring expression, the learner identifies five heuristics, which apply to the 63 participants as follows:

- For 37 participants it learned to always use colour, irrespective of the context (this corresponds to the baseline rule learned above).
- For the rest of the participants it always uses colour if the target and the landmark are of the same type (which again is intuitively quite appropriate).
- When the target and the landmark are not of the same type, we see more variation in learned behaviour:
  - for 19 participants colour simply doesn't get used;
  - in scenes where the target is on top of the landmark, six participants use colour if the target and the distractor have the same size, with two of these six always using colour in scenes where the target is in front of the landmark, and the other four using colour only if target and distractor do not share size; and
  - one participant is characterised as using colour if the target and distractor do not share colour.

Participants showed least commonality when it came to heuristics for the inclusion of a *relation*. For 15 participants, the model predicts that they always use a relation; 29 are predicted never to include a relation. Of the remaining participants, eight share a heuristic with one other participant, while the remaining 11 have a unique decision pattern. In total, 17 different rules were learned to account for the inclusion of relations in the descriptions.

Landmark colour is predicted to be used in all relational descriptions by 18 participants and in none by eight participants. For four participants, the use of the landmark's colour is predicted if the landmark shares its type with the target object; one of these four also includes the landmark colour for all cases where the target is in front of the landmark, and two participants use the landmark colour if target and distractor share size. The remaining seven participants each have their own heuristic. This results in 15 different heuristics.

As indicated by the lack of change in accuracy of prediction when participant ID is included (see Table 4), all participants are predicted to share the same heuristics for the inclusion of the *target's size* and the *landmark's size*, as shown in Figure 4 for these two properties. Size is therefore the property with the highest degree of commonality between the participants who contributed to this corpus.

The specific content of the rules mentioned above may appear idiosyncratic; they are just what the limited data in the corpus supports, and some elements of the rules may be due to artefacts of the specific stimuli used in the data gathering. We would require a more diverse set of stimuli to determine whether this is the case, but the basic point stands: *we can find correlations between characteristics of the scenes and the presence or absence of a particular attribute in referring expressions, even if we cannot predict so well the exact combinations of these correlations that a given speaker will use.* Of course, this is in some sense what all referring expression generation algorithms aim to capture; our claim here is that an attribute-centric model is more able to explain the human data.

From the behaviour of the 63 participants that contributed to the data set used in this study, we learn 30 different combinations of these attribute-inclusion heuristics. One way to think of this is that each combination of attribute-inclusion heuristics corresponds to a *speaker profile*. We have avoided listing the details of all 30 speaker profiles here; however, the sets of heuristics that were each used by more than one participant are shown in Table 5.

Table 5. The most common speaker profiles. The first column indicates the number of speakers sharing each profile.

#	tg_col	tg_size	tg_size	rel	lm_size
13	TgCol-T	TgSize-1	Rel-F	n/a	n/a
10	TgCol-T	TgSize-1	Rel-T	LmCol-T	LmSize-1
9	TgCol-1	TgSize-1	Rel-F	n/a	n/a
2	TgCol-3	TgSize-1	Rel-4	LmCol-F	LmSize-1
2	TgCol-T	TgSize-1	Rel-2	LmCol-T	LmSize-1
2	TgCol-1	TgSize-1	Rel-T	LmCol-1	LmSize-1

TgCol-T = always include tg colour
TgCol-1 = include tg colour if tg and lm share type.
TgCol-3 = include tg colour if tg and lm share type
          or if the tg is in front of the lm
          or if tg and dr share size.
TgSize-1 = include tg size if tg and dr share type.
Rel-F = never include a relation.
Rel-T = always include a relation.
Rel-2 = include a relation if tg shares type with dr but not lm.
Rel-4 = include a relation if the tg is on top of the lm
        and lm and dr share size
LmCol-T = if a relation is present, include the lm colour.
LmCol-F = never include lm colour.
LmCol-1 = include lm colour if tg and lm share type.
LmSize-1 = if a relation is present, include the lm size
           if tg and lm share colour.

So, for example, our data contains 13 people who automatically include the colour of an intended referent, never use a relation to a landmark, and make the inclusion of the referent's size dependent on the similarity of the referent to another object. We also have two people who make the inclusion of all attributes dependent on contextual factors, with the exception of the landmark's colour, which they never mention; and so on. Grouping people according to the collection of individual attribute-heuristics they use can be seen as a more fine-grained alternative to predicting a particular content pattern for each participant–scene combination, which also more easily generalises to new scenes.

5   What's Missing?

As we have argued above, and as captured schematically in Figure 3, existing approaches to the generation of referring expressions are primarily focussed on the development of algorithms whose main function is to control the serial incorporation of attribute values into a developing referring expression. This focus has either meant that all attributes are considered a priori equal (as in algorithms which compute discriminatory power, and use this as the sole determinant for attribute selection), or are ordered on the basis of some pre-defined but unexplained preferences (as in the Incremental Algorithm and its variants, as well as Krahmer et al.'s [16] graph-based algorithm).

We have proposed here a rather more bottom-up way of thinking about referring expression generation: rather than focus on algorithms whose purpose is to control the combination of individual attributes, we argue instead that we should focus on the individual attributes themselves, and explore what it is that makes them appropriate for inclusion in a developing referring expression in a given situation. The picture that emerges is one where we can think of an individual speaker's approach to reference as consisting of a collection of attribute-specific heuristics. These individual heuristics are shared across speakers to a greater or lesser degree, and the combinations of heuristics vary by speaker. A speaker may even use different heuristics depending upon features orthogonal to the referential context (such as who they are talking to, or the mission-critical nature of getting it right first time).

As we have suggested above, this makes for a much richer story of how humans produce referring expressions, and provides a basis for the development of algorithms that enable us to incorporate much more sophistication: depending on the circumstances, a referring expression might be generated almost without thinking, simply on the basis of general properties of a scene; or, in mission-critical situations, a more cautious, reflective and reasoned approach might be used.[4]

Clearly the model we have sketched here is not a complete picture of how referring expressions might be generated. What we have argued is that, at least in simple scenes, effective referring expressions almost 'jump out' at the speaker,

---

[4] The distinction here is deliberately reminiscent of Carletta's risky vs cautious distinction [2] in regard to referential behaviour in the HCRC Map Task Corpus.

without any need for complex reasoning or retrospective analysis to determine whether they are indeed successful distinguishing descriptions. Particularly as scenes get more complex, it is perhaps unlikely that referring expressions which just happen to be distinguishing can be 'read off' the scene, and more plausible that some reasoning is required by the hearer in order to determine whether the expression constructed so far is sufficient for the identification task at hand. Van der Sluis [18] cites an example of a referring expression that appears in an episode of the television series *Twin Peaks*, where a character by the name of Lucy is attempting to identify a referent for a speaker:

> Uhm, I'm gonna transfer to the phone on the table by the red chair ... [points in the direction of the phone] the ... the red chair, against the wall, uh the little table, with the lamp on it, the lamp that we moved from the corner? ... the black phone, not the brown phone ...

Examples like this suggest that, in appropriately complex referential situations, what we have is a repeated iteration between 'reading off' attributes from the scene, and reflective analysis that indicates more work is needed to achieve successful reference; however, unlike existing algorithms, there is no need for the success check to be carried out after each and every attribute. To model this kind of behaviour, we need to develop algorithms which, although they may construct a first-pass referring expression by means of a kind of 'parallel gestalt', are then open to a process of selective extension whereby a monitoring process (somewhat akin to the anticipation-feedback loop of [12]) decides what needs to be added in a more reasoned way. Our existing data does not allow us to explore these kinds of questions, and we are not aware of any data sets that do; but the framework we have sketched is suggestive of experiments that might be carried out to probe these kinds of phenomena more delicately.

6   Conclusions

In this chapter, we have suggested that the bulk of the existing work on referring expression generation, including our own previous work, has mistakenly placed the focus on the development of algorithms for combining attributes in a serial fashion to produce distinguishing descriptions. We have suggested an alternative way of thinking about the problem, where the focus is instead placed on the specific attributes that make up a referring expression. By doing this, we have shown how it is possible, despite the apparent wide variation in the forms of reference that speakers produce, to identify a number of component strategies that are common across many speakers. Under this analysis, the apparently broad variation that we see is a result of different speakers using different combinations of component strategies which themselves vary in the extent to which they are shared across speakers.

There are many questions left unresolved here. As we suggested earlier, the practical application of NLG techniques for generating referring expressions—as might be required in object location in 'omniscient room' scenarios or landmark description in navigation systems—is probably best served by identifying the component strategies that are most effective, and we have not touched on that question here. Also, as discussed in Section 5, the picture we have presented here is only a part of the story, and a complete algorithm still requires the integration of these component strategies into a model which is capable of extending a referring expression appropriately when it is inadequate for the task. However, we believe the framework described here provides a new perspective on what is involved in referring expression generation, and points the way to a range of experimental studies that will provide greater insight.

References

1. Belz, A., Kow, E., Viethen, J., Gatt, A.: The GREC challenge 2008: Overview and evaluation results. In: Proceedings of the 5th International Natural Language Generation Conference, Salt Fork OH, USA, pp. 183–191 (2008)
2. Carletta, J.C.: Risk-taking and Recovery in Task-Oriented Dialogue. Ph.D. thesis, University of Edinburgh (1992)
3. Dale, R.: Cooking up referring expressions. In: Proceedings of the 27th Annual Meeting of the Association for Computational Linguistics, Vancouver BC, Canada (1989)
4. Dale, R., Haddock, N.: Content determination in the generation of referring expressions. Computational Intelligence 7(4), 252–265 (1991)
5. Dale, R., Reiter, E.: Computational interpretations of the Gricean maxims in the generation of referring expressions. Cognitive Science 19(2), 233–263 (1995)
6. van Deemter, K.: Generating referring expressions: Boolean extensions of the Incremental Algorithm. Computational Linguistics 28(1), 37–52 (2002)
7. van Deemter, K.: Generating referring expressions that involve gradable properties. Computational Linguistics 32(2), 195–222 (2006)
8. Gardent, C.: Generating minimal definite descriptions. In: Proceedings of the 40th Annual Meeting of the Association for Computational Linguistics, Philadelphia PA, USA (2002)
9. Gatt, A., Belz, A., Kow, E.: The TUNA challenge 2008: Overview and evaluation results. In: Proceedings of the 5th International Natural Language Generation Conference, Salt Fork OH, USA, pp. 198–206 (2008)
10. Gatt, A., van Deemter, K.: Conceptual coherence in the generation of referring expressions. In: Proceedings of the 21st COLING and the 44th ACL Conference, Sydney, Australia (2006)
11. Horacek, H.: On referring to sets of objects naturally. In: Proceedings of the 3rd International Conference on Natural Language Generation, Brockenhurst, UK, pp. 70–79 (2004)
12. Jameson, A., Wahlster, W.: User modelling in anaphora generation: ellipsis and definite description. In: Proceedings of the 5th European Conference on Artificial Intelligence, Orsay, France, pp. 222–227 (1982)
13. Jordan, P.W.: Contextual influences on attribute selection for repeated descriptions. In: van Deemter, K., Kibble, R. (eds.) Information Sharing: Reference and Presupposition in Language Generation and Interpretation. CSLI Publications, Stanford (2002)

14. Kelleher, J., Kruijff, G.J.M.: Incremental generation of spatial referring expressions in situated dialog. In: Proceedings of the 21st COLING and the 44th ACL Conference, Sydney, Australia (2006)
15. Krahmer, E., Theune, M.: Efficient context-sensitive generation of referring expressions. In: van Deemter, K., Kibble, R. (eds.) Information Sharing: Reference and Presupposition in Language Generation and Interpretation, pp. 223–264. CSLI Publications, Stanford (2002)
16. Krahmer, E., van Erk, S., Verleg, A.: Graph-based generation of referring expressions. Computational Lingustics 29(1), 53–72 (2003)
17. Quinlan, J.R.: C4.5: Programs for Machine Learning. Morgan Kaufmann, San Francisco (1993)
18. van der Sluis, I.: Multimodal Reference: Studies in Automatic Generation of Multimodal Referring Expressions. Ph.D. thesis, Tilburg University, The Netherlands (2005)
19. Viethen, J., Dale, R.: Algorithms for generating referring expressions: Do they do what people do? In: Proceedings of the 4th International Conference on Natural Language Generation, Sydney, Australia, pp. 63–70 (2006)
20. Viethen, J., Dale, R.: Generating referring expressions: What makes a difference? In: Australasian Language Technology Association Workshop 2008, Hobart, Australia, pp. 160–168 (2008)
21. Viethen, J., Dale, R.: The use of spatial relations in referring expression generation. In: Proceedings of the 5th International Conference on Natural Language Generation, Salt Fork OH, USA (2008)
22. Witten, I.H., Frank, E.: Data Mining: Practical machine learning tools and techniques. Morgan Kaufmann, San Francisco (2005)

Assessing the Trade-Off between System Building Cost and Output Quality in Data-to-Text Generation

Anja Belz and Eric Kow

Natural Language Technology Group
School of Computing, Mathematical and Information Sciences
University of Brighton
Brighton BN2 3PB, UK
http://www.nltg.brighton.ac.uk/

Abstract. Data-to-text generation systems tend to be knowledge-based and manually built, which limits their reusability and makes them time and cost-intensive to create and maintain. Methods for automating (part of) the system building process exist, but do such methods risk a loss in output quality? In this paper, we investigate the cost/quality trade-off in generation system building. We compare six data-to-text systems which were created by predominantly automatic techniques against six systems for the same domain which were created by predominantly manual techniques. We evaluate the systems using intrinsic automatic metrics and human quality ratings. We find that there is some correlation between degree of automation in the system-building process and output quality (more automation tending to mean lower evaluation scores). We also find that there are discrepancies between the results of the automatic evaluation metrics and the human-assessed evaluation experiments. We discuss caveats in assessing system-building cost and implications of the discrepancies in automatic and human evaluation.

1 Introduction

Traditional Natural Language Generation (NLG) systems tend to be handcrafted know-ledge-based systems. Such systems tend to be brittle, expensive to create and hard to adapt to new domains or applications. Over the last decade or so, in particular following Knight and Langkilde's work on n-gram-based generate-and-select surface realisation [13,16], NLG researchers have become increasingly interested in systems that are automatically trainable from data. Systems that have a trainable component tend to be easier to adapt to new domains and applications, and increased automation in system building is often taken as self-evidently a good thing. The question is, however, whether reduced system building cost and increased adaptability are achieved at the price of a reduction in output quality, and if so, how great the price is. This in turn raises the question of how to evaluate output quality so that a potential decrease can be detected and quantified.

E. Krahmer, M. Theune (Eds.): Empirical Methods in NLG, LNAI 5790, pp. 180–200, 2010.

In this paper we set about trying to find answers to these questions. We start, in the following section, by briefly describing the corpus of weather forecasts which we used in our experiments. In the next section (Section 3), we outline four different approaches to building data-to-text generation systems which involve different combinations of manual and automatic techniques. Next (Section 4) we describe ten systems in the four categories which we used in the original set of experiments. In Section 5 we describe the human-assessed and automatically computed evaluation methods we used to comparatively evaluate the quality of the outputs of the ten systems. We then present the evaluation results and discuss implications of discrepancies we found between the results of the human and automatic evaluations (Section 6).

In material not previously reported in Belz & Kow 2009 [5], we also present the results of some follow-up experiments involving two additional, newly built systems, which we carried out in order to look into the impact of alternative input representations on the fully automatically trainable systems (Section 7), and present some further discussion of (i) issues in assessing the extent to which a system has been built manually, and (ii) the implications of the discrepancies we see between automatic and human-assessed evaluations (Section 8).

2   Data

The main component of the Prodigy-METEO corpus is a set of pairs of wind data (input) and corresponding wind forecast text (output).[1] The Prodigy-METEO outputs were extracted from the SumTime-METEO corpus [26]. We used only those forecasts from SumTime-METEO that are for the period 06:00–24:00 GMT. These are issued in the a.m. and make up roughly half of SumTime-METEO. An extract from an example forecast file (for 5Oct2000_03) is shown in Figure 2. From this subset of forecasts, we then extracted all 'wind statements' except the one for the long range outlook (i.e. all wind forecasts statements under points 2, 3 and 4, for 10m and 50m. Only the first wind statement from each a.m. forecast was included in Prodigy-METEO (for full details of the corpus creation process, see the corpus release notes [2]).

The Prodigy-METEO inputs (an example of such an input can be found at the top of Figure 3) are vectors of time stamps and wind parameters, and were 'reverse-engineered', by automatically aligning wind speeds and wind directions in the forecasts with time-stamps in the wind data file (an extract from the wind data file for 5 Oct 2000 is shown in Figure 1). In order to do this, wind speed and directions in the data file have to be matched with those in the forecast. This was not entirely straightforward, because often there is no exact match in the data file for the wind speeds and directions in the forecast. The strategy adopted was the

---

[1] In response to requests for our data, we have now released a pre-alpha version of the Prodigy-METEO Corpus which also includes outputs from the 10 systems described in Section 4 that were built with this data, as well as some additional human-authored wind forecasts. The corpus can be downloaded here: http://www.nltg.brighton.ac.uk/home/Anja.Belz/. For full details of the corpus contents, see Belz, 2009b [2].

```
Oil1/Oil2/Oil3_FIELDS
05-10-00

05/06  SSW  18  22  27   3.0  4.8  SSW  2.59
05/09  S    16  20  25   2.7  4.3  SSW  2.39
05/12  S    14  17  21   2.5  4.0  SSW  2.29
05/15  S    14  17  21   2.3  3.7  SSW  2.28
05/18  SSE  12  15  18   2.4  3.8  SSW  2.38
05/21  SSE  10  12  15   2.4  3.8  SSW  2.48
06/00  VAR  6   7   8    2.4  3.8  SSW  2.48
...
```

Fig. 1. Meteorological data file for 05-10-2000, a.m. (names of oil fields anonymised).

```
FORECAST FOR:-
Oil1/Oil2/Oil3 FIELDS
...

2. FORECAST 06-24 GMT, THURSDAY, 05-Oct 2000

=====WARNINGS: RISK THUNDERSTORM.    =======

WIND(KTS)  CONFIDENCE: HIGH
  10M:   SSW 16-20 GRADUALLY BACKING SSE THEN
         FALLING VARIABLE 04-08 BY LATE EVENING
  50M:   SSW 20-26 GRADUALLY BACKING SSE THEN
         FALLING VARIABLE 08-12 BY LATE EVENING
...
```

Fig. 2. Wind forecast for 05-10-2000, a.m. (names of oil fields anonymised).

same as in the SUMTIME work, in order to make the systems comparable.[2] From each of these alignments, numerical data vectors were automatically created; e.g. the example at the top of Figure 3 is the input vector for 5Oct2000_03. The input vector is a sequence of 7-tuples $\langle i, d, s_{min}, s_{max}, g_{min}, g_{max}, t \rangle$ where i is the tuple's ID, d is the wind direction, s_{min} and s_{max} are the minimum and maximum wind speeds, g_{min} and g_{max} are the minimum and maximum gust speeds, and t is a time stamp (indicating for what time of the day the data is valid).

In order to obtain these input vectors, we chunked the forecast texts into adjacent phrases, each of which realises one 7-tuple. Each forecast corresponds to at least one 7-tuple. One or more parts of a 7-tuple may not be realised in a given forecast. A -1 value for a timestamp t means that the procedure described above failed to identify a time for a segment. A '-' value means that the corresponding wind information is not included in the forecast text.

The Prodigy-METEO corpus consists of 483 wind data/forecast pairs. For the purposes of the experiments reported in this and previous papers we created 5 randomly determined subdivisions of the corpus into training and test data, for 5-fold cross-validation. For all generation methods that involve training from data, we repeated the training and testing process 5 times, once for each fold, as is standard under cross-validation. For the automatically computed evaluation methods, we therefore compute scores for each of the 5 test sets, and then average

---

[2] Reiter *et al.* selected the time stamp of the data in the table that most closely matched the data in the forecast, and if there was not a close enough match, they derived a time stamp from the time expression in the forecast, and finally, if that could not be done with enough confidence, then time was left unspecified.

Input	`[[1,SSW,16,20,-,-,0600],[2,SSE,-,-,-,-,NOTIME],[3,VAR,04,08,-,-,2400]]`
Corpus	SSW 16-20 GRADUALLY BACKING SSE THEN FALLING VARIABLE 4-8 BY LATE EVENING
SUMTIME-Hybrid	SSW 16-20 GRADUALLY BACKING SSE THEN BECOMING VARIABLE 10 OR LESS BY MIDNIGHT
PCFG-greedy	SSW 16-20 BACKING SSE FOR A TIME THEN FALLING VARIABLE 4-8 BY LATE EVENING
PCFG-roulette	SSW 16-20 GRADUALLY BACKING SSE AND VARIABLE 4-8
PCFG-viterbi	SSW 16-20 BACKING SSE VARIABLE 4-8 LATER
PCFG-2gram	SSW 16-20 BACKING SSE VARIABLE 4-8 LATER
PCFG-random	SSW 16-20 AT FIRST FROM MIDDAY BECOMING SSE DURING THE AFTERNOON THEN VARIABLE 4-8
PSCFG-semantic	SSW 16-20 BACKING SSE THEN FALLING VARIABLE 04-08 BY LATE EVENING
PSCFG-unstructured	SSW 16-20 GRADUALLY BACKING SSE THEN FALLING VARIABLE 04-08 BY LATE EVENING
PBSMT-unstructured	LESS SSW 16-20 SOON BACKING SSE BY END OF THEN FALLING VARIABLE 04-08 BY LATE EVENING
PBSMT-structured	GUSTS SSW 16-20 BY EVENING STEADILY LESS GUSTS GRADUALLY BACKING SSE BY LATE EVENING MINONE BY MIDDAY THEN AND FALLING UNKNOWN VARIABLE 04-08 LATER GUSTS

Fig. 3. Example input with corresponding outputs by all systems and from the corpus (for 5 Oct 2000)

over them. In order to be able to compare the forecasts produced by our systems directly to those produced by the original SUMTIME system,[3] we had to remove a small number of forecasts from each test set for which we did not have SumTime system outputs.

In the human-assessed evaluation experiment (see Section 5.2), where we used 22 randomly selected forecast dates, we needed exactly one output from each system for each of the 22 dates. For the trainable systems we therefore randomly selected a fold for each date to obtain an output from.

3   Four Ways to Build an NLG System

In this section, we describe four approaches to building language generators involving different combinations of automatic and manual techniques: traditional handcrafted systems (Section 3.1); handcrafted but trainable probabilistic context-free grammar (PCFG) generators (Section 3.2); partly automatically constructed and trainable probabilistic synchronous context-free grammar (PSCFG) generators; and generators automatically built with phrase-based statistical

---

[3] The SUMTIME-Hybrid system to be precise (see Section 4.1).

machine translation (PBSMT) methods (Section 3.4). In Section 4 we explain how we used these techniques to build the systems in our evaluation.

3.1   Rule-Based NLG

Traditional NLG systems are handcrafted as rule-based deterministic decision-makers that make decisions locally, at each step in the generation process. Decisions are encoded as generation rules with conditions for rule application (often in the form of if-then rules or rules with parameters to be matched), usually on the basis of corpus analysis and expert consultation. Reiter and Dale's influential 1997 paper [23] recommended that NLG systems be built largely "by careful analysis of the target text corpus, and by talking to domain experts" (p. 74, and reiterated on pp. 58, 61, 72 and 73).

Handcrafted generation tools have always formed the mainstay of NLG research, a situation virtually unchanged by the statistical revolution that swept through other NLP fields in the 1990s. Well-known examples include the surface realisers Penman, FUF/SURGE and RealPro, the referring expression generation components created by Dale and Reiter, and content-to-text generators built in the M-PIRO and PLAN-Doc projects, to name but a very few.[4]

3.2   PCFG Generation

Context-free grammars (CFG) are non-directional, and can be used for generation as well as for analysis (parsing). One approach that uses CFGs for generation is Probabilistic Context-free Representationally Underspecified (pCRU) language generation [1]. As mentioned above, traditional NLG systems tend to be composed of generation rules that apply transformations to representations. The basic idea in pCRU is that as long as the generation rules are all of the form $relation(arg_1, ...arg_n) \rightarrow relation_1(arg_1, ...arg_p) ... relation_m(arg_1, ...arg_q)$, $m \geq 1, n, p, q \geq 0$, then the set of all generation rules can be seen as defining a context-free language and a single probabilistic model can be estimated from raw or annotated text to guide generation processes.

In this approach, a CFG is created by hand that encodes the space of all possible generation processes from inputs to outputs, and has no decision-making ability. A probability distribution over this base CFG is estimated from a corpus, and this is what enables decisions between alternative generation rules to be made. The pCRU package permits this distribution to be used in one of the following three modes to drive generation processes: (i) greedy – apply only the most likely rule at each choice point; (ii) Viterbi – apply all expansion rules to each nonterminal to create the generation forest for the input, then do a Viterbi search of the generation forest; (iii) greedy roulette-wheel – select a rule to expand a nonterminal according to a non-uniform random distribution proportional to the likelihoods of expansion rules.

---

[4] See `http://www.nlg-wiki.org/` for information about all these systems and their creators.

In addition there are two baseline modes: (i) random – where generation rules are randomly selected at each choice point; and (ii) n-gram – where all alternatives are generated and the most likely is selected according to an n-gram language model (as in NITROGEN [13] and its successor HALOGEN [16]).

For the linguistically constrained weather forecasting domain, pCRU generators trained on raw corpora have been shown to perform well [1], but for more complex domains it is likely that manually annotated corpora will be needed for training the CFG base generator. As this is in addition to the manually constructed CFG base generator, the manual component in PCFG generator building is potentially substantial.

3.3 PSCFG Generation

Synchronous context-free grammars (SCFGs) are mostly used in machine translation [10], but have also been used for simple content-to-text generation [28]. The simplest form of SCFG can be viewed as a pair of CFGs G_1, G_2 with paired production rules such that for each rule in G_1 there is a rule in G_2 with the same left-hand side, and the same non-terminals in the right-hand side (RHS). The order of non-terminals on the RHS may differ, and each RHS may additionally contain any number of terminals in any order. An SCFG can equivalently be seen as a single grammar G encoding a set of pairs of strings. A probabilistic SCFG is defined by the 6-tuple $G = \langle \mathcal{N}, \mathcal{T}_e, \mathcal{T}_f, L, S, \lambda \rangle$, where $\mathcal{N}$ is a finite set of non-terminals, $\mathcal{T}_e$, $\mathcal{T}_f$ are finite sets of terminal symbols, L is a set of paired production rules, S is a start symbol $\in \mathcal{N}$, and λ is a set of parameters that define a probability distribution of derivations under G. Each rule in L has the form $A \to \langle \alpha; \beta \rangle$, where $A \in \mathcal{N}$, $\alpha \in N \cup \mathcal{T}_e^+$, $\beta \in N \cup \mathcal{T}_f^+$, and $N \subseteq \mathcal{N}$. SCFGs can be trained from aligned corpora to produce probabilistic (or 'weighted') SCFGs.

In MT the two CFGs that make up an SCFG are used to encode (the structure of) the two languages which the MT system translates between. Translation with an SCFG then consists of (i) parsing the input string with the source language CFG to produce a derivation tree, and then (ii) generating along the same derivation tree, but using the target language CFG to produce the output string.

When using SCFGs for content-to-text generation one of the paired CFGs encodes the meaning representation language, and the other the (natural) language in which text is supposed to be generated. A generation process then consists of (i) 'parsing' the meaning representation (MR) into its constituent structure, and, in the opposite direction, (ii) assembling strings of words corresponding to constituent parts of the input MR into a sentence or text that realises the entire MR.

We used the WASP^{-1} method [27,28] which provides a way in which a probabilistic SCFG can be constructed for the most part automatically. The training process requires two resources as input: a CFG of MRs and a set of sentences paired with their MRs. As output, it produces a probabilistic SCFG. The training process works in two phases, producing a (non-probabilistic) SCFG in the *lexical acquisition phase*, and associating the rules with probabilities in the *parameter estimation phase*.

The lexical acquisition phase uses the GIZA++ word-alignment tool, an implementation [17] of IBM Model 5 [8] to construct an alignment of MRs with NL strings. An SCFG is then constructed by using the MR CFG as a skeleton and inferring the NL grammar from the alignment.

For the parameter estimation phase, WASP^{-1} uses a log-linear model from Koehn et al. [15] which defines a conditional probability distribution over derivations D given an input MR f as

$$P_\lambda(D|f) \propto P(e(D))^{\lambda_1} \prod_{d \in D} w_\lambda(r(d))$$

where $e(D)$ is the output sentence that the derivation D yields and $w_\lambda(r(d))$ is the weight an individual rule used in a derivation, defined as

$$w_\lambda(A \to \langle \alpha, \beta \rangle) =$$
$$P(\beta|\alpha)^{\lambda_2} P(\alpha|\beta)^{\lambda_3} P_w(\beta|\alpha)^{\lambda_4} P_w(\alpha|\beta)^{\lambda_5} \exp(-|\alpha|)^{\lambda_6}$$

where $P(\beta|\alpha)$ and $P(\alpha|\beta)$ are the relative frequencies of β and α, $P_w(\beta|\alpha)$ and $P_w(\alpha|\beta)$ are lexical weights, and $\exp(-|\alpha|)$ is a word penalty to control output sentence length. The model parameters λ_i are trained using minimum error rate training.

Compared to probabilistic CFGs, probabilistic SCFGs trained with WASP^{-1} have a much reduced manual component in system building. In the latter, the NL grammar for the output language, the mapping from MRs to word strings and the rule probabilities are all created automatically, moreover from raw corpora, whereas in PCFGs, only the rule probabilities are obtained automatically.

3.4 SMT Methods

A Statistical Machine Translation (SMT) system is essentially composed of a translation model and a language model, where the former translates source language substrings into target language substrings, and the language model determines the most likely linearisation of the translated substrings. The currently most popular phrase-based SMT (PBSMT) approach translates phrases (arbitrary sequences of words, rather than phrases in the linguistic sense), whereas the original 'IBM models' translated words. Different PBSMT methods differ in how they construct the phrase translation table.

We used the phrase-based translation model included in the MOSES toolkit [14] which is based on the noisy channel model, where Bayes's rule is used to reformulate the task of translating a source language string f into a target language string $\hat{e}$ as finding the sentence $\hat{e}$ such that $\hat{e} = \text{argmax}_e\, P(e)P(f|e)$. The translation model (which gives $P(f|e)$) is obtained from a parallel corpus of source and target language texts, where the first step is automatic alignment using the GIZA++ word-level aligner. Word-level alignments are used to obtain phrase translation pairs using a set of heuristics. A 3-gram language model (which gives $P(e)$) for the target language is trained either on the same or a different corpus. For full details refer to Koehn et al. [14,15].

PBSMT offers a completely automatic method for constructing generators from given corpora of paired MRs and realisations, on the basis of which the PBSMT approach constructs a mapping from MRs to realisations.

4   Ten Weather Forecast Text Generators

4.1   SUMTIME-Hybrid

We included the original SUMTIME system [24] in our evaluations. This rule-based system has two modules: a content-determination module and a microplanning and realisation module. It can be run without the content-determination module, taking content representations (7-tuple sequences as described in Section 2) as inputs, and is then called SUMTIME-Hybrid. SUMTIME-Hybrid is a traditional deterministic rule-based generation system. Figure 3 shows an example forecast from the SUMTIME-Hybrid system (and corresponding outputs from the other systems, described below).

4.2   PCFG Generators

We also included five pCRU generators for the METEO domain created previously [1]. The pCRU base grammar for the Prodigy-METEO data is a set of generation rules with atomic arguments that convert an input into a set of forecast texts. To create inputs for the pCRU generators from the corpus input vectors (Section 2), first information is added (automatically) to each of the 7-tuples encoding whether the change in wind direction compared to the preceding 7-tuple is clockwise or anti-clockwise; whether change in wind speed is an increase or a decrease; and whether a 7-tuple is the last in the vector. Then, the augmented 7-tuples are converted into nonterminals with arguments.

A probability distribution over the base generator was obtained by the multi-treeban-king method [1] from the (un-annotated) Prodigy-METEO corpus. This method first parses the corpus with the base CFG and then obtains rule-application frequency counts from the parsed corpus which are used to obtain a probability distribution by straightforward maximum likelihood estimation. If there is more than one parse for a sentence then the frequency count increment is equally split over rules in alternative parses.

4.3   PSCFG Generators

We created two probabilistic synchronous CFG (PSCFG) generators for the METEO domain using WASP^{-1}. The main task here was to create a CFG for wind data representations. We used two different grammars (resulting in two different generators). The 'unstructured' grammar encodes raw corpus input vectors augmented as described in Section 4.2, whereas the 'semantic' grammar encodes representations with recursive predicate-argument structure that more resemble semantic forms. These were produced automatically from the raw input vectors.

Both the PSCFG-unstructured and the PSCFG-semantic generators were built in the same way, by feeding the CFG for wind data representations and the corpus of paired wind data representations and forecasts to WASP^{-1} which then created PSCFGs from it.

4.4 PBSMT Generators

We also created two generators with the MOSES toolkit. The main question here was how to represent the 'source language' inputs. While SMT methods are often applied with no linguistic knowledge at all (and are therefore blind as to whether paired inputs and outputs are NL strings or something else), it was not clear how well they would cope with the task of mapping from number/symbol vectors to NL strings. We tested two different input representations, one of which was simply the augmented corpus input vectors as described above (PBSMT-unstructured), and another in which the individual 7-tuples of which the vectors are composed are explicitly marked by predicate-argument structure (PBSMT-structured). As in Wong & Mooney's content-to-text generation work [28] we wanted to test the effect of treating the structure markers as tokens.

We built two different generators by feeding the two different versions of the paired corpus to MOSES. We did not use a factored translation model (the words used in weather forecasts did not vary sufficiently), and we did not tune our parameters to optimise for BLEU (or other) scores, a method common in SMT in [7,18].

5   Evaluation Methods

5.1   Automatic Evaluation Methods

The two automatic metrics used in the evaluations, NIST[5] and BLEU,[6] have been shown to correlate well with expert judgments (Pearson's $r = 0.82$ and 0.79 respectively) in the METEO domain [3]. BLEU-x is an n-gram based string comparison measure, originally proposed by Papineni et al. [19] for evaluation of MT systems. It computes the proportion of word n-grams of length x and less that a system output shares with several reference outputs, and ranges from 0 to 1. Setting $x = 4$ (i.e. considering all n-grams of length ≤ 4) is standard. NIST [11] is a version of BLEU, but where BLEU gives equal weight to all n-grams, NIST gives more importance to less frequent (hence more informative) n-grams, and the range of NIST scores depends on the size of the test set. Some research has shown NIST to correlate with human judgments more highly than BLEU [3,11,25].

5.2   Human Evaluation

We designed an experiment in which participants were asked to rate forecast texts for Clarity and Readability on scales of 1–7. Clarity was explained as indicating how understandable a forecast was, and Readability as indicating how fluent and readable it was. After an introduction and detailed explanations, participants carried out the evaluations over the web. They were able to interrupt and resume the evaluation at any time.

---

[5] http://cio.nist.gov/esd/emaildir/lists/mt_list/bin00000.bin
[6] ftp://jaguar.ncsl.nist.gov/mt/resources/mteval-v11b.pl

Table 1. Mean forecast-level NIST scores and homogeneous subsets (Tukey HSD, alpha = .05) for test data

System	NIST	Homogeneous subsets						
corpus	4.062	A						
PCFG-greedy	3.361		B					
PSCFG-semantic	3.303		B					
PSCFG-unstructured	3.191		B	C				
PCFG-roulette	3.033			C	D			
PBSMT-unstructured	2.924				D			
PCFG-viterbi	2.854				D	E		
PCFG-2gram	2.854				D	E		
SUMTIME-Hybrid	2.707					E	F	
PCFG-random	2.540						F	
PBSMT-structured	2.331							G

Table 2. Mean forecast-level BLEU scores and homogeneous subsets (Tukey HSD, alpha = .05) for test data

System	BLEU	Homogeneous subsets						
corpus	1.00	A						
PCFG-greedy	.65		B					
PSCFG-semantic	.637		B					
PSCFG-unstructured	.617		B	C				
PCFG-viterbi	.57			C	D			
PCFG-2gram	.561				D			
PCFG-roulette	.516				D	E		
PBSMT-unstructured	.500					E		
SUMTIME-Hybrid	.437						F	
PBSMT-structured	.338							G
PCFG-random	.269								H

We randomly selected 22 forecast dates and used outputs from the 10 systems described in Section 4 for those dates (as well as the corresponding forecasts in the corpus) in the evaluation, i.e. a total of 242 forecast texts. We used a repeated Latin squares design where each combination of forecast date and system is assigned two trials. As there were 2 evaluation criteria, there were a total of 968 individual ratings in this experiment. An evaluation session started with three training examples; the real trials were then presented in random order.

We recruited 22 participants from among our university colleagues whose first language was English and who had no experience of NLP. We did not try to recruit master mariners as in earlier experiments reported by Belz and Reiter [3], because these experiments also demonstrated that the correlation between the ratings by such expert evaluators and lay-people is very strong in the METEO domain (Pearson's $r = 0.845$).

Table 3. Mean Clarity and Readability ratings from human evaluation; NIST and BLEU scores on same 22 forecasts as used in human evaluation

	Clarity	Readability	NIST	BLEU
SUMTIME-Hybrid	6.06	6.18	5.71	0.52
PSCFG-semantic	5.79	5.70	6.76	0.65
corpus	5.79	5.93	8.45	1
PCFG-greedy	5.79	5.63	6.73	0.67
PSCFG-unstruc	5.72	5.84	6.61	0.64
PCFG-roulette	5.29	5.56	6.07	0.52
PCFG-2gram	5.29	5.29	5.23	0.52
PCFG-viterbi	4.90	5.34	5.15	0.51
PCFG-random	4.43	4.52	4.52	0.25
PBSMT-unstruc	3.70	3.93	5.38	0.49
PBSMT-struc	2.79	2.77	4.21	0.33

6   Results

For each evaluation method, we carried out a one-way ANOVA with 'System' as the fixed factor, and the evaluation measure as the dependent variable. In each case we report the main effect of System on the measure and (if it is significant) we also report significant differences between pairs of systems in the form of homogeneous subsets obtained with a post-hoc Tukey HSD analysis.

Tables 1 and 2 display the results for the BLEU and NIST evaluations, where scores were calculated on test data, using 5-fold cross-validation. System names are shown in the first column, mean forecast-level scores in the second, and the remaining columns indicate significant differences between systems. The way to read the homogeneous subsets is that two systems which do not have a letter in common are significantly different with $p < .05$.

For the BLEU evaluation, the main effect of System on BLEU score was $F_{(10,2420)} = 248.274$, at $p < .001$. PCFG-greedy, PSCFG-semantic and PSCFG-unstructured come out top, although only the first two are significantly better than all other systems. SUMTIME-Hybrid, PBSMT-structured and PCFG-random bring up the rear, with the remaining systems distributed over the middle ground. A striking result is that the handcrafted SUMTIME-Hybrid system comes out near the bottom, being significantly worse than all other systems except PCFG-structured and PBSMT-random.

For the NIST evaluation, the main effect of System on BLEU score was $F_{(10,2420)} = 108.086$, at $p < .001$. The systems were ranked in the same way as in the BLEU evaluation except for the systems in the D subset. The correlation between the NIST and BLEU scores is Pearson's $r = .739, p < .001$.

The main results from the automatic evaluations are that the two PSCFG systems and the PCFG system with the greedy generation algorithm are best overall. However, the human evaluations produced rather different results.

Figure 4 is a series of bar charts representing the results of the human evaluation for Clarity. For each system (indicated by the labels on the x-axis), there

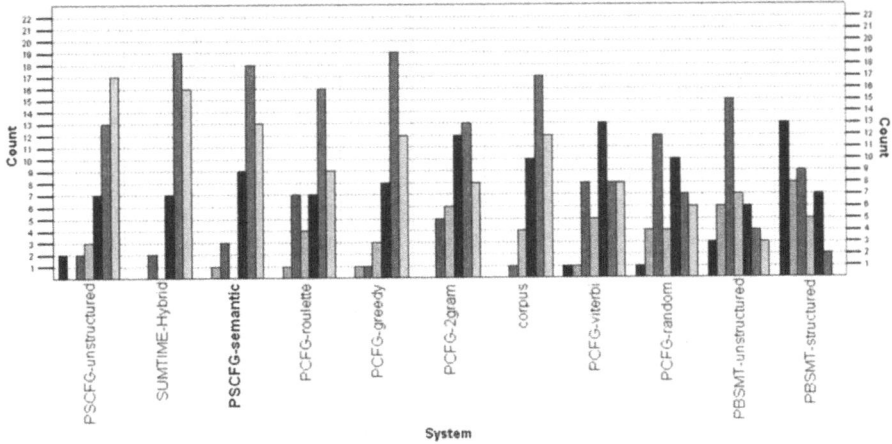

Fig. 4. Clarity ratings: Number of times each system was rated 1, 2, 3, 4, 5, 6, and 7 on Clarity. Systems in descending order of mode (most frequent rating).

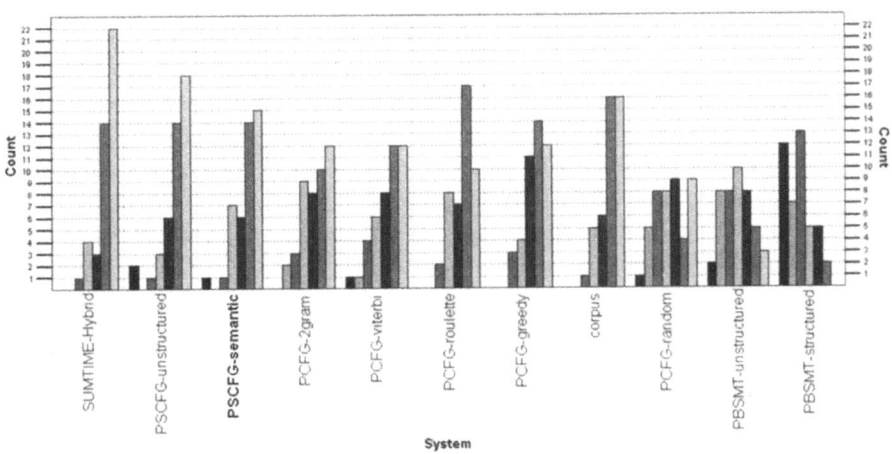

Fig. 5. Readability ratings: Number of times each system was rated 1, 2, 3, 4, 5, 6, and 7 on Readability. Systems in descending order of mode (most frequent rating).

are 7 bars, showing how many ratings of 1, 2, 3, 4, 5, 6 and 7 (7 being the best) a system was given. So the left-most bar for a system shows how many ratings of 1 a system was given, the second bar how many ratings of 2, etc. Systems are shown in descending order of mode (the value of the most frequently assigned rating, e.g. 7 for PSCFG-unstructured on the left, and 1 for PBSMT-structured on the right). The PSCFG-unstructured and SUMTIME-Hybrid systems come out top in this evaluation, with PSCFG-semantic, PCFG-roulette and PCFG-greedy close behind. Conversely, PBSMT-structured clearly came out worst, with no ratings of 7 and a mode of 1 (=completely unclear).

Figure 5 consists of the same kind of bar charts, for the Readability ratings. Here the SUMTIME-Hybrid system is the clear winner, with no ratings of 1 or 2, and 22 ratings of 7 (=excellent, all parts read well). It is closely followed by PSCFG-unstructured, the corpus forecasts and PSCFG-semantic. Again, PBSMT-structured is clearly worst with no ratings of 7, although this time the mode is 3 (=fairly bad).

We also looked at the means of the ratings, and these are shown in the second and third columns of Table 3. The means have to be treated with some caution, because ratings are ordinal data and it is not clear how meaningful it is to compute means. However, it is a simple way of obtaining a system ranking for comparison with the two automatic scores (shown in the remaining two columns of Table 3, computed over the system outputs used in the human evaluation). In terms of means, SUMTIME-Hybrid comes out top for both Clarity and Readability. In Clarity, it is followed by the two PSCFG systems, the corpus files (the only forecasts actually written by humans), and PCFG-greedy which have virtually the same means. For Readability, corpus and PSCFG-unstructured are ahead of PSCFG-semantic and PCFG-greedy (in this order). Bringing up the rear for both Clarity and Readability, as in the NIST evaluations, is PBSMT-structured, with PCFG-random and and PBSMT-unstructured faring somewhat better.

There are some striking differences between the automatic and human evaluations. For one, the human evaluators rank the SUMTIME-Hybrid system very high, whereas both automatic metrics rank it very low, just above PCFG-random and PBSMT-structured. Furthermore, the metrics rank PBSMT-unstructured more highly than the human evaluators, placing it above the SUMTIME-Hybrid system and in the case of NIST, also above two of the PCFG systems (Table 1). In fact, the human and the automatic evaluations agree only that the PSCFG systems and PCFG-greedy are equally good.

7   Issues Around Input Representations

The different input representations required by our systems are all derived from the basic sequence of 7-tuples described in Section 2, augmented by information regarding segment position, changes in wind speed (increase/decrease) and direction (clockwise/ anti-clockwise). The exact nature of each system's input representations, e.g. whether they are structured or flat, is in part determined by system type.

In Figure 6, we give examples of the input representations required by each system. For the PCFG and PSCFG systems input representation is determined entirely by the specific grammar used by the given system (the same is true of the SUMTIME-Hybrid system). As all four PCFG systems use the same base grammar, they all take the same predicate-argument sequences as inputs (see PCFG-* input representation in Figure 6). The two PSCFG systems have different content representation grammars, and their input representations therefore also differ (PSCFG-* in Figure 6).

The PBSMT systems differ from the others in that the entire mapping is automatically constructed, and in principle any input representations could therefore

System	Input Representation
Basic input rep.	`[[1,SSW,16,20,-,-,0600], [2,SSE,-,-,-,-,NOTIME],[3,VAR,04,08,-,-,2400]]`
PCFG-*	`Segment_Init(-1,_SSW,16,20,n,n,six)`
	`Segment(2,-1,same,ctrclock,_SSE,n,n,n,n,minone)`
	`Segment(3,1,down,unknown,_VAR,04,08,n,n,zero)`
PSCFG-semantic	`segment(1,false,nowindchange,wind(dir(ssw,x),16,20),nogust,six)`
	`segment(2,false,windchange(same,ctrclock),wind(dir(sse,x),n,n),nogust,`
	`minone)`
	`segment(3,true,windchange(down,unknown),wind(dir(var,x),04,08),nogust,`
	`zero)`
PSCFG-unstruct.	`-1 _SSW 16 20 n n six`
	`2 -1 same ctrclock _SSE n n n n minone`
	`3 1 down unknown _VAR 04 08 n n zero`
PBSMT-struct.	`segment_init ( -1 , _ssw , 16 , 20 , n , n , six )`
	`segment ( 2 , -1 , same , ctrclock , _sse , n , n , n , n , minone )`
	`segment ( 3 , 1 , down , unknown , _var , 04 , 08 , n , n , zero )`
PBSMT-unstruct.	`-1 _ssw 16 20 n n six`
	`2 -1 same ctrclock _sse n n n n minone`
	`3 1 down unknown _var 04 08 n n zero`
Corpus	`SSW 16-20 GRADUALLY BACKING SSE THEN FALLING VARIABLE 4-8 BY LATE EVENING`

Fig. 6. Example input representations for different generators (for 5 Oct 2000)

be used in the corpus of paired inputs/outputs that are fed to the system building component (as long as inputs are sequences of tokens). In our previous experiments, we tried out two alternatives (see PBSMT-* in Figure 6). As mentioned above, we first used input representations with structure markers (predicate names, brackets, commas) in place (PBSMT-structured). However, as can be seen from our results (Tables 1, 2 and 3), simply removing these structure markers (as in PBSMT-unstructured) results in a significant improvement in terms of all evaluation methods we used.

In order to investigate the difference further we created PBSMT-structured-2, a minor variation derived from the PBSMT-structured representation by removing only commas and brackets (leaving the predicate names in).

We also experimented with a rather more different input representation. By looking at the phrase tables used in the other PBSMT generators, we observed that the PBSMT systems often run into problems, because numbers are used with different meanings (e.g. a number can be a segment index or a wind speed), and the phrase tables have no way of distinguishing between one use of the same number and another. The information about which numbers are indices and which are wind speeds is there in the input representations—the first and second elements of each 7-tuple are segment indices, whereas the sixth and seventh elements are

Table 4. PBSMT systems: Mean forecast-level NIST scores and homogeneous subsets (Tukey HSD, alpha = .05) for test set (cross-validated)

System	NIST	Homogeneous subsets			
corpus	4.062	A			
PBSMT-tagged	3.173		B, C		
PBSMT-unstructured	2.924			D	
PBSMT-structured-2	2.831			D	
PBSMT-structured	2.331				G

wind speeds—but the PBSMT training mechanism has no way of utilising positional information. So for the fourth input representation, PBSMT-tagged, we inserted (in a fully automatic process) the 'missing' positional information explicitly in the form of 'tags', i.e. short words that precede each relevant token in the input (tags in bold):

PBSMT-tagged `-1 dir _ssw dir2 wind 16 wind2 20 n n six`

`windchange same ctrclock dir _sse dir2 wind n wind2 n n n minone`

`1 windchange down unknown dir _var wind 04 wind2 08 n n zero`

The NIST and BLEU scores for the two new PBSMT systems are shown in Table 4 and 5, respectively, along with scores for the other two PBSMT systems and the corpus texts, for ease of comparison. We are using the same letters to indicate the homogeneous subsets as in Tables 1 and 2.

The results for PBSMT-structured-2 indicate that removing commas and tags leads to the bigger part of the improvement from PBSMT-structured to PBSMT-unstructured, but the latter still outperforms it.

PBSMT-tagged is the best performing of the PBSMT systems; in terms of NIST scores there is no significant difference between it and the top-performing system (PCFG-greedy). This shows that in the case of data-to-text generation systems built using PBSMT methodology, very substantial improvements can be achieved by rewriting input representations. Of course, such improvements are achieved by additional manual work—error analysis and redesigning representations—and we cannot claim that the PBSMT-tagged system was created entirely by automatic techniques. although this took no more than three hours, and if the new tagging technique can be usefully applied to other generation tasks, we can treat the cost as one-off.

8   Discussion

In Section 3 we effectively ranked our four system-building methods in terms of how much manual effort they involve, which gave us the following ranking, in decreasing order: SUMTIME, PCFG-*, PSCFG-*, PBSMT-*. Not one of our four evaluation methods produced exactly the same rank order for these four groups (not if we base the comparison on the best system score in each group nor if we

Table 5. PBSMT systems: Mean forecast-level BLEU scores and homogeneous subsets (Tukey HSD, alpha = .05) for test set (cross-validated)

System	BLEU	Homogeneous subsets		
corpus	1.00	A		
PBSMT-tagged	.576		C, D	
PBSMT-unstructured	.500			E
PBSMT-structured-2	.481			E
PBSMT-structured	.338				G

base it on the average), i.e. we could not have predicted output quality rankings with complete accuracy from the degree of automation of the system-building process. Nevertheless, while there is no significant correlation between degree of automation and either BLEU or NIST, there is a significant correlation between degree of automation and both Clarity and Readability.

To establish this, we ranked our original 10 systems in order of degree of automation as follows: SUMTIME, {PCFG-greedy, PCFG-viterbi, PCFG-roulette}, {PCFG-2gram, PCFG-random}, {PSCFG-semantic, PSCFG-unstructured}, PBSMT-unstructured, PBSMT-structured (curly brackets indicating joint rank). We then computed Spearman's Rho between the automation ranks and the system-level Readability and Clarity scores, resulting in $\rho = -.59$, Sig. (1-tailed) $= .036$ for Clarity, and $\rho = -.549$, Sig. (1-tailed) $= .05$ for Readability.[7] Rho values for BLEU and NIST were close to 0, and not significant.

Our main finding in this research is thus that degree of automation in system building can (to some extent) predict human-assessed intrinsic scores, but automatic metric scores cannot be predicted with it. This conclusion relies on the estimation of degree of automation being accurate; it also raises the question of how to interpret the discrepancy between human-assessed and metric scores. Below we look at each of these issues in slightly more detail.

Assessing System Building Cost: In order to arrive at our estimation of degree of automation, we made the following assumptions: cost was defined as person-months spent in system building from the point at which the system-builder decided to build a wind forecast text generator in the METEO domain up to the point at which the system existed in the form in which we evaluated it; resources available to the system builder from the beginning of this process were the METEO corpus in either the SUMTIME or the PRODIGY version, and *any existing software*. Given these assumptions, the SUMTIME-Hybrid system took on the order of a year to build,[8] the PCFG systems on the order of 1 month, the

---

[7] We are using 1-tailed significance tests here, because we started out with a 1-tailed (or directional) hypothesis, namely that increasing automation would lead to decreasing scores, and we were testing the strength of the correlation, rather than the direction.

[8] Belz [1], estimated on the basis of personal communication with E. Reiter and S. Sripada.

PSCFG systems 1 day, and the PBSMT systems 1 hour; within some of the groups we were able to draw finer distinctions (e.g. the PCFG-random and PCFG-ngram systems use off-the-shelf tools for selection). None of the systems used existing, reusable generator modules.[9]

The above system ranking is thus appropriate and accurate in our case. In the more general case, however, it is not necessarily the case that handcrafted systems rank highest and PBSMT systems lowest in terms of system-building cost (as defined in the preceding paragraph). For example, not all parts of a completely manually constructed system will have necessarily been built for a given system (i.e. they might have been reused), while use of a fully automatic system-building method can hide substantial manual effort. The cost of building handcrafted systems could very well be reduced with a good set of reusable libraries, or a well designed domain-specific programming language (where "domain" here refers to "natural language generation"). Handcrafted systems could also be partly expressed in terms of some formal grammar which in turn can be made cheaper through judicious use of automation (for example, using a metagrammar compiler such as XMG [20]). On the other hand, as we saw in Section 7, it is possible to invest manual effort in improving fully automatic system-building methods, in order to try and improve their evaluation scores. In our case this was no more than a few hours, but clearly this could be an open-ended process.

The main point here is that system-building cost must be established on a case-by-case basis, not generically on the basis, say, of system type or generation methodology. While manual effort may well be amortised in the future because (part of) a system is reusable, or, conversely, maintenance costs may accrue in the future, these do not fall under the building-cost heading.

Implications for Evaluation Methods: Given the discrepancy between our human-assessed and automatic evaluation methods, one might wonder which are 'more right'? BLEU in particular has come in for some hefty criticism over recent years, with some papers (e.g. [9]) exposing individual examples of high BLEU scores for patently bad-quality outputs, and vice versa. We can identify such examples in our data too, e.g.:

System	Output	BLEU
corpus	N-NNE 16-20 RISING 20-24 FOR A TIME THIS AFTERNOON	1
SUMTIME-Hybrid	N-NNE 16-20 INCREASING 20-24 BY MID AFTERNOON	0.43
PCFG	N-NNE 16-20 INCREASING 20-24 FOR A TIME MID PERIOD	0.65

There is really nothing wrong with the above SUMTIME-Hybrid output, in fact there is evidence that master mariners prefer more precise statements of time [22], but its BLEU score is low (whereas the human evaluators give it clear 7s).

Yet human evaluation does not always produce ideal scores either, something that is perhaps not discussed frequently enough (but see e.g. [4]). Human

---

[9] The systems only used reusable interpretors and compilers in addition to the training tools.

evaluators are notorious for their lack of agreement with other evaluators and even themselves [21]; and individual scores can be just as unintuitive as in the case of BLEU. Again, we can find examples in our data (scores from two evaluators for each of the two forecasts):

System	Output	Clar.	Read.
PCFG-2gram	SSW 26 - 30 VEERING SW 32 - 36 VEERING WSW 20 - 24	3	2
	BACKING SW 28 - 32	5	6
PBSMT-unstruct.	OR LESS S'LY 05 - 10 THIS EVENING INCREASING	7	7
	10 - 14 BY LATE EVENING	1	1

In the first example above, the high readability score by the second evaluator does not seem justified, because of the repetition of *veering* and the absence of 'link words' between segments such as *then* and *and*; in the second example, clearly the two evaluators who gave these scores could not disagree more, and again it is the high scores that are unjustified—a forecast cannot start with *or less*.

Both automatic metrics and human evaluators are fallible in individual cases. What matters more is how they perform on the whole, on a set of outputs. If inappropriate scores are noise that disappears 'in the statistical wash', then all is well, otherwise (and only then) there is a problem.

Without a third, objective point of reference, we cannot be sure which of our evaluation methods is 'more right'. What we have found again and again in our evaluation experiments (see also Belz et al. [6], and Gatt & Belz [12], both in this volume), is that different types of evaluation do not necessarily agree with each other, and that we should not regard any single one of them as an objective measure of quality, but rather as assessing one particular aspect of systems. If we want our wind forecasts to be similar to the corpus forecasts, then BLEU and NIST can probably give us a fair indication of that type of similarity; if we are interested in how readable and clear human readers (think they) find our forecasts, then we should look at the Clarity and Readability scores.

9   Conclusions

Reports of research on automating (part of) system building often take it as read that such automation is a good thing. The resulting systems are not often compared to handcrafted alternatives in terms of output quality or other quality criteria, and little is therefore known about the loss of system quality that results from automation. The existence of several independently developed systems for the METEO domain of weather forecasts, to which we have added six new systems in the research reported in this paper, provides a unique opportunity to examine the system building cost vs. system quality trade-off in data-to-text generation.

We investigated 12 systems which fall into four categories in terms of the manual work involved in creating them, ranging from completely manual to completely automatic system building. We used two automatically assessed corpus-similarity metrics and two human assessed quality criteria to evaluate systems.

We found some significant correlation between degree of automation of system building and the human-assessed Clarity and Readability criteria, but no correlation between the former and either of the automatic metrics.

We found striking differences between the results from tests of human acceptability and measurements of corpus similarity. Relative to the human ratings, the automatic metrics underestimated the quality of the handcrafted SUMTIME-Hybrid system, but overestimated the quality of the automatically constructed SMT systems. This will not come as a surprise to those familiar with the machine translation evaluation literature where this is a major complaint about BLEU [9]. From our research (see also e.g. [22] it seems clear that when the quality of diverse types of systems is compared, automatic metrics such as BLEU on their own do not give a complete and reliable picture, and carrying out additional evaluations is crucial.

Increased reusability and adaptability of systems and components have cost and time benefits, and methods for automatically training systems from data offer advantages in both these respects. However, careful evaluation is needed to ensure that these advantages are not achieved at the price of a reduction in system quality that renders systems unacceptable to human users.

Acknowledgments. The research reported in this paper was supported under EPSRC grant EP/E029116/1 (the Prodigy Project). We thank Emiel Krahmer and the anonymous ENLG'09 reviewers for their helpful comments.

References

1. Belz, A.: Automatic generation of weather forecast texts using comprehensive probabilistic generation-space models. Natural Language Engineering 14(4), 431–455 (2008)
2. Belz, A.: Prodigy-METEO: Pre-alpha release notes (Nov 2009). Tech. Rep. NLTG-09-01, Natural Language Technology Group, CMIS, University of Brighton (2009)
3. Belz, A., Reiter, E.: Comparing automatic and human evaluation of NLG systems. In: Proceedings of the 11th Conference of the European Chapter of the Association for Computational Linguistics (EACL 2006), pp. 313–320 (2006)
4. Belz, A.: That's nice.. what can you do with it? Computational Linguistics 35(1), 111–118 (2009)
5. Belz, A., Kow, E.: System building cost vs. output quality in data-to-text generation. In: Proceedings of the 12th European Workshop on Natural Language Generation (2009)
6. Belz, A., Kow, E., Viethen, J., Gatt, A.: Generating referring expressions in context: The GREC task evaluation challenges. In: Krahmer, E., Theune, M. (eds.) Empirical Methods in NLG. LNCS (LNAI), vol. 5790, pp. 294–328. Springer, Heidelberg (2010)
7. Bertoldi, N., Haddow, B., Fouet, J.: Improved Minimum Error Rate Training in Moses. The Prague Bulletin of Mathematical Linguistics 91, 7–16 (2009)
8. Brown, P.F., Della Pietra, V.J., Della Pietra, S.A., Mercer, R.L.: The mathematics of statistical machine translation: parameter estimation. Computational Linguistics 19(2), 263–311 (1993)

9. Callison-Burch, C., Osborne, M., Koehn, P.: Re-evaluating the role of BLEU in machine translation research. In: Proceedings of EACL 2006 (2006)
10. Chiang, D.: An introduction to synchronous grammars (part of the course materials for the ACL 2006 tutorial on synchronous grammars) (2006)
11. Doddington, G.: Automatic evaluation of machine translation quality using n-gram co-occurrence statistics. In: Proceedings of the ARPA Workshop on Human Language Technology (2002)
12. Gatt, A., Belz, A.: Introducing Shared Tasks to NLG: The TUNA Shared Task Evaluation Challenges. In: Krahmer, E., Theune, M. (eds.) Empirical Methods in NLG. LNCS (LNAI), vol. 5790, pp. 264–293. Springer, Heidelberg (2010)
13. Knight, K., Langkilde, I.: Generation that exploits corpus-based statistical knowledge. In: Proceedings of the 36th Annual Meeting of the Association for Computational Linguistics and 17th International Conference on Computational Linguistics (COLING-ACL 1998), pp. 704–710 (1998)
14. Koehn, P., Hoang, H., Birch, A., Callison-Burch, C., Federico, M., Bertoldi, N., Cowan, B., Shen, W., Moran, C., Zens, R., Dyer, C., Bojar, O., Constantin, A., Herbst, E.: Moses: Open source toolkit for statistical machine translation. In: Proceedings of the 45th Annual Meeting of the Association for Computational Linguistics (ACL 2007), pp. 177–180 (2007)
15. Koehn, P., Och, F.J., Marcu, D.: Statistical phrase-based translation. In: Proceedings of Human Language Technologies: The Annual Conference of the North American Chapter of the Association for Computational Linguistics on Human Language Technology (HLT-NAACL 2003), pp. 48–54 (2003)
16. Langkilde, I.: Forest-based statistical sentence generation. In: Proceedings of the 6th Applied Natural Language Processing Conference and the 1st Meeting of the North American Chapter of the Association of Computational Linguistics (ANLP-NAACL 2000), pp. 170–177 (2000)
17. Och, F.J., Ney, H.: A Systematic Comparison of Various Statistical Alignment Models. Computational Linguistics 29(1), 19–51 (2003)
18. Och, F.: Minimum error rate training in statistical machine translation. In: Proceedings of the 41st Annual Meeting on Association for Computational Linguistics, vol. 1, p. 167. Association for Computational Linguistics (2003)
19. Papineni, K., Roukos, S., Ward, T., Zhu, W.J.: BLEU: A method for automatic evaluation of machine translation. IBM research report, IBM Research Division (2001)
20. Parmentier, Y., Le Roux, J.: XMG: a Multi-formalism Metagrammatical Framework. In: 17th European Summer School in Logic, Language and Information - ESSLLI 2005, Edinburgh/Scotland (August 2005)
21. Reidsma, D., Op den Akker, R.: Exploiting 'subjective' annotations. In: Proceedings of the COLING 2008 Workshop on Human Judgements in Computational Linguistics, pp. 8–16 (2008)
22. Reiter, E., Belz, A.: An investigation into the validity of some metrics for automatically evaluating NLG systems. Computational Linguistics 35(4) (2009)
23. Reiter, E., Dale, R.: Building applied natural language generation systems. Natural Langauge Engineering 3(1), 57–87 (1997)
24. Reiter, E., Sripada, S., Hunter, J., Yu, J.: Choosing words in computer-generated weather forecasts. Artificial Intelligence 167, 137–169 (2005)
25. Riezler, S., Maxwell, J.T.: On some pitfalls in automatic evaluation and significance testing for MT. In: Proceedings of the ACL 2005 Workshop on Intrinsic and Extrinsic Evaluation Measures for MT and/or Summarization, pp. 57–64 (2005)

26. Sripada, S., Reiter, E., Hunter, J., Yu, J.: SumTime-Meteo: A parallel corpus of naturally occurring forecast texts and weather data. Tech. Rep. AUCS/TR0201, Computing Science Department, University of Aberdeen (2002)
27. Wong, Y.W., Mooney, R.: Learning for semantic parsing with statistical machine translation. In: Proceedings of Human Language Technologies: The Annual Conference of the North American Chapter of the Association for Computational Linguistics on Human Language Technology (HLT-NAACL 2006), pp. 439–446 (2006)
28. Wong, Y.W., Mooney, R.: Generation by inverting a semantic parser that uses statistical machine translation. In: Proceedings of Human Language Technologies: The Annual Conference of the North American Chapter of the Association for Computational Linguistics on Human Language Technology (HLT-NAACL 2007), pp. 172–179 (2007)

Human Evaluation of a German Surface Realisation Ranker

Aoife Cahill[1] and Martin Forst[2]

[1] Institut für Maschinelle Sprachverarbeitung (IMS)
University of Stuttgart
70174 Stuttgart, Germany
`aoife.cahill@ims.uni-stuttgart.de`
[2] Powerset/Microsoft
475 Brannan Street
San Francisco, CA 94304, USA
`martin.forst@microsoft.com`

Abstract. In this chapter we present a human-based evaluation of surface realisation alternatives. We examine the relative rankings of naturally occurring corpus sentences and automatically generated strings chosen by statistical models (language model, log-linear model), as well as the naturalness of the strings chosen by the log-linear model. We also investigate to what extent preceding context has an effect on choice. We show that native speakers do accept quite some variation in word order, but that there are clearly also factors that make certain realisation alternatives more natural than others. We then examine correlations between native speaker judgements of automatically generated German text and automatic evaluation metrics. We look at a number of metrics from the MT and Summarisation communities and find that for a relative ranking task, most automatic metrics perform equally well and have fairly strong correlations to the human judgements. In contrast, on a naturalness judgement task, the correlation between the human judgements and the automatic metrics was quite weak, the General Text Matcher (GTM) tool providing the only metric that correlates with the human judgements at a statistically significant level.

Keywords: generation evaluation, surface realisation, human evaluation, German, human judgements, automatic metrics, correlation.

1   Introduction

An important component of research on surface realisation (the task of generating strings for a given abstract representation) is evaluation, especially if we want to be able to compare across systems. There is consensus that exact match with respect to an actually observed corpus sentence is too strict a metric and that BLEU score measured against corpus sentences can only give a rough impression of the quality of the system output. It is unclear, however, what kind of metric would be most suitable for the evaluation of string realisations, so that,

E. Krahmer, M. Theune (Eds.): Empirical Methods in NLG, LNAI 5790, pp. 201–221, 2010.

as a result, there have been a range of automatic metrics applied, including *inter alia* exact match, string edit distance, NIST SSA, BLEU, NIST, ROUGE, generation string accuracy, generation tree accuracy, word accuracy [1,3,8,16,26].

It is not always clear how appropriate these metrics are, especially at the level of individual sentences. Automatic evaluation metrics are an indispensable tool for rapid experiment turnover, but ideally, a metric for the evaluation of realisation rankers would rank alternative realisations in the same way as native speakers of the language for which the surface realisation system is developed, and not only globally, but also for individual sentences.

Another major consideration in evaluation is what to take as the gold standard. The easiest option is to take the original corpus string that was used to produce the abstract representation from which we generate. However, there may well be other realisations of the same input that are as suitable in the given context. Reiter and Sripada [21] argue that while we should take advantage of large corpora in NLG, we also need to take care not to introduce errors by learning from incorrect data present in corpora.

In order to better understand what makes good evaluation data and metrics, we designed and implemented an experiment in which human judges evaluated German string realisations. The main aims of this experiment were: (i) to establish how much variation in German word order is acceptable for human judges, (ii) to find an automatic evaluation metric that mirrors the findings of the human evaluation, (iii) to provide detailed feedback for the designers of the surface realisation ranking model and (iv) to establish what effect preceding context has on the choice of realisation. In this chapter, we concentrate on points (i), (ii) and (iv).

The remainder of the chapter is structured as follows: In Section 2, we outline the realisation ranking system that provided the data for the experiment. In Section 3, we outline the design of the experiment, and in Section 4, we present our findings. Section 5 outlines the correlation between the human judgements and automatic evaluation metrics. In Section 6, we relate this to other work, and finally we conclude in Section 7.

2   A Realisation Ranking System for German

We take the realisation ranking system for German described in Cahill et al. [6] and present the output to human judges. One goal of this series of experiments is to examine whether the results based on automatic evaluation metrics published in that paper are confirmed in an evaluation by humans. Another goal is to collect data that will allow us and other researchers[1] to explore more fine-grained and reliable automatic evaluation metrics for realisation ranking.

The system presented by Cahill et al. [6] ranks the strings generated by a hand-crafted broad-coverage Lexical Functional Grammar [5] for German [22] on the basis of a given input f-structure. Their system was trained and evaluated on data derived from the TIGER Corpus [4], a treebank comprising 50,472 sentences

---

[1] The data is available for download from
http://www.ims.uni-stuttgart.de/projekte/pargram/geneval/data/

"Williams war in der britischen Politik äußerst umstritten."

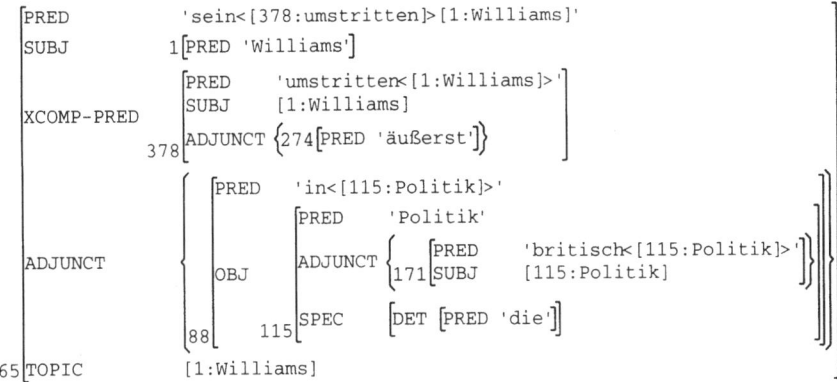

Williams war in der britischen Politik äußerst umstritten.
In der britischen Politik war Williams äußerst umstritten.
Äußerst umstritten war Williams in der britischen Politik.
Äußerst umstritten war in der britischen Politik Williams.

Fig. 1. F-structure associated with (1) and strings generated from it

of German newspaper text. In the experiments presented in this article, we use f-structures from their held-out and test sets, of which 96% can be associated with surface realisations by the grammar. F-structures are attribute-value matrices representing grammatical functions and morphosyntactic features; roughly speaking, they are predicate-argument structures. In LFG, f-structures are assumed to be a crosslinguistically relatively parallel syntactic representation level, alongside the more surface-oriented c-structures, which are context-free trees. Figure 1 shows the f-structure[2] associated with TIGER Corpus sentence 8609, glossed in (1), as well as the 4 string realisations that the German LFG generates from this f-structure. The LFG is reversible, i.e. the same grammar is used for parsing as for generation. It is a hand-crafted grammar, and has been carefully constructed to only parse (and therefore generate) grammatical strings.[3] This

---

[2] Note that only grammatical functions are displayed; morphosyntactic features are omitted due to space constraints. Also note that the discourse function TOPIC was ignored in generation.

[3] A ranking mechanism based on so-called optimality marks can lead to a certain "asymmetry" between parsing and generation in the sense that not all sentences that can be associated with a certain f-structure are necessarily generated from this same f-structure. E.g. the sentence *Williams war äußerst umstritten in der britischen Politik.* can be parsed into the f-structure in Figure 1, but it is not generated because an optimality mark penalises the extraposition of PPs to the right of a clause. OT marks were introduced to the generation grammar for efficiency reasons when generating certain unlikely word orders had a negative impact on performance. Only few optimality marks were used in the process of generating the data for our experiments, so that the bias they introduce should not be too noticeable.

being said, the grammar used for our experiments is a bit too permissive in terms of constituent order, probably because it had previously been developed mostly for parsing, so that, occasionally, highly marked or even ungrammatical constituent orders are generated.

(1)   Williams war in der britischen Politik äußerst   umstritten.
       Williams was in the British   politics extremely controversial.
       'Williams was extremely controversial in British politics.'

The ranker consists of a log-linear model that is based on linguistically informed structural features as well as a trigram language model, whose score is integrated into the model simply as an additional feature.[4] The log-linear model is trained on corpus data, in this case sentences from the TIGER Corpus, for which f-structures are available; the observed corpus sentences are considered as references whose probability is to be maximised during the training process.

The output of the realisation ranker is evaluated in terms of exact match and BLEU score, both measured against the actually observed corpus sentences. In addition to the figures achieved by the ranker, the corresponding figures achieved by the employed trigram language model on its own are given as a baseline, and the exact match figure of the best possible string selection is given as an upper bound.[5] We summarise these figures in Table 1.

Table 1. Results achieved by trigram LM ranker and log-linear model ranker in Cahill et al. [6]

	Exact Match	BLEU score
Language model	27%	0.7306
Log-linear model	37%	0.7939
Upper bound	62%	–

By means of these figures, Cahill et al. [6] show that a log-linear model based on structural features and a language model score performs considerably better realisation ranking than just a language model. In our experiments, presented in

---

[4] One category of features in the log-linear model is based on C-structural features. These are very similar to the ones described by Nenkova et al. [17] (this volume) for English in their work on finding features that could be useful for distinguishing fluent sentences from disfluent ones. They show that simple shallow features based on the output of the Charniak parser can be helpful for this task.

[5] The observed corpus sentence can be (re)generated from the corresponding f-structure for only 62% of the sentences used, usually because of differences in punctuation. Hence this exact match upper bound. Similarly, the upper bound BLEU score is not 1.0, since we do not always have the original corpus string in the list of alternatives. The exact upper bound BLEU score is unknown because BLEU is calculated on entire corpora rather than individual sentences and it is technically impractical to generate all candidate corpora.

detail in the following section, we examine whether human judges confirm this and how natural and/or acceptable the selection performed by the realisation ranker under consideration is for German native speakers.

3   Experiment Design

The experiment was divided into three major parts, with Part 1 and Part 3 each being subdivided further into two tasks. Each major part took between 30 and 45 minutes to complete, and participants were asked to leave some time (e.g. a week) between parts. In total, 24 participants completed the experiment. All were native German speakers (mostly from South-Western Germany[6]) and almost all had a linguistic background. Table 2 gives a breakdown of the items in each part of the experiment. Note that Tasks 3a and 3b contained the same items as Tasks 1a and 1b.

Table 2. Statistics for each experiment part

	Task 1a	Task 1b	Part 2
Num. items	44	52	41
Avg. sent length	14.4	12.1	9.4

3.1   Part 1

The aim of Part 1 of the experiment was twofold. First, to identify the relative rankings of the systems evaluated in Cahill et al. [6] according to the human judges, and second to evaluate the quality of the strings as chosen by the log-linear model of Cahill et al. [6]. To these ends, Part 1 was further subdivided into two tasks:

Task 1a: During the first task, participants were presented with alternative realisations for an input f-structure (but not shown the original f-structure) and asked to rank them in order of how natural sounding they were, 1 being the best and 3 being the worst.[7] Each item contained three alternatives, (i) the original string found in TIGER, (ii) the string chosen as most likely by the trigram language model, and (iii) the string chosen as most likely by the log-linear model.

---

[6] It is known that word order preferences differ between different dialects of German, so the geographic origin of our participants may have had an effect on our results. However, since the task involved written language and since most speakers identify their variety as either "neutral" standard German or "South-Western" standard German rather than the actual South-Western dialect, we assume that this effect is small. In any case, our sample of German speakers is too small to investigate the effects of dialectal influences on word order preferences.

[7] Joint rankings were not allowed, i.e. the participants were forced to make strict ranking decisions, and in hindsight this may have introduced some noise into the data.

5 (von 52)

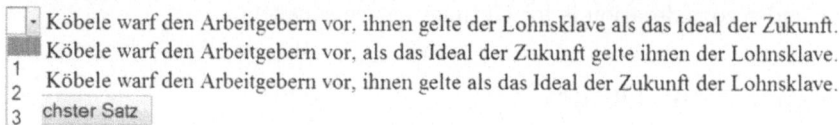

- Köbele warf den Arbeitgebern vor, ihnen gelte der Lohnsklave als das Ideal der Zukunft.
Köbele warf den Arbeitgebern vor, als das Ideal der Zukunft gelte ihnen der Lohnsklave.
1
2   Köbele warf den Arbeitgebern vor, ihnen gelte als das Ideal der Zukunft der Lohnsklave.
3   chster Satz

Fig. 2. Screenshot of Task 1a of the Experiment

A random subset of the items where each system chose a different alternative were chosen from the evaluation data of Cahill et al. [6]. The three alternatives were presented in random order for each item, and the items were presented in random order for each participant. Some items were presented randomly to participants more than once as a sanity check, and in total for Part 1a, participants made 52 ranking judgements on 44 items. Figure 2 shows a screen shot of what the participant was presented with for this task.

Task 1b: In the second task of Part 1, participants were presented with the string chosen by the log-linear model as being the most likely while not being identical to the original corpus string and asked to evaluate it on a scale from 1 to 5 on how natural sounding it was, 1 being very unnatural or marked and 5 being completely natural. Figure 3 shows a screen shot of what the participant saw during the experiment. Again some random items were presented to the participants more than once, and the items themselves were presented in random order. In total, the participants made 59 judgements on 52 items.

21 (von 59)

Die Beschäftigungs-politische Prognose fällt trostlos aus.

unnatürlich bzw. stark markiert ○ 1 ○ 2 ○ 3 ○ 4 ○ 5 vollkommen natürlich

Nächster Satz

Fig. 3. Screenshot of Task 1b of the Experiment

3.2   Part 2

The aim of the second part of the experiment was to determine to what degree the human judges would select the original corpus string among all the generated alternatives as the most natural surface realisation in a given context. In this part, participants were hence presented between 4 and 8 alternative surface realisations for an input f-structure, as well as some preceding context.[8] This preceding context was automatically determined using information from the export

---

[8] The preceding context usually comprised 2 sentences, but sometimes more if the total length of the preceding 2 sentences was less than 240 characters.

release of the TIGER treebank and was not hand-checked for relevance.[9] The participants were then asked to choose the realisation that they felt fit best given the context defined by the preceding sentences.

The items used in this part were selected on the basis of the criteria that (i) the original corpus string could be generated from the input f-structure and that (ii) the total number of generated surface realisations was between 4 and 8. The items were presented in random order, and the list of alternatives were presented in random order to each participant. Some items were randomly presented more than once, resulting in 50 judgements on 41 items. Figure 4 shows a screen shot of what the participant saw.[10]

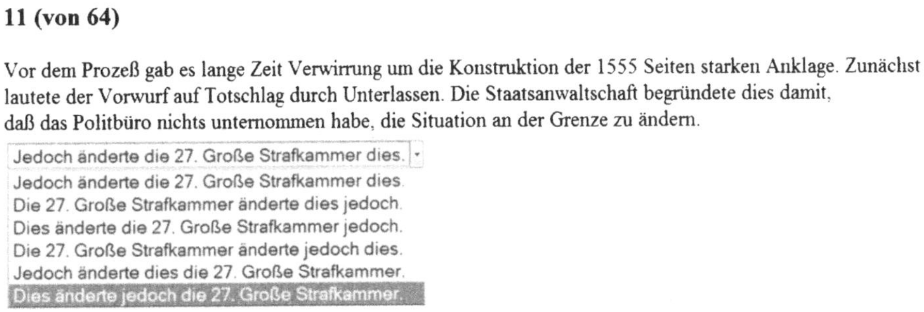

11 (von 64)

Vor dem Prozeß gab es lange Zeit Verwirrung um die Konstruktion der 1555 Seiten starken Anklage. Zunächst lautete der Vorwurf auf Totschlag durch Unterlassen. Die Staatsanwaltschaft begründete dies damit, daß das Politbüro nichts unternommen habe, die Situation an der Grenze zu ändern.

Jedoch änderte die 27. Große Strafkammer dies.
Jedoch änderte die 27. Große Strafkammer dies.
Die 27. Große Strafkammer änderte dies jedoch.
Dies änderte die 27. Große Strafkammer jedoch.
Die 27. Große Strafkammer änderte jedoch dies.
Jedoch änderte dies die 27. Große Strafkammer.
Dies änderte jedoch die 27. Große Strafkammer.

Fig. 4. Screenshot of Part 2 of the Experiment

3.3   Part 3

Part 3 of the experiment was identical to Part 1, except that now, rather than being presented with sentences in isolation, the participants were given some preceding context. The context was determined automatically, in the same way as in Part 2. The items themselves were the same as in Part 1. The aim of this part of the experiment was to see what effect preceding context had on judgements.

4   Results

In this section we present the results and analysis of the experiments outlined above.

4.1   How Good Were the Strings?

The data collected in Task 1a showed the overall human relative ranking of the three systems. We calculate the total numbers of each rank for each system.

---

[9] The export release of the TIGER treebank includes an article ID for each sentence. Unfortunately, this is not completely reliable for determining relevant context, since an article can also contain several short news snippets which are completely unrelated. In addition, paragraph boundaries within articles are not marked. This leads to some noise, which unfortunately is difficult to measure objectively.

[10] We had the participants make 64 judgments, but could only use 50 of them because the other 14 concerned items for which the original corpus string was not generated.

Table 3 summarises the results. The original string is the string found in the TIGER Corpus, the LL String is the string chosen as being most likely by the log-linear model, and the LM String is the string chosen as being most likely by the trigram language model.

Table 3 confirms the overall relative rankings of the three systems as determined using BLEU scores. The original TIGER strings are ranked best (average 1.4), the strings chosen by the log-linear model are ranked better than the strings chosen by the language model (average 2.56 vs 2.04).

Table 3. Task 1a: Ranks for each system

	Rank 1	Total Rank 2	Rank 3	Average Rank
Original String	817	366	65	1.40
LL String	303	593	352	2.04
LM String	128	289	831	2.56

In Task 1b, the aim was to find out how acceptable the strings chosen by the log-linear model were, although they were not the same as the original string. Figure 5 summarises the data. The graph shows that the majority of strings chosen by the log-linear model ranked very highly on the naturalness scale.

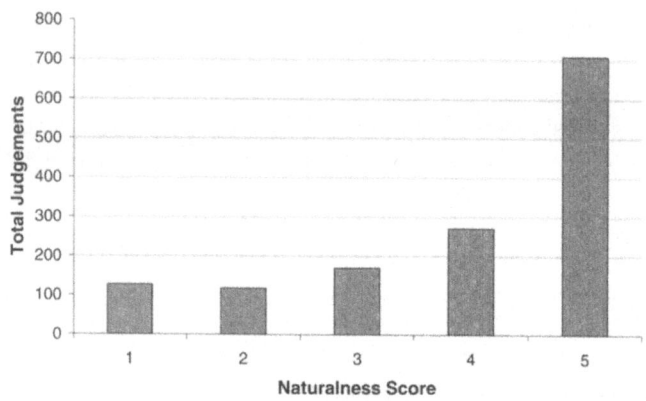

Fig. 5. Task 1b: Naturalness scores for strings chosen by log-linear model, 1=worst

4.2   Did the Human Judges Agree with the Original Authors?

In Part 2, the aim was to find out how often the human judges chose the same string as the original author (given alternatives generated by the LFG grammar). Most items had between 4 and 6 alternative strings. In 70% of all items, the human judges chose the same string as the original author. However, the remaining 30% of the time, the human judges picked an alternative as being the

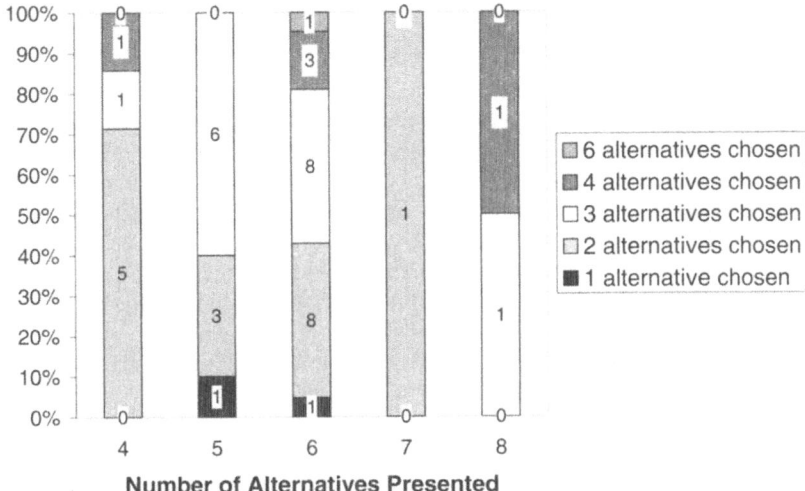

Fig. 6. Part 2: Number of Alternatives Chosen

most fitting in the given context.[11] This suggests on the one hand that there is quite some variation in what native German speakers will accept, but on the other hand the 70% of choices being the same string as the original author's are far beyond what would result from a random selection.

Figure 6 shows, for each bin of possible alternatives, the percentage of items with a given number of choices made. For example, for the items with 4 possible alternatives, over 70% of the time, the judges chose between only 2 of them. For the items with 5 possible alternatives, in 10% of those items the human judges chose only 1 of those alternatives; in 30% of cases, the human judges all chose the same 2 solutions, and for the remaining 60% they chose between only 3 of the 5 possible alternatives. These figures indicate that, although judges could not always agree on one best string, they were often choosing between only 2 or 3 of the possible alternatives. This again suggests that, on the one hand, native speakers do accept quite some variation, but that, on the other hand, there are clearly factors, probably of a stylistic or discourse-related nature, that make certain realisation alternatives more preferable than others.

Table 4. The 4 alternatives given by the grammar for (2) and their frequencies

Alternative	Freq.
Zunächst blieb die Brandursache unbekannt.	2
Die Brandursache blieb zunächst unbekannt.	24
Unbekannt blieb die Brandursache zunächst.	1
Unbekannt blieb zunächst die Brandursache.	1

---

[11] Recall that almost all strings presented to the judges were grammatical.

Table 5. The 6 alternatives given by the grammar for (3) and their frequencies

Alternative	Freq.
Mit einem sechsstelligen Betrag fördert die Unternehmensgruppe Tengelmann die Arbeit im brandenburgischen Biosphärenreservat Schorfheide.	7
Mit einem sechsstelligen Betrag fördert die Arbeit im brandenburgischen Biosphärenreservat Schorfheide die Unternehmensgruppe Tengelmann.	1
Die Arbeit im brandenburgischen Biosphärenreservat Schorfheide fördert die Unternehmensgruppe Tengelmann mit einem sechsstelligen Betrag.	4
Die Arbeit im brandenburgischen Biosphärenreservat Schorfheide fördert mit einem sechsstelligen Betrag die Unternehmensgruppe Tengelmann.	5
Die Unternehmensgruppe Tengelmann fördert die Arbeit im brandenburgischen Biosphärenreservat Schorfheide mit einem sechsstelligen Betrag.	5
Die Unternehmensgruppe Tengelmann fördert mit einem sechsstelligen Betrag die Arbeit im brandenburgischen Biosphärenreservat Schorfheide.	5

The graph in Figure 6 also shows that only in two cases did the human judges choose from among all possible alternatives. In one case, there were 4 possible alternatives, and in the other, 6. The original sentence that had 4 alternatives is given in (2). The four alternatives that participants were asked to choose from are given in Table 4, with the frequency of each choice. The original sentence that had 6 alternatives is given in (3). The six alternatives generated by the grammar and the frequencies with which they were chosen is given in Table 5.

(2)   Die Brandursache blieb   zunächst unbekannt.
       The cause of fire   remained initially unknown.
       'The cause of the fire remained unknown initially.'

Tables 4 and 5 tell different stories. On the one hand, although each of the 4 alternatives was chosen at least once from Table 4, there is a clear preference for one string (and this is also the original string from the TIGER Corpus). On the other hand, there is no clear preference[12] for any one of the alternatives in Table 5, and, in fact, the alternative that was selected most frequently by the participants is not the original string. Interestingly, out of the 41 items presented to participants, the original string was chosen by the majority of participants in 36 cases. Again, this confirms the hypothesis that there is a certain amount of acceptable variation for native speakers but there are clear preferences for certain strings over others. It also indicates that corpus sentences are not necessarily the only valid surface realisation for a given abstract representation, and caution should be exercised in their use as gold standard strings.

(3)   Die Unternehmensgruppe Tengelmann fördert mit einem sechsstelligen Betrag
       The group of companies   Tengelmann assists with a   6-figure   sum
       die Arbeit im brandenburgischen Biosphärenreservat Schorfheide.
       the work   in of-Brandenburg   biosphere reserve   Schorfheide.
       'The Tengelmann group of companies is supporting the work at the biosphere reserve in Schorfheide, Brandenburg, with a 6-figure sum.'

---

[12] Although it is clear that alternative 2 is dispreferred.

4.3   Effects of Context

As explained in Section 3.1, Part 3 of our experiment was identical to Part 1, except that the participants could see some preceding context. The aim of this part was to investigate to what extent discourse factors influence the way in which human judges evaluate the output of the realisation ranker. In Task 3a, we expected the original strings to be ranked (even) higher in context than out of context; consequently, the ranks of the realisations selected by the log-linear and the language model would have to go down. With respect to Task 3b, we had no particular expectation, but were just interested in seeing whether some preceding context would affect the evaluation results for the strings selected as most probable by the log-linear model ranker in any way.

Table 6 summarises the results of Task 3a. It shows that, at least overall, our expectation that the original corpus sentences would be ranked higher within context than out of context was not borne out. Actually, they were ranked a bit lower than they were when presented in isolation, and the only realisations that are ranked slightly higher overall are the ones selected by the trigram LM.

Table 6. Task 3a: Ranks for each system (compared to ranks in Task 1a)

	Rank 1	Total Rank 2	Rank 3	Average Rank
Original String	810 (-7)	365 (-1)	71 (+6)	1.41 (+0.01)
LL String	274 (-29)	615 (+22)	357 (+5)	2.07 (+0.03)
LM String	162 (+34)	266 (-23)	818 (-13)	2.53 (-0.03)

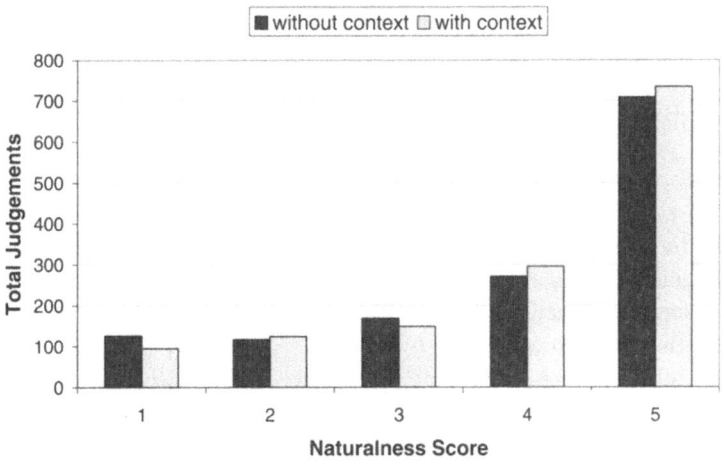

Fig. 7. Tasks 1b and 3b: Naturalness scores for strings chosen by log-linear model, presented without and with context

The overall results of Task 3b are presented in Figure 7. Interestingly, although we did not expect any particular effect of preceding context on the way the participants would rate the realisations selected by the log-linear model, the naturalness scores were higher in the condition with context (Task 3b) than in the one without context (Task 1b). One explanation might be that sentences, regardless of the contextual adequacy of their constituent order, are generally rated higher in context than out of context, simply because the context makes the content conveyed less surprising.

Since, contrary to our expectations, we could not detect a clear effect of context in the overall results of Task 3a, we investigated how the average ranks of the three alternatives presented for individual items differ between Task 1a and Task 3a. An example of an original corpus sentence which many participants ranked higher in context than in isolation is given in (4a.). The realisations selected by the the log-linear model and the trigram LM are given in (4b.) and (4c.) respectively, and the context shown to the participants is given above these alternatives. We believe that the context has this effect because it prepares the reader for the structure with the sentence-initial predicative participle *entscheidend*; usually, these elements tend to appear in clause-final position.

(4)   -2 Betroffen sind die Antibabypillen   Femovan, Lovelle, [...] und Dimirel.
        Concerned are the contraceptive pills Femovan, Lovelle, [...], and Dimirel.
        -1 Das Bundesinstitut schließt nicht aus, daß sich die Thrombose-Warnung
        The federal institute excludes not   that   the thrombosis warning
        als grundlos   erweisen könnte.
        as unfounded turn out could.
        a. Entscheidend sei die [...] abschließende Bewertung, sagte Jürgen Beckmann
        Decisive   is the [...] final   evaluation, said Jürgen Beckmann
        vom   Institut dem ZDF.
        of the institute the ZDF.
        b. Die [...] abschließende Bewertung sei entscheidend, sagte Jürgen Beckmann
        vom Institut dem ZDF.
        c. Die [...] abschließende Bewertung sei entscheidend, sagte dem ZDF Jürgen
        Beckmann vom Institut.

In contrast, (5a.) is an example of a corpus sentence which our participants tended to rank lower in context than in isolation. Actually, the human judges preferred the realisation selected by the trigram LM (5c.) to the original sentence and the realisation chosen by the log-linear model (5b.) in both conditions, but this preference was even reinforced when context was available. One explanation might be that the two preceding sentences are precisely about the decision to which the initial phrase of variant (5c.) refers, which ensures a smooth flow of the discourse.

(5)   -2 Im   konkreten Fall darf der Kurde allerdings trotz   der Entscheidung
        In the concrete   case may the Kurd however   despite the decision
        der   Bundesrichter nicht in die Türkei abgeschoben werden, weil   ihm
        of the federal judges not   to the Turkey deported   be   because him

dort nach   den Feststellungen der   Vorinstanz   politische
there according to the conclusions   of the court of lower instance political
Verfolgung droht.
persecution threatens.

-1 Es besteht Abschiebeschutz   nach   dem Ausländergesetz.
   It exists   deportation protection according to the foreigner law.

a. Der 9. Senat [...] äußerte   sich in seiner Entscheidung nicht zur
   The 9th senate [...] expressed itself in its   decision   not   to the
   Verfassungsgemäßheit der   Drittstaatenregelung.
   constitutionality   of the third-country rule.

b. Der 9. Senat [...] äußerte sich in seiner Entscheidung zur Verfassungsgemäßheit
   der Drittstaatenregelung nicht.

c. In seiner Entscheidung äußerte sich der 9. Senat [...] nicht zur Verfassungsgemäßheit der Drittstaatenregelung.

4.4   Annotator Agreement

We measure two types of annotator agreement. First we measure how well each annotator agrees with him/herself. This is done by evaluating what percentage of the time an annotator made the same choice when presented with the same item choices (recall that, as described in Section 3, a number of items were presented randomly more than once to each participant). The results are given in Table 7. The results show that in between 70% and 77% of cases, judges make the same decision when presented with the same data. We found this to be a surprisingly low number and think that it is most likely due to the acceptable variation in word order for speakers. Another measure of agreement is how well the individual participants agree with each other. In order to establish this, we calculate an average Spearman's correlation coefficient (non-parametric Pearson's correlation coefficient) between all pairs of participants for each experiment. The results are summarised in Table 8. Although these figures indicate a high level of inter-annotator agreement, more tests are required to establish exactly what these figures mean for each experiment.

Table 7. How often did a participant make the same choice?

Experiment	Agreement (%)
Task 1a	77.4
Task 1b	71.0
Part 2	74.3
Task 3a	72.6
Task 3b	70.9

Table 8. Inter-Annotator Agreement for each experiment

Experiment	Spearman coefficient
Task 1a	0.62
Task 1b	0.60
Part 2	0.58
Task 3a	0.61
Task 3b	0.51

5   Correlation with Automatic Metrics

In this section, we examine the correlation between the human judgements obtained in Tasks 1a and 1b and a number of automatic metrics in order to determine which one is most suitable for use in the development of a surface realisation ranker. We looked at automatic metrics that are purely string-based as well as automatic metrics that rely on syntactic analysis.

5.1   String-Based Metrics

The following is the list of string-based metrics that we considered:

BLEU [20] calculates the number of n-grams a solution shares with a reference, adjusted by a brevity penalty. Usually the geometric mean for scores up to 4-gram are reported.

ROUGE [14] is an evaluation metric designed to evaluate automatically generated summaries. It comprises a number of string comparison methods including n-gram matching and skip n-grams. We use the default ROUGE-L longest common subsequence f-score measure. Preliminary experiments with the skip n-grams performed worse than the default parameters.

GTM General Text Matching [15] calculates word overlap between a reference and a solution, without double counting duplicate words. It places less importance on word order than BLEU.

SED Levenshtein (String Edit) distance is the minimum number of insert, delete and substitute operations needed to transform one string into the other.

WER Word Error Rate is derived from the Levenshtein distance and is the number of insertions, deletions and substitutions divided by the total number of words in the reference.

TER Translation Error Rate [23] computes the number of insertions, deletions, substitutions and shifts needed to match a solution to a reference.

Most of these metrics come from the Machine Translation field, where the task is arguably very different. In the evaluation of a surface realisation system (as opposed to a complete generation system), typically the choice of vocabulary is limited and often the task is closer to word re-ordering. Many of the MT metrics have methods for attempting to account for different but equivalent translations of a given source word, typically by integrating a lexical resource

Table 9. Average scores of each metric for Part 1 data

		Task 1a		Task 1b
	GOLD	LL	LM	LL
human A (rank 1–3)	1.4	2.05	2.55	
human B (scale 1–5)				3.92
BLEU	1.0	0.72	0.67	0.79
ROUGE-L	1.0	0.78	0.85	0.85
GTM	1.0	0.60	0.55	0.74
SED	1.0	0.61	0.54	0.71
WER	0.0	39.88	48.04	28.83
TER	0.0	0.14	0.16	0.11

such as WordNet. Also, these metrics were mostly designed to evaluate English output, so it is not clear that they will be equally appropriate for other languages, especially freer word order ones such as German.

The scores given by each metric for the data used in both experiments are presented in Table 9. For the Task 1a data, we use the Spearman rank correlation coefficient to measure the correlation between the human judgements and the automatic scorers. The results are presented in Table 10 for both the sentence and the corpus level correlations;[13] we also present p-values for statistical significance. Since we only have judgements on three systems, the corpus correlation is not that informative. Interestingly, the ROUGE-L metric is the only one that does not rank the output of the three systems in the same order as the judges, since it ranks the strings chosen by the language model higher than the strings chosen by the log-linear model. However, at the level of the individual sentence, the ROUGE-L metric correlates best with the human judgements. The GTM metric correlates at about the same level, but in general there seems to be little difference between the metrics. In general, it should be kept in mind in the interpretation of Table 10 that the best surface realisation got rank 1 and the worst got rank 3, so that the similarity metrics (BLEU, etc.) are correlated negatively with ranks whereas the the dissimilarity metrics (WER, TER) are correlated positively.

For the Task 1b data, we use the Pearson correlation coefficient to measure the correlation between the human judgements and the automatic metrics. The results are given in Table 11. Here we only look at the correlation at the individual sentence level, since we are looking at data from only one system. For this data, the GTM metric clearly correlates most closely with the human judgements, and it is the only metric that exhibits a statistically significant correlation. BLEU and TER correlate particularly poorly with the human judgements, with correlation coefficients very close to zero.

---

[13] Calculating BLEU on a sentence level may lead to score of 0 if there are no matching 4-grams, so the sentence-level BLEU correlations cannot be directly compared. This should be kept in mind when interpreting the results in Tables 10 and 11.

Table 10. Spearman correlation coefficients between human judgements for task 1a (rank 1–3) and automatic metrics

	Sentence		Corpus	
	corr	p-value	corr	p-value
BLEU	-0.615	<0.001	-1	0.3333
ROUGE-L	-0.644	<0.001	-0.5	1
GTM	-0.643	<0.001	-1	0.3333
SED	-0.628	<0.001	-1	0.3333
WER	0.623	<0.001	1	0.3333
TER	0.608	<0.001	1	0.3333

Table 11. Pearson correlation coefficients between human judgements for task 1b (naturalness scale 1–5) and automatic metrics

	Sentence	
	corr	p-value
BLEU	0.095	0.5048
ROUGE-L	0.207	0.1417
GTM	0.424	0.0017
SED	0.168	0.2344
WER	-0.188	0.1817
TER	-0.024	0.8646

The results show that there is little difference in the performance of the metrics in the relative ranking task, and that all systems correlate well with the human judgements. The ability of the metrics to tell whether (relatively speaking) one string is more natural than another is in itself an encouraging result, and suggests that in certain situations most automatic metrics will perform adequately. However, if what is required is an absolute judgement on a particular string, it seems that none of the automatic metrics are performing particularly well. One possible explanation for this is the lack of linguistic insight into the differences between alternatives. In a relatively free word-order language like German, there are certain discourse-related constraints that influence the "naturalness" of a sentence. These constraints apply predominantly to constituent order, whereas the order within constituents is generally fixed.[14] All of the automatic metrics view constituent reorderings as equally likely, since they do not integrate any linguistic knowledge, and this limits their ability to give an absolute score to sentences where the constituents have been reordered.

5.2   Syntactic Metrics

Recently, there has been a move towards more syntactic, rather than purely string-based, evaluation of MT output and summarisation [13,19]. The idea is

---

[14] See also Filippova and Strube [10].

to go beyond simple string comparisons and evaluate at a deeper linguistic level. Since most of the work in this direction has only been carried out for English so far, we apply the idea rather than a specific tool to the data. We parse the data from both experiments with a German dependency parser [12] trained on the TIGER Treebank (with sentences 8000-10000 heldout for testing). This parser achieves 91.23% labelled accuracy on the 2000-sentence test set.

To calculate the correlation between the human judgements and the dependency parser, we parse the original strings as well as the strings chosen by the log-linear and language models. The standard procedure for dependency parser evaluation relies on both strings being identical to calculate (un-)labelled dependency accuracy, and so we map the dependencies produced by the parser into sets of triples as used in the evaluation software of Crouch et al. [9] where each dependency is represented as `deprel(head,word)` and each word is indexed with its position in the original string.[15] We compare the parses for both experiments against the parses of the original strings. We calculate both a weighted and unweighted dependency f-score, as given in Table 12.

Table 12. Average scores of the syntactic metric for Part 1 data

| | Task 1a | | | Task 1b |
	GOLD	LL	LM	LL
DEP	1.0	0.88	0.83	0.93
UNWEIGHTED-DEP	1.0	0.82	0.70	0.90

Table 13. Correlation between dependency-based evaluation and human judgements

| | Task 1a | | Task 1b | |
	corr	p-value	corr	p-value
DEP	-0.640	<0.001	0.186	0.1860
UNWEIGHTED DEP	-0.657	<0.001	0.290	0.03686

The unweighted f-score is calculated by taking the average of the scores for each dependency type, while the weighted f-score weighs each average score by its frequency in the test corpus. We calculate the Spearman and Pearson correlation coefficients as before; the results are given in Table 13. The results show that the unweighted dependencies correlate more closely (and statistically significantly) with the human judgements than the weighted ones. This suggests that the frequency of a dependency type does not matter as much as its overall correctness.

The large discrepancy between the absolute correlation coefficients for Task 1a and 1b can be explained by the fact that they are different tasks. Task 1a

---

[15] This is a 1-1 mapping, and the indexing ensures that duplicate words in a sentence are not confused.

ranks 3 strings relative to one another, while Task 1b measures the naturalness of the string. We would expect automatic metrics to be better at the first task than the second, as it is easier to rank systems relative to each other than to give a system an absolute score.

6 Related Work

The work that is most closely related to what is presented in this chapter is that of Velldal's thesis [25], where several models of realisation ranking are presented and evaluated against the original corpus text. Chapter 8 of his thesis describes a small human-based experiment, where 7 native English speakers rank the output of 4 systems. One system is the original text, another is a randomly chosen baseline, the third is a string chosen by a log-linear model and the fourth is one chosen by a language model. Joint rankings were allowed. The results presented in Velldal's thesis [25] mirror our findings in Tasks 1a and 3a, that native speakers rank the original strings higher than the log-linear model strings which are ranked higher than the language model strings. In both cases, the log-linear models include the language model score as a feature in the log-linear model. Nakanishi et al. [16] report that they achieve the best BLEU scores when they do not include the language model score in their log-linear model, but they also admit that their language model was not trained on enough data.

Belz and Reiter [3] carry out a comparison of automatic evaluation metrics against human domain experts and human non-experts in the domain of weather forecast statements. In their evaluations, the NIST score correlated more closely than BLEU or ROUGE to the human judgements. They conclude that more than 4 reference texts are needed for automatic evaluation of NLG systems. Gatt and Belz [11] (this volume) use the string edit distance, the BLEU and the NIST metrics as well as human assessments of adequacy and fluency in the evaluation of string realisations of referring expressions. They also carry out a task-based evaluation of the realisation candidates. They then correlate the intrinsic metrics among each other and with the task-based metric, but cannot find any significant correlation between any of the automatic metrics and any of the human-assessed ones. The significant correlations they do observe are between human-assessed intrinsic metrics and task-based ones. Belz and Kow [2] (this volume) show that the automatic evaluation metrics under-estimate the performance of hand-crafted systems and over-estimate the performance of automatically created systems, further emphasising the need for additional evaluation of any NLG system.

The correlation between the dependency parsing metric and the human judgements was no higher than those of the simple string-based metrics GTM and ROUGE (although it did outperform all other automatic metrics). This does not correspond to related work on English Summarisation evaluation [18] which shows that a metric based on an automatically induced LFG parser for English achieves comparable or higher correlation with human judgements than ROUGE and Basic Elements (BE).[16] Parsers of German typically do not achieve as high

---

[16] The GTM metric was not compared in that paper.

performance as their English counterparts, and further experiments including alternative parsers are needed to see if we can improve performance of this metric. Summarisation and Surface Realisation are also different tasks, so it is not clear that evaluation metrics that work well in Summarisation should also work well for surface realisation.

The data used in our experiments was almost always grammatically correct. Therefore the task of an evaluation system is to score more natural sounding strings higher than marked or unnatural ones. In this respect, our findings mirror those of Stent et al. [24] for English data, that the automatic metrics do not correlate well with human judges on syntactic correctness.

7 Conclusion

In this chapter, we have presented a human-based experiment to evaluate the output of a realisation ranking system for German. We evaluated the original corpus text, and strings chosen by a log-linear model and a language model. We found that, at a global level, the human judgements mirrored the relative rankings of the three system according to the BLEU score. In terms of naturalness, the strings chosen by the log-linear model were generally given 4 or 5, indicating that although the log-linear model might not choose the same string as the original author had written, the strings it chose were mostly very natural strings.

When presented with all alternatives generated by the grammar for a given input f-structure, the human judges chose the same string as the original author 70% of the time. In 5 out of 41 cases, the majority of judges chose a string other than the original string. These figures show that native speakers accept some variation in word order, and so caution should be exercised when using corpus-derived reference data. The observed acceptable variation was often linked to information-structural considerations. Cahill and Riester [7] show how generation ranking can be further improved by automatically learning morphosyntactic approximations of information status patterns.

In examining the effect of preceding context, we found that overall context had very little effect. At the level of individual sentences, however, clear tendencies were observed, but some sentences were judged better in context and others were ranked lower. This may be because, given the corpus annotation, it was impossible to guarantee that the automatically extracted preceding context was relevant.

We presented data that examined the correlation between native speaker judgements and automatic evaluation metrics on automatically generated German text. We found that for our first task, all metrics were correlated to roughly the same degree, with ROUGE-L achieving the highest correlation at an individual sentence level and the GTM tool right behind. At the corpus level, all except ROUGE were in agreement with the human judgements. In the second task, the GTM tool had the strongest correlation. We carried out an experiment to test whether a more sophisticated syntax-based evaluation metric performed better than the more simple string-based ones. For the more difficult naturalness task, we found that while the unweighted dependency evaluation metric correlated with the human judgements more strongly than almost all other metrics, it did not outperform the GTM tool. The correlation between the human judgements and the automatic evaluation metrics was much higher for the relative ranking task than for the naturalness task.

Acknowledgments. We are extremely grateful to all of our participants for their generosity in terms of time and patience, without which we could not have carried out this experiment. Furthermore, we would like to thank Johan Hall, Joakim Nivre and Yannick Versely for their help in retraining the MALT dependency parser with our data set. This work was partly funded by the Collaborative Research Centre (SFB 732) at the University of Stuttgart.

References

1. Bangalore, S., Rambow, O., Whittaker, S.: Evaluation metrics for generation. In: Proceedings of the First International Natural Language Generation Conference (INLG 2000), Mitzpe Ramon, Israel, pp. 1–8 (2000)
2. Belz, A., Kow, E.: Assessing the trade-off between system building cost and output quality in data-to-text generation. In: Krahmer, E., Theune, M. (eds.) Empirical Methods in NLG. LNCS (LNAI), vol. 5790, pp. 180–200. Springer, Heidelberg (2010)
3. Belz, A., Reiter, E.: Comparing automatic and human evaluation of NLG systems. In: Proceedings of the 11th Conference of the European Chapter of the Association for Computational Linguistics, Trento, Italy, pp. 313–320 (2006)
4. Brants, S., Dipper, S., Hansen, S., Lezius, W., Smith, G.: The TIGER treebank. In: Proceedings of the Workshop on Treebanks and Linguistic Theories, Sozopol, Bulgaria, pp. 24–41 (2002)
5. Bresnan, J.: Lexical-Functional Syntax. Blackwell, Oxford (2001)
6. Cahill, A., Forst, M., Rohrer, C.: Stochastic realisation ranking for a free word order language. In: Proceedings of the Eleventh European Workshop on Natural Language Generation, DFKI GmbH, Saarbrücken, Germany, June 2007, pp. 17–24 (2007) (document D-07-01)
7. Cahill, A., Riester, A.: Incorporating information status into generation ranking. In: Proceedings of the Joint Conference of the 47th Annual Meeting of the ACL and the 4th International Joint Conference on Natural Language Processing of the AFNLP, August 2009, pp. 817–825. Association for Computational Linguistics, Suntec (August 2009)
8. Callaway, C.: Evaluating coverage for large symbolic NLG grammars. In: Proceedings of the 18th International Joint Conference on Artificial Intelligence (IJCAI 2003), Acapulco, Mexico, pp. 811–817 (2003)
9. Crouch, R., Kaplan, R., King, T.H., Riezler, S.: A comparison of evaluation metrics for a broad coverage parser. In: Proceedings of the LREC Workshop: Beyond PARSEVAL – Towards Improved Evaluation Measures for Parsing Systems, Las Palmas, Canary Islands, Spain, pp. 67–74 (2002)
10. Filippova, K., Strube, M.: Generating constituent order in German clauses. In: Proceedings of the 45th Annual Meeting of the Association of Computational Linguistics, pp. 320–327. Association for Computational Linguistics, Prague (June 2007)

11. Gatt, A., Belz, A.: Introducing shared task evaluation to NLG: The TUNA shared task evaluation challenges. In: Krahmer, E., Theune, M. (eds.) Empirical Methods in NLG. LNCS (LNAI), vol. 5790, pp. 264–293. Springer, Heidelberg (2010)
12. Hall, J., Nivre, J.: A dependency-driven parser for German dependency and constituency representations. In: Proceedings of the Workshop on Parsing German, pp. 47–54. Association for Computational Linguistics, Columbus (June 2008)
13. Hovy, E., Yew Lin, C., Zhou, L.: Evaluating duc 2005 using basic elements. In: Proceedings of DUC 2005 (2005)
14. Lin, C.Y.: ROUGE: A package for automatic evaluation of summaries. In: Marie-Francine Moens, S.S. (ed.) Text Summarization Branches Out: Proceedings of the ACL 2004 Workshop, pp. 74–81. Association for Computational Linguistics, Barcelona (July 2004)
15. Melamed, I.D., Green, R., Turian, J.P.: Precision and recall of machine translation. In: NAACL 2003: Proceedings of the 2003 Conference of the North American Chapter of the Association for Computational Linguistics on Human Language Technology, pp. 61–63. Association for Computational Linguistics, Morristown (2003)
16. Nakanishi, H., Miyao, Y., Tsujii, J.: Probabilistic models for disambiguation of an HPSG-based chart generator. In: Proceedings of the Ninth International Workshop on Parsing Technology, pp. 93–102. Association for Computational Linguistics, Vancouver (October 2005)
17. Nenkova, A., Chae, J., Louis, A., Pitler, E.: Structural features for predicting the linguistic quality of text: Applications to machine translation, automatic summarization and human-authored text. In: Krahmer, E., Theune, M. (eds.) Empirical Methods in NLG. LNCS (LNAI), vol. 5790, pp. 222–241. Springer, Heidelberg (2010)
18. Owczarzak, K.: DEPEVAL(summ): Dependency-based evaluation for automatic summaries. In: Proceedings of the Joint Conference of the 47th Annual Meeting of the ACL and the 4th International Joint Conference on Natural Language Processing of the AFNLP, pp. 190–198. Association for Computational Linguistics, Suntec (August 2009)
19. Owczarzak, K., van Genabith, J., Way, A.: Evaluating machine translation with LFG dependencies. Machine Translation 21, 95–119 (2008)
20. Papineni, K., Roukos, S., Ward, T., Zhu, W.J.: Bleu: a method for automatic evaluation of machine translation. In: Proceedings of 40th Annual Meeting of the Association for Computational Linguistics, pp. 311–318. Association for Computational Linguistics, Philadelphia (July 2002)
21. Reiter, E., Sripada, S.: Should corpora texts be gold standards for NLG? In: Proceedings of INLG 2002, Harriman, NY, pp. 97–104 (2002)
22. Rohrer, C., Forst, M.: Improving coverage and parsing quality of a large-scale LFG for German. In: Proceedings of the Language Resources and Evaluation Conference (LREC 2006), Genoa, Italy, pp. 2206–2211 (2006)
23. Snover, M., Dorr, B., Schwartz, R., Micciulla, L., Weischedel, R.: A study of translation error rate with targeted human annotation. In: Proceedings of the Association for Machine Translation in the Americas Conference 2006, pp. 223–231 (2006)
24. Stent, A., Marge, M., Singhai, M.: Evaluating evaluation methods for generation in the presence of variation. In: Proceedings of CICLING, pp. 341–351 (2005)
25. Velldal, E.: Empirical Realization Ranking. Ph.D. thesis, University of Oslo (2008)
26. Velldal, E., Oepen, S.: Statistical ranking in tactical generation. In: Proceedings of the 2006 Conference on Empirical Methods in Natural Language Processing, pp. 517–525. Association for Computational Linguistics, Sydney (July 2006)

Structural Features for Predicting the Linguistic Quality of Text
Applications to Machine Translation, Automatic Summarization and Human-Authored Text

Ani Nenkova, Jieun Chae, Annie Louis, and Emily Pitler

University of Pennsylvania
{nenkova,chaeji,lannie,epitler}@seas.upenn.edu

Abstract. Sentence structure is considered to be an important component of the overall linguistic quality of text. Yet few empirical studies have sought to characterize how and to what extent structural features determine fluency and linguistic quality. We report the results of experiments on the predictive power of syntactic phrasing statistics and other structural features for these aspects of text. Manual assessments of sentence fluency for machine translation evaluation and text quality for summarization evaluation are used as gold-standard. We find that many structural features related to phrase length are weakly but significantly correlated with fluency and classifiers based on the entire suite of structural features can achieve high accuracy in pairwise comparison of sentence fluency and in distinguishing machine translations from human translations. We also test the hypothesis that the learned models capture general fluency properties applicable to human-authored text. The results from our experiments do not support the hypothesis. At the same time structural features and models based on them prove to be robust for automatic evaluation of the linguistic quality of multi-document summaries.

1 Introduction

Numerous natural language applications involve the task of producing fluent text. This is a core problem for surface realization in natural language generation [29,2], as well as an important step in machine translation (MT). Considerations of sentence fluency are also key in sentence simplification [42], sentence compression [24,28,11,34,46,18], text re-generation for summarization [6,48] and headline generation [4,49,43]. Despite the popularity of these applications, the factors contributing to sentence fluency have not been researched in depth. Much more attention has been devoted to discourse-level constraints on adjacent sentences indicative of coherence and good text flow [30,5,27]. But the development of fully automatic measures of fluency will make it possible to evaluate system output without the involvement of human assessors, which in turn will facilitate system development.

E. Krahmer, M. Theune (Eds.): Empirical Methods in NLG, LNAI 5790, pp. 222–241, 2010.
© Springer-Verlag Berlin Heidelberg 2010

In many applications fluency is assessed in combination with other qualities and the assessment is performed in comparison with a human model. For example, in machine translation evaluation, automatic evaluation methods such as BLEU [37] use n-gram overlap comparisons with a model to judge the overall translation quality, with higher n-grams meant to capture fluency considerations. More sophisticated ways to compare a system production and a model involve the use of syntax, but even in these cases fluency is only indirectly assessed and the main advantage of the use of syntax is better estimation of the *semantic* overlap between a model and an output. Similarly, the metrics proposed for text generation by [3] (simple accuracy, generation accuracy) are based on string-edit distance from an ideal output.

In contrast, the work of [48] and [35] directly sets as a goal the assessment of sentence-level fluency, regardless of content and without any human gold-standard. In [48] the main premise is that syntactic information from a parser can more robustly capture fluency than language models, giving more direct indications of the degree of ungrammaticality of a sentence. The idea is extended in [35], where features derived from four different parsers are shown to lead to impressive success in the assessment of fluency of artificially generated sentences with varying level of fluency. Their fluency models hold promise for actual improvements in machine translation output quality [50].

Syntactic tree features that capture common parse configurations and that are used in discriminative parsing [12,9,23] are expected to be beneficial for predicting sentence fluency as well. Indeed, early work has demonstrated that syntactic features, and branching properties in particular, are helpful features for automatically distinguishing human translations from machine translations [15]. The exploration of branching properties of human and machine translations was motivated by the observations during failure analysis that MT system output tends to favor right-branching structures over noun compounding. Branching preference mismatches manifest themselves in the English output when translating from languages whose branching properties are radically different from English. Accuracy close to 80% was achieved for distinguishing human translations from machine translations.

Structural features have also been used for ranking different surface realizations corresponding to the same input semantics, for example in the work of [47] and [8]. In these prior studies, a corpus of English and German sentences respectively are parsed into HPSG/LFG structures. Then all possible surface realizations for the structures are generated and a log-linear model ranker is trained to recognize the original sentence which is considered to be the best realization. Structural features lead to better models than n-gram language model features for both languages. In a follow-up work on human assessment of surface realization variability, Cahill and Forst [7] (this volume) present findings that further motivate the need for automatic objective metric for sentence fluency evaluation. In their experiments, they found that subjects agreed with their own ranking of surface realizations only 70% of the time. A suitable automatic model of fluency will not only be cheaper than manual evaluation but will also remove noise due to human judgement variability.

In our work we continue the investigation of sentence level fluency based on features that capture surface statistics of the syntactic structure in a sentence. We define the features in Sect. 2.1. We revisit the task of distinguishing machine translations from human translations (Sect. 2.3) , but also further our understanding of fluency by providing a comprehensive analysis of the association between fluency assessments of translations and structural features (Sect. 2.2 and Sect. 2.5). We also demonstrate that based on the same class of features, it is possible to distinguish fluent machine translations from non-fluent machine translations (Sect. 2.4). Finally, we test the models on human written text in order to verify if the classifiers trained on data coming from machine translation evaluations can be used for general predictions of fluency and readability (Sect. 3.1 and Sect. 3.2). The results indicate that the models do not generalize well for the different type of data.

Given the findings that fluency models trained on machine translation data do not perform well on human-authored text, we conducted a study where training in testing is performed over the same domain. Specifically, we test the feasibility of performing automatic evaluation of linguistic quality of multi-document summaries using the same structural features (Sect. 4). To ensure that findings are not specific to a given dataset, we train and test the model on consecutive years of evaluations of summarization systems.

2   Sentence Fluency and Machine Translation

For our experiments we use the evaluations of Chinese to English translations distributed by the Linguistic Data Consortium (catalog number LDC2003T17), for which both machine and human translations are available. Machine translations have been assessed by evaluators for fluency on a five point scale (5: flawless English; 4: good English; 3: non-native English; 2: disfluent English; 1: incomprehensible). Assessments by different annotators were averaged to assign overall fluency assessment for each machine-translated sentence. For each segment (sentence), there are four human and three machine translations.

In this setting we address four tasks with increasing difficulty:

- Distinguish human and machine translations.
- Distinguish fluent machine translations from poor machine translations.
- Distinguish the better (in terms of fluency) translation among two translations of the same input segment. This task corresponds to input-level automatic evaluation of fluency.[1]
- Use the models trained on data from MT evaluations to predict potential fluency problems of human-written texts from the Wall Street Journal.

---

[1] Our data is not suitable for experiments with system-level evaluation where the task is to predict which system is better than others over an entire test suite because there are only three systems. We will address this task for multi-document summarization, where we have summaries produced by 30 or more participating systems.

It is important to note that the purpose of our study is not evaluation of machine translation per se. Our goal is more general and the interest is in finding predictors of sentence fluency. There are no corpora with fluency assessments collected for human-authored text, so it seems advantageous to use the assessments done in the context of machine translation for preliminary investigations of fluency. Nevertheless, our findings are also potentially beneficial for sentence-level evaluation of machine translation.

2.1   Features

Perceived sentence fluency is influenced by many factors. The way the sentence fits in the context of surrounding sentences is one obvious factor [5]. Another well-known factor is vocabulary use: the presence of uncommon difficult words is known to pose problems to readers and to render text less readable [13,41]. But these discourse- and vocabulary-level features measure properties at granularities different from the sentence level.

Structural sentence level features have not been investigated as a stand-alone class, as has been done for the other types of features. This is why we constrain our study to syntactic features alone, and do not initially discuss discourse and language model features in our experiments with machine translation data. For our experiments on evaluation of the linguistic quality of multi-sentential summaries, we do compare several classes of features.

In our work, instead of looking at the syntactic structures present in the sentences, e.g. the syntactic rules used, we use surface statistics of phrase length and types of modification. The sentences were parsed with Charniak's parser [10] in order to calculate these features.

In order to facilitate later reference to features that turn out to be significant in correlation analysis with fluency ratings, we denote some of the Feature Classes by FC_n.

Sentence length is the number of words in a sentence. Evaluation metrics such as BLEU [37] have a built-in preference for shorter translations. In general one would expect that shorter sentences are easier to read and thus are perceived as more fluent. We added this feature in order to test directly the hypothesis for brevity preference.

Parse tree depth and the number of subordinating conjunctions (*SBAR count*) are considered to be a measure of sentence complexity, as well as the number of noun phrases, verb phrases and prepositional phrases [38]. Generally, longer sentences are syntactically more complex but when sentences are approximately the same length parse tree depth can be indicative of increased complexity that can slow processing and lead to lower perceived fluency of the sentence.

Number of fragment tags in the sentence parse. Fragments occur without necessarily causing fluency problems in headlines (e.g. "Cheney willing to hold bilateral talks if Arafat observes U.S. cease-fire arrangement") but in machine translation the presence of fragments can signal a more serious problem.

Phrase type proportion was computed for prepositional phrases (PP), noun phrases (NP) and verb phrases (VP). The length in number of words of each phrase type was counted, then divided by the sentence length. Embedded phrases were also included in the calculation: for example a noun phrase (NP1 ... (NP2)) would contribute $length(NP1) + length(NP2)$ to the phrase length count.

Average phrase length is the number of words comprising a given type of phrase, divided by the number of phrases of this type. It was computed for PP, NP, VP, ADJP, ADVP. Two versions of the features were computed— (FC_1) one with embedded phrases included in the calculation and (FC_2) one just for the largest phrases of a given type; the average length of *any* phrase type in a sentence was also calculated. *Normalized average phrase length* (FC_3) is computed for PP, NP and VP and is equal to the average phrase length of given type divided by the sentence length. These were computed only for the largest phrases.

Phrase type rate was also computed for PPs, VPs and NPs and is equal to the number of phrases of the given type that appeared in the sentence, divided by the sentence length. For example, the sentence "The boy caught a huge fish this morning" will have NP phrase number equal to 3/8 and VP phrase number equal to 1/8.

Phrase length. (FC_4) The number of words in a PP, NP, VP, without any normalization; it is computed only for the largest phrases. *Normalized phrase length* is the average phrase length (for VPs, NPs, PPs) divided by the sentence length. This was computed both for (FC_5) longest phrase where embedded phrases of the same type were counted only once and (FC_6) for each phrase regardless of embedding.

Length of NPs/PPs contained in a VP. The average number of words that constitute a NP or PP within a verb phrase, divided by the length of the verb phrase. Similarly, the *length of PP in NP* was computed.

Head noun modifiers. Noun phrases can be very complex, and the head noun can be modified in a variety of ways—pre-modifiers, prepositional phrase modifiers, apposition. The length in words of these modifiers was calculated. Each feature also had a variant in which the modifier length was divided by the sentence length. Finally, two more features on total modification were computed: one was the sum of all modifier lengths, the other the sum of normalized modifier length.

2.2   Feature Analysis

In this section, we analyze the association of the features that we described above and fluency. Note that the purpose of the analysis is not feature selection—all features will be used in the later experiments. Rather, the analysis is performed in order to better understand which factors are predictive of good fluency.

The distribution of fluency scores in the dataset is rather skewed, with the majority of the sentences rated as being of average fluency 3 as can be seen in Table 1.

Table 1. Distribution of fluency scores

Fluency score	Number of sentences
$1 \leq$ fluency < 2	7
$1 \leq$ fluency < 2	295
$2 \leq$ fluency < 3	1789
$3 \leq$ fluency < 4	521
$4 \leq$ fluency < 5	22

Table 2 lists the features for which Pearson's correlation coefficient between the fluency ratings and the values of features was highest.

First of all, fluency and adequacy as given by MT evaluators are highly correlated (0.7). This is surprisingly high, given that separate fluency and adequacy assessments were elicited with the idea that these are qualities of the translations that are independent of each other. Fluency was judged directly by the assessors, while adequacy was meant to assess the content of the sentence compared to a human gold-standard. Yet, the assessments of the two aspects were often the same—readability/fluency of the sentence is important for understanding the sentence. Only after the assessor has understood the sentence can (s)he judge how it compares to the human model. One can conclude then that a model of fluency/readability that will allow systems to produce fluent text is key for developing a successful machine translation system.

The next feature most strongly associated with fluency is sentence length. Shorter sentences are easier and perceived as more fluent than longer ones, which is not surprising. Such preference for brevity has been empirically validated in computational linguistics work both for written text [39] and for utterances in dialog [40] (this volume). Note though that the correlation is actually rather weak. It is only one of various fluency factors and has to be accommodated alongside the possibly conflicting requirements shown by the other features. Still, length considerations reappear at sub-sentential (phrasal) levels as well.

Noun phrase length for example has almost the same correlation with fluency as sentence length does. The longer the noun phrases, the less fluent the sentence is. Long noun phrases take longer to interpret and reduce sentence fluency/readability.

Consider the following example:

- *[The dog]* jumped over the fence and fetched the ball.
- *[The big dog in the corner]* fetched the ball.

The long noun phrase is more difficult to read, especially in subject position. Similarly the length of the verb phrases signals potential fluency problems as can be seen from the examples of human translation in our corpus:[2]

---

[2] Human translations were not rated for fluency and were considered ideal, as if rated 5. Such assumptions might be too strong. As we will see later, summaries written by people were occasionally rated as being of poor quality by assessors different from the original writer.

- Most of the US allies in Europe publicly [object to invading Iraq]$_{VP}$.
- But this [is dealing against some recent remarks of Japanese financial minister, Masajuro Shiokawa]$_{VP}$.

VP distance (the average number of words separating two verb phrases) is also negatively correlated with sentence fluency. In machine translations there is the obvious problem that they might not include a verb for long stretches of text. But even in human written text, the presence of more verbs can make a difference in fluency [1]. Consider the following two sentences:

- In his state of the Union address, Putin also **talked** about the national development plan for this fiscal year and the domestic and foreign policies.
- Inside the courtyard of the television station, a reception team of 25 people **was formed to attend** to those who **came to make** donations in person.

The next strongest correlation is with unnormalized verb phrase length. In fact in terms of correlations, in turned out that it was best not to normalize the phrase length features at all. The normalized versions were also correlated with fluency, but the association was lower than for the direct count without normalization.

Parse tree depth is the final feature correlated with fluency with correlation above 0.1.

Table 2. Pearson's correlation coefficient between fluency and different features. P-values are given in parenthesis.

adequacy	sentence length	FC$_4$ for NP
0.701 (0.00)	-0.132 (0.00)	-0.124 (0.00)
VP distance	**FC$_4$ for VP**	**max tree depth**
-0.116 (0.00)	-0.109 (0.00)	-0.106 (0.00)
FC$_2$ any phrase	**FC$_1$ for NP**	**FC$_1$ for VP**
-0.105 (0.00)	-0.097 (0.00)	-0.094 (0.00)
SBAR length	**FC$_2$ for NP**	**FC$_4$ for PP**
-0.086 (0.00)	-0.084 (0.00)	-0.082 (0.00)
FC$_1$ for PP	**SBAR count**	**PP length in VP**
-0.070 (0.00)	-0.069 (0.001)	-0.066 (0.001)
FC$_5$ for PP	**NP length in VP**	**FC$_6$ PP**
0.065 (0.001)	-0.058 (0.003)	-0.054 (0.006)
FC$_6$ for VP	**PP length in NP**	**Fragment**
0.054 (0.005)	0.053 (0.006)	-0.049(0.011)

None of the features related to noun modification—apposition length, number of appositions, number of pre-modifiers, etc—were significantly correlated with fluency at the 0.95 confidence level.

2.3 Distinguishing Human from Machine Translations

In this section we use all the features introduced in Section 2.1 for several classification tasks. Note that while we discussed the high correlation between fluency and adequacy, we do not use adequacy in the experiments that we report from here on.

For all experiments we used four of the classifiers in the WEKA machine learning toolkit [22]: decision tree (J48), logistic regression, support vector machines (SMO), and multi-layer perceptron. All results are for 10-fold cross validation.

We extracted the 300 sentences with highest fluency scores, 300 sentences with lowest fluency scores among machine translations and 300 randomly chosen human translations. We then tried the classification task of distinguishing human and machine translations with different fluency quality (highest and lowest fluency score). We expect that low fluency MT will be more easily distinguished from human translation in comparison with machine translations rated as having high fluency. We also ran experiments with the entire dataset, including all human translations and all machine translations regardless of fluency level.

Results are shown in Table 3. Overall the best classifier is the multi-layer perceptron. On the task using all available data of machine and human translations, the classification accuracy is 86.99%. We expected that distinguishing the machine translations from the human ones will be harder when the best translations are used, compared to the worse translations, but this expectation is fulfilled only for the support vector machine classifier.

The high accuracies shown in Table 3 give convincing evidence that the surface structural statistics can distinguish very well between fluent and non-fluent sentences when the examples come from human and machine-produced text respectively. If this is the case, will it be possible to distinguish between good and bad machine translations as well? In order to answer this question, we ran one more binary classification task. The two classes were the 300 machine translations with highest and lowest fluency respectively. The results are not as good as those for distinguishing machine and human translation, but still significantly outperform a random baseline. All classifiers performed similarly on the task, and achieved accuracy close to 61%.

Table 3. Accuracy for the task of distinguishing machine and human translations

Classifier	worst 300 MT	best 300 MT	all MT
SMO	86.00%	78.33%	82.68%
Logistic reg.	77.16%	79.33%	82.68%
MLP	78.00%	82%	86.99%
Decision Tree(J48)	71.67 %	81.33%	86.11%

2.4   Pairwise Fluency Comparisons

We also considered the possibility of pairwise comparisons for fluency: given two sentences, can we distinguish which is the one scored more highly for fluency. The feature vector for each pair of sentences is obtained as the difference of features of the individual sentences.

There are two ways this task can be set up. First, we can use all assessed translations and make pairings for every two sentences with different fluency assessment. In this setting, the question being addressed is *Can sentences with differing fluency be distinguished?*, without regard to the sources of the sentence. The harder question is *Can a more fluent translation be distinguished from a less fluent translation of the same sentence?*

The results from these experiments can be seen in Table 4. When any two sentences with different fluency assessments are paired, the prediction accuracy is very high: 91.34% for the multi-layer perceptron classifier. In fact all classifiers have accuracy higher than 80% for this task. The surface statistics of syntactic form are powerful enough to distinguishing sentences of varying fluency.

The task of pairwise comparison for translations of the same input is more difficult: doing well on this task would be equivalent to having a reliable measure for ranking different possible translation variants.

Table 4. Accuracy for pairwise fluency comparison. "Same sentence" are comparisons constrained between different translations of the same sentences, "Any pair" contains comparisons of sentences with different fluency over the entire dataset.

Task	J48	Logistic Regression	SMO	MLP
Any pair	89.73%	82.35%	82.38%	91.34%
Same Sentence	67.11%	70.91%	71.23%	69.18%

In fact, the problem is *much* more difficult as can be seen in the second row of Table 4, and the performance for all classifiers is more than 10% lower than those for comparisons not constrained to be translations of the same sentence. Logistic regression, support vector machines and multi-layer perceptron perform similarly, with support vector machine giving the best accuracy of 71.23%. This number is still impressively high, and significantly higher than baseline performance.

2.5   Feature Analysis: Differences among Tasks

In the previous sections we presented three variations involving fluency predictions based on syntactic phrasing features: distinguishing human from machine translations, distinguishing good machine translations from bad machine translations, and pairwise ranking of sentences with different fluency. The results differ considerably and it is interesting to know whether the same kind of features are useful in making the three distinctions.

In Table 5 we show the five features with largest weight in the support vector machine model for each task. In many cases, certain features appear to be important only for particular tasks. For example the number of prepositional phrases is an important feature only for ranking different versions of *the same sentence* but is not important for other distinctions. The number of appositions is helpful in distinguishing human translations from machine translations, but is

Table 5. The five features with highest weights in the support vector machine model for the different tasks

MT vs HT	good MT vs Bad MT	Ranking	Same sentence Ranking
FC$_4$ for PP	# of SBARs	FC$_2$ for NP	FC$_5$ for NP
PP length in VP	FC$_4$ for VP	FC$_3$ for PP	# of PP
FC$_2$ for NP	post modification length	# of NP	FC$_6$ for NP
# of appositions	# of VP	FC$_3$ for NP	max tree depth
SBAR length	sentence length	FC$_3$ for VP	FC$_2$ any

not that useful in the other tasks. So the predictive power of the features is very directly related to the variant of fluency distinctions one is interested in making.

3   Applications to Human-Authored Text

3.1   Identifying Hard-to-Read Sentences in Wall Street Journal Texts

The goal we set out in the beginning of this paper was to derive a predictive model of sentence fluency from data coming from MT evaluations. In the previous sections, we demonstrated that indeed structural features can enable us to perform this task very accurately *in the context of machine translation*. But will the models conveniently trained on data from MT evaluation be at all capable to identify sentences in human-written text that are not fluent and are difficult to understand?

To answer this question, we performed an additional experiment on 30 Wall Street Journal articles from the Penn Treebank that were previously used in experiments for assessing overall text quality [39]. The articles were chosen at random and comprised a total of 290 sentences. One human assessor was asked to read each sentence and mark the ones that seemed disfluent because they were hard to comprehend. These were sentences that needed to be read more than once in order to fully understand the information conveyed in them. There were 52 such sentences. The assessments served as a gold-standard against which the predictions of the fluency models were compared.

Two models trained on machine translation data were used to predict the status of each sentence in the WSJ articles. One of the models was that for distinguishing human translations from machine translations (human vs. MT), the other was the model for distinguishing the 300 best from the 300 worst machine translations (good MT vs. bad MT). The classifiers used were decision trees for human vs. machine distinction and support vector machines for good MT vs. bad MT. For the first model sentences predicted to belong to the "human translation" class are considered fluent; for the second model fluent sentences are the ones predicted to be in the "good MT" class.

The results are shown in Table 6. The two models differ in performance considerably. The model for distinguishing machine translations from human translations is the better one, with accuracy of 57%. For both, prediction accuracy is much lower than when tested on data from MT evaluations. These findings indicate that building a new corpus for the finer fluency distinctions present in human-written text is likely to be more beneficial than trying to leverage data from existing MT evaluations.

Table 6. Accuracy, precision and recall (for fluent class) for each model when test on WSJ sentences

Model	Accuracy	Precision	Recall
human vs machine trans.	57%	0.79	0.58
good MT vs bad MT	44%	0.57	0.44

Below, we show several example sentences on which the assessor and the model for distinguishing human and machine translations (dis)agreed.

1. Model and assessor agree that sentence is problematic.
   (a) The Soviet legislature approved a 1990 budget yesterday that halves its huge deficit with cuts in defense spending and capital outlays while striving to improve supplies to frustrated consumers.
   (b) Officials proposed a cut in the defense budget this year to 70.9 billion rubles (US$114.3 billion) from 77.3 billion rubles (US$125 billion) as well as large cuts in outlays for new factories and equipment.
   (c) Rather, the two closely linked exchanges have been drifting apart for some years, with a nearly five-year-old moratorium on new dual listings, separate and different listing requirements, differing trading and settlement guidelines and diverging national-policy aims.
2. The model predicts the sentence is good, but the assessor finds it problematic.
   (a) Moody's Investors Service Inc. said it lowered the ratings of some $145 million of Pinnacle debt because of "accelerating deficiency in liquidity," which it said was evidenced by Pinnacle's elimination of dividend payments.
   (b) Sales were higher in all of the company's business categories, with the biggest growth coming in sales of foodstuffs such as margarine, coffee and frozen food, which rose 6.3%.
   (c) Ajinomoto predicted sales in the current fiscal year ending next March 31 of 480 billion yen, compared with 460.05 billion yen in fiscal 1989.
3. The model predicts the sentences are bad, but the assessor considered them fluent.
   (a) The sense grows that modern public bureaucracies simply don't perform their assigned functions well.
   (b) Amstrad PLC, a British maker of computer hardware and communications equipment, posted a 52% plunge in pretax profit for the latest year.
   (c) At current allocations, that means EPA will be spending $300 billion on itself.

3.2   Correlation with Overall Text Quality

Here we focus on the relationship between sentence fluency and overall text quality. We would expect that the presence of disfluent sentences in text will make it appear less well written. Five annotators had previously assessed the overall text quality of each of the WSJ articles on a scale from 1 to 5 [39]. The average of the assessments was taken as a single number describing the linguistic quality article. The correlation between this number and the percentage of fluent sentences in the article according to the different models is shown in Table 7.

The correlation between the percentage of fluent sentences in the article as given by the human assessor and the overall text quality is rather low, 0.127. Correlation with the percentage of fluent sentences predicted by the two automatic models are even closer to zero. Note that none of the correlations are actually significant for the small dataset of 30 points.

Table 7. Correlations between text quality assessment of the articles and the percentage of fluent sentences according to different models

Fluency given by	Correlation	p-value
human	0.127	0.504
human vs machine trans. model	-0.055	0.772
good MT vs bad MT model	0.076	0.69

The low correlations indicate that binary decisions on sentence level fluency are not likely to be helpful for determining the overall quality of text. A question that remains unanswered from the experiments presented so far is whether structural features can be used to predict overall text quality directly. A dataset larger than the 30 WSJ documents is necessary for this purpose. So, in the next section we turn to a large collection of multi-document summaries evaluated for linguistic quality.

4   Predicting Linguistic Quality for Multi-document Summarization

Efforts for the development of automatic text summarizers have focused almost exclusively on improving content selection capabilities of systems, ignoring the linguistic quality of the system output. Part of the reason for this imbalance is the existence of ROUGE [32,33], the system for automatic evaluation of content selection, which allows for frequent system evaluation during system development and for reporting results of experiments performed outside of the annual NIST-led evaluations (DUC[3] and TAC[4]). Few metrics, however, have been proposed

---

[3] http://duc.nist.gov/
[4] http://www.nist.gov/tac/

[31] for evaluating linguistic quality and none have been tested for correlation with the manual metrics used by NIST.

So here we use the same structural features described in the experiments on sentence level fluency in order to directly predict the linguistic quality of summaries. We compare their performance with that of several other metrics of text quality. We evaluate the predictive power of the linguistic quality metrics by training and testing models on consecutive years of NIST evaluations, showing the robustness of each class and their abilities to reproduce human rankings of systems and summaries with high accuracy.

4.1 Summarization Data

We use a large corpus of system- and human-authored summaries from the Document Understanding Conference (DUC) workshops [36] from years 2006 and 2007. These summaries were produced for inputs consisting of a set of 25 related documents on a topic. The length of the summary was constrained to be 250 or fewer words. In DUC 2006, there were 50 inputs to be summarized and 35 summarization systems which participated in the evaluation. In DUC 2007, there were 45 inputs and 32 different summarization systems. Four human summaries are also available for each input.

All summaries were manually evaluated for several aspects of linguistic quality, including (a) referential clarify, (b) focus and (c) structure and coherence. For each of the questions, Summaries were rated on a scale from 1 to 5, in which 5 is the best separately for each of these aspects.

Judging from the 2006 scores, systems are currently the worst at structure (mean=2.4, median=2), middling at referential clarity (mean=3.1, median=3), and relatively better at focus (mean=3.6, median=4). Structure is the aspect of linguistic quality where there is the most room for improvement. Excluding the baseline system, which simply extracts the leading sentences from the most recent article in the input and therefore has well-formed summaries, all of the other systems have average structure scores below 3.5 in DUC 2006. Human summaries were predominantly scored 5, but some scores of 4 and 3 also occur.

4.2 Predictors of Linguistic Quality

Structural features. The structural features we described in Sect. 2.1 apply for individual sentences. In order to apply they to summaries which consist of more than one sentence, we simply take the average value of features for the sentences in the summary.

Coh-Metrix. The Coh-Metrix tool[5] provides an implementation of 54 features known in the psycholinguistic literature to correlate with the coherence of human-written texts [19]. These include for example commonly used readability metrics based on sentence length and number of syllables in constituent words.

---

[5] http://cohmetrix.memphis.edu/

Other measures implemented in the system are surface text properties known to contribute to text processing difficulty such as the number of words before the main verb, the prevalence of pronouns and low frequency content words. Also included are measures of cohesion between adjacent sentences such as similarity under a latent semantic analysis model [16], stem and content word overlap, and syntactic similarity between adjacent sentences. In addition, the presence in a text of different types of discourse connectives such as *causal (e.g. 'because',* *'consequently')* and *temporal (e.g. 'after', 'until')* are also recorded. Coh-Metrix has been designed with the goal of capturing properties of coherent text and has been used for grade level assessment, predicting student essay grades, identifying differences between spoken and written texts, authorship identification, and various other tasks.

Vocabulary: language models. Psycholinguistic studies have shown that people read frequent words and phrases more quickly [26,21], so the words that appear in a text might influence people's perception of its quality. Language models are a way of computing how familiar the words in a text are to readers by using the distribution of words and phrases from a large background corpus. We built unigram, bigram, and trigram language models with Good-Turing smoothing over the New York Times section of the English GigaWord corpus (over 900 million words). We used the SRI Language Modeling Toolkit [45] for this purpose. For each of the three n-gram language models, we include the *min,* *max,* and *average* log probability of the sentences contained in a summary, as well as the *overall log probability* of the entire summary.

Word coherence. Word co-occurrence patterns across adjacent sentences provide a way of measuring local coherence which can be easily computed using large amounts of unannotated text [30,44]. Specifically, we used the two features introduced by [44]. [44] make an analogy to machine translation: in translation, two words are likely to be translations of each other if they often appear in *parallel* sentences (a sentence and its translation); in texts, two words are likely to signal local coherence if they often appear in *adjacent* sentences. The two features of word coherence are the *forward likelihood,* the likelihood of observing the words in sentence s_i conditioned on s_{i-1}, and the *backward likelihood,* the likelihood of observing the words in sentence s_i conditioned on sentence s_{i+1}. "Parallel texts" of 5 million adjacent sentences were extracted from the New York Times section of the English GigaWord corpus. We used the GIZA++[6] implementation of IBM Model 1 to align the words in adjacent sentences and obtain all relevant probabilities.

The equation for the forward likelihood of a text T containing n sentences is below:

$$P_F(T) = \prod_{i=1}^{n-1} \prod_{j=1}^{|s_{i+1}|} \frac{\epsilon}{|s_i| + 1} \sum_{k=0}^{|s_i|} t(s_{i+1}^j | s_i^k) \qquad (1)$$

---

[6] http://www.fjoch.com/GIZA++.html

Here, sentence s_{i+1} is assumed to be generated from events (words) in sentence s_i. The events in s_i include a special *NULL* word.

The backward likelihood is identical, with s_i and s_{i+1} interchanged.

Entity coherence. Linguistic theories, and Centering theory [20] in particular, have hypothesized that the transition of attention between entities from one sentence to the next plays a major role in the determination of local coherence. [5], inspired by Centering, proposed an easily computable representation for sequences of entity mentions across a text. In their Entity Grid model, a text is represented by a matrix with rows corresponding to each sentence in a text, and columns to each entity mentioned anywhere in the text. The value of a cell in the grid is the entity's grammatical role in that sentence (Subject, Object, Neither, or Absent). This representation captures the pattern of entities across sentences in terms of entity transitions. For example, if an entity that occurs in a subject position in sentence s_i is an object in s_{i+1}, the text would have a transition *SO*. One would expect that coherent texts would contain a certain distribution of entity transitions which would differ from those in incoherent sequences.

We use the Brown Coherence Toolkit[7] [17] to construct the grids. The tool does not perform full coreference resolution. Instead, noun phrases are considered to refer to the same entity if their heads are identical.

The actual entity coherence features are the probabilities of local entity transitions (SS, SO, etc), computed as the fraction of each type of transition in the entire entity grid for the text.

4.3   Experimental Setup

We used the summaries from DUC 2006 for training and feature development and DUC 2007 served as the test set. Validating the results on consecutive years of evaluation is important, as results that hold for the data in one year might not carry over to the next, as happened for example in [14]'s work.

We experiment with the predictive power of the linguistic quality classes of our features in two settings. In *system-level* evaluation, we would like to rank all participating systems according to their performance on the entire test set. In *input-level* evaluation, we would like to rank all summaries produced for a single given input.

We use a Ranking SVM (SVM^{light} [25]) to learn how to rank summaries using our features. Just as in a SVM used for classification, the Ranking SVM learns a weight vector from the training data. The output of the Ranking SVM is the dot product of the weight vector and the feature values, which is a real number. However, rather than optimizing for this score to be as close as possible to the true score, as in regression, the Ranking SVM instead seeks to minimize the number of discordant pairs (pairs in which the gold standard has x_1 ranked strictly higher than x_2, but the learner ranks x_2 strictly higher than x_1). The default regularization parameter was used.

---

[7] http://www.cs.brown.edu/~melsner/manual.html

Following [5], we report summary ranking accuracy as the fraction of correct pairwise rankings in the test set.

For input-level evaluation, the pairs are formed from summaries of the *same input*. Pairs in which the gold standard ratings are tied are not included. After removing the ties, the test set thus consists of 51 pairs for human referential clarity; 15,736 pairs for system referential clarity; 57 pairs for human focus; 13,660 pairs for system focus; 88 pairs for human structure; and 14,398 pairs for system structure.

For system-level evaluation, we treat the real-valued output of the SVM ranker for each summary as the linguistic quality score. The 45 individual scores for summaries produced by a given system are averaged to obtain an overall score for the system. The gold-standard system-level quality rating is equal to the average *human ratings* for the system's summaries over the 45 inputs. Again, we compare all pairs of systems with non-tied gold-standard scores and compute the prediction accuracy for these pairs. At the system level, there are 491 pairs for referential clarity, 492 pairs for focus, and 490 pairs for structure in the test set.

For both evaluation settings, a random baseline which ranked the summaries in a random order would have an expected pairwise accuracy of 50%.

4.4   Results

The performance of each class of features is shown in Table 8. The best result in each colum is given in bold, and the rank of the structural features class is noted in brackets.

Structural and language model features are the best predictors of input-level evaluation of human summaries. The pairwise ranking prediction accuracy of structural features is 80% for referential clarity and lower 70s for focus and structure. For system evaluation structural features do reasonably—accuracies of low 60s for input-level and around 85% for system-level for each of the three quality aspects.

No class of predictors stand out as the overall best because the performance differs considerably across tasks. Structural features are very good for input-level human summaries, middle of the range for input level system summaries and about the worst class of features for system-level evaluation of automatic summaries.

Table 8. Pairwise ranking prediction accuracy

Features	Input-level; Systems			Input-level; Humans			System-level		
	Refs	Focus	Struct.	Refs	Focus	Struct.	Refs	Focus	Struct.
LM	62.2	60.5	62.5	76.5	**71.9**	**78.4**	**91.2**	**85.2**	86.3
Coh-metrix	**67.9**	63.0	62.4	68.6	59.6	67.0	88.6	83.9	86.3
Entity coh.	64.3	**64.2**	**63.6**	54.9	52.6	56.8	89.6	85.0	**87.1**
Word coh.	53.3	53.2	53.7	62.7	70.2	60.2	87.8	81.7	79.0
Structural	64.4 [2]	61.9 [3]	62.6 [2]	**80.4** [1]	**71.9** [1]	72.7 [2]	87.6 [5]	82.3 [4]	84.9 [4]

The language model and entity coherence classes seem to be the two classes that tend to perform uniformly well for the three tasks.

System-level accuracies are high for all classes of features, above 85% which suggest that using the trained ranker can be a practical substitute of manual evaluation.

5   Conclusion

We presented a study of sentence fluency based on data from machine translation evaluations. These data allow for two types of comparisons: human (fluent) text and (not so good) machine-generated text, and levels of fluency in the automatically produced text. The distinctions were possible even when based solely on features describing syntactic phrasing in the sentences.

Correlation analysis reveals that the structural features are significantly but weakly correlated with fluency. Interestingly, the features correlated with fluency levels in machine-produced text are not the same as those that distinguish between human and machine translations. Such results raise the need for caution when using assessments for machine produced text to build a general model of fluency. The captured phenomena in this case might not be the same as these from comparing human texts with differing fluency. For future research it will be beneficial to build a dedicated corpus in which *human-produced* sentences are assessed for fluency.

Our experiments show that basic fluency distinctions can be made with high accuracy. Machine translations can be distinguished from human translations with accuracy of 87%; machine translations with low fluency can be distinguished from machine translations with high fluency with accuracy of 61%. In pairwise comparison of sentences with different fluency, accuracy of predicting which of the two is better is 90%. Results are not as high but still promising for comparisons in fluency of translations of the same text.

We also demonstrated that while fluency models based on structural features learned on machine translation data do not generalize well to human texts, the models of overall text quality for summarization are robust and can be used for automatic evaluation of linguistic quality. Structural features compare favorably to other classes of predictors of linguistic quality for input-level ranking of human summaries particularly, but also for input-level evaluation of automatic summaries.

References

1. Bailin, A., Grafstein, A.: The linguistic assumptions underlying readability formulae: a critique. Language and Communication 21, 285–301 (2001)
2. Bangalore, S., Rambow, O.: Exploiting a probabilistic hierarchical model for generation. In: Proceedings of the 18th International Conference on Computational Linguistics (COLING 2000), pp. 42–48 (2000)

3. Bangalore, S., Rambow, O., Whittaker, S.: Evaluation metrics for generation. In: Proceedings of the First International Conference on Natural Language Generation (INLG 2000), pp. 1–8 (2000)
4. Banko, M., Mittal, V., Witbrock, M.: Headline generation based on statistical translation. In: Proceedings of the 38th Annual Meeting of the Association for Computational Linguistics (ACL 2000), pp. 318–325 (2000)
5. Barzilay, R., Lapata, M.: Modeling local coherence: An entity-based approach. Computational Linguistics 34(1), 1–34 (2008)
6. Barzilay, R., McKeown, K.R.: Sentence fusion for multidocument news summarization. Computational Linguistics 31(3), 297–328 (2005)
7. Cahill, A., Forst, M.: Human evaluation of a German surface realisation ranker. In: Krahmer, E., Theune, M. (eds.) Empirical Methods in NLG. LNCS (LNAI), vol. 5790, pp. 201–221. Springer, Heidelberg (2010)
8. Cahill, A., Forst, M., Rohrer, C.: Stochastic realisation ranking for a free word order language. In: Proceedings of the Eleventh European Workshop on Natural Language Generation (ENLG 2007), pp. 17–24 (2007)
9. Charniak, E., Johnson, M.: Coarse-to-fine n-best parsing and maxent discriminative reranking. In: Proceedings of the 43rd Annual Meeting of the Association for Computational Linguistics (ACL 2005), pp. 173–180 (2005)
10. Charniak, E.: A maximum-entropy-inspired parser. In: Proceedings of the 1st North American chapter of the Association for Computational Linguistics Conference (NAACL 2000), pp. 132–139 (2000)
11. Clarke, J., Lapata, M.: Models for sentence compression: A comparison across domains, training requirements and evaluation measures. In: Proceedings of the 21st International Conference on Computational Linguistics and 44th Annual Meeting of the Association for Computational Linguistics (COLING/ACL 2006), pp. 377–384 (2006)
12. Collins, M., Koo, T.: Discriminative reranking for natural language parsing. Computational Linguistics 31(1), 25–70 (2005)
13. Collins-Thompson, K., Callan, J.P.: A language modeling approach to predicting reading difficulty. In: Proceedings of the Human Language Technology Conference of the North American Chapter of the Association for Computational Linguistics: HLT-NAACL 2004, pp. 193–200 (2004)
14. Conroy, J., Dang, H.: Mind the gap: dangers of divorcing evaluations of summary content from linguistic quality. In: Proceedings of the 22nd International Conference on Computational Linguistics (COLING 2008), pp. 145–152 (2008)
15. Corston-Oliver, S., Gamon, M., Brockett, C.: A machine learning approach to the automatic evaluation of machine translation. In: Proceedings of 39th Annual Meeting of the Association for Computational Linguistics (ACL 2001), pp. 148–155 (2001)
16. Deerwester, S., Dumais, S., Furnas, G., Landauer, T., Harshman, R.: Indexing by latent semantic analysis. Journal of the American Society for Information Science 41, 391–407 (1990)
17. Elsner, M., Austerweil, J., Charniak, E.: A unified local and global model for discourse coherence. In: Human Language Technologies 2007: The Conference of the North American Chapter of the Association for Computational Linguistics; Proceedings of the Main Conference, pp. 436–443 (2007)
18. Galley, M., McKeown, K.: Lexicalized Markov grammars for sentence compression. In: Human Language Technologies 2007: The Conference of the North American Chapter of the Association for Computational Linguistics; Proceedings of the Main Conference, pp. 180–187 (2007)

19. Graesser, A., McNamara, D., Louwerse, M., Cai, Z.: Coh-Metrix: Analysis of text on cohesion and language. Behavior Research Methods Instruments and Computers 36(2), 193–202 (2004)

20. Grosz, B., Joshi, A., Weinstein, S.: Centering: a framework for modelling the local coherence of discourse. Computational Linguistics 21(2), 203–226 (1995)

21. Haberlandt, K., Graesser, A.: Component processes in text comprehension and some of their interactions. Journal of Experimental Psychology: General 114(3), 357–374 (1985)

22. Holmes, G., Donkin, A., Witten, I.: Weka: A machine learning workbench. In: Second Australian and New Zealand Conference on Intelligent Information Systems, pp. 357–361 (1994)

23. Huang, L.: Forest reranking: Discriminative parsing with non-local features. In: Proceedings of the 46th Annual Meeting of the Association for Computational Linguistics: Human Language Technologies (ACL 2008: HLT), pp. 586–594 (2008)

24. Jing, H.: Sentence reduction for automatic text summarization. In: Proceedings of the Sixth Conference on Applied Natural Language Processing (ANLP 2000), pp. 310–315 (2000)

25. Joachims, T.: Optimizing search engines using clickthrough data. In: Proceedings of the Eighth ACM SIGKDD International Conference on Knowledge Discovery and Data Mining, pp. 133–142 (2002)

26. Just, M., Carpenter, P.: The psychology of reading and language comprehension, Allyn, Bacon (1987)

27. Karamanis, N., Mellish, C., Poesio, M., Oberlander, J.: Evaluating centering for information ordering using corpora. Computational Linguististics 35(1), 29–46 (2009)

28. Knight, K., Marcu, D.: Summarization beyond sentence extraction: a probabilistic approach to sentence compression. Artificial Intelligence 139(1), 91–107 (2002)

29. Langkilde, I., Knight, K.: Generation that exploits corpus-based statistical knowledge. In: Proceedings of the 36th Annual Meeting of the Association for Computational Linguistics and the 17th International Conference on Computational Linguistics (COLING-ACL 1998), pp. 704–710 (1998)

30. Lapata, M.: Probabilistic text structuring: Experiments with sentence ordering. In: Proceedings of the 41st Annual Meeting of the Association for Computational Linguistics (ACL 2003), pp. 545–552 (2003)

31. Lapata, M., Barzilay, R.: Automatic evaluation of text coherence: models and representations. In: Proceedings of the 19th International Joint Conference on Artificial Intelligence (IJCAI 2005), pp. 1085–1090 (2005)

32. Lin, C.Y., Hovy, E.: Automatic evaluation of summaries using n-gram co-occurrence statistics. In: Proceedings of the 2003 Conference of the North American Chapter of the Association for Computational Linguistics on Human Language Technology (NAACL 2003), pp. 71–78 (2003)

33. Lin, C.: Rouge: A package for automatic evaluation of summaries. In: Proceedings of the Workshop on Text Summarization Branches Out (WAS 2004), pp. 25–26 (2004)

34. McDonald, R.: Discriminative sentence compression with soft syntactic evidence. In: Proceedings of the 11th Conference of the European Chapter of the Association for Computational Linguistics (EACL 2006), pp. 297–304 (2006)

35. Mutton, A., Dras, M., Wan, S., Dale, R.: GLEU: Automatic evaluation of sentence-level fluency. In: Proceedings of the 45th Annual Meeting of the Association of Computational Linguistics (ACL 2007), pp. 344–351 (2007)

36. Over, P., Dang, H., Harman, D.: DUC in context. Information Processing Management 43(6), 1506–1520 (2007)

37. Papineni, K., Roukos, S., Ward, T., Zhu, W.J.: BLEU: a method for automatic evaluation of machine translation. In: Proceedings of the 40th Annual Meeting of the Association for Computational Linguistics (ACL 2002), pp. 311–318 (2002)
38. Petersen, S.E., Ostendorf, M.: A machine learning approach to reading level assessment. Computer Speech and Language 23(1), 89–106 (2009)
39. Pitler, E., Nenkova, A.: Revisiting readability: a unified framework for predicting text quality. In: Proceedings of the Conference on Empirical Methods in Natural Language Processing (EMNLP 2008), pp. 186–195 (2008)
40. Rieser, V., Lemon, O.: Natural language generation as planning under uncertainty for spoken dialogue systems. In: Krahmer, E., Theune, M. (eds.) Empirical Methods in NLG. LNCS (LNAI), vol. 5790, pp. 105–120. Springer, Heidelberg (2010)
41. Schwarm, S., Ostendorf, M.: Reading level assessment using support vector machines and statistical language models. In: Proceedings of the 43rd Annual Meeting of the Association for Computational Linguistics (ACL 2005), pp. 523–530 (2005)
42. Siddharthan, A.: Syntactic simplification and Text Cohesion. Ph.D. thesis, University of Cambridge, UK (2003)
43. Soricut, R., Marcu, D.: Abstractive headline generation using WIDL-expressions. Information Processing and Management 43(6), 1536–1548 (2007)
44. Soricut, R., Marcu, D.: Discourse generation using utility-trained coherence models. In: Proceedings of the COLING/ACL 2006 Main Conference Poster Sessions, pp. 803–810 (2006)
45. Stolcke, A.: SRILM – an extensible language modeling toolkit. In: Seventh International Conference on Spoken Language Processing (ICSLP 2002), vol. 3 (2002)
46. Turner, J., Charniak, E.: Supervised and unsupervised learning for sentence compression. In: Proceedings of the 43rd Annual Meeting on Association for Computational Linguistics (ACL 2005), pp. 290–297 (2005)
47. Velldal, E., Oepen, S.: Maximum entropy models for realization ranking. In: Proceedings of the 10th Machine Translation Summit, pp. 109–116 (2005)
48. Wan, S., Dale, R., Dras, M.: Searching for grammaticality: Propagating dependencies in the Viterbi algorithm. In: Proceedings of the Tenth European Workshop on Natural Language Generation (ENLG 2005), pp. 211–216 (2005)
49. Zajic, D., Dorr, B., Lin, J., Schwartz, R.: Multi-candidate reduction: Sentence compression as a tool for document summarization tasks. Information Processing Management 43(6), 1549–1570 (2007)
50. Zwarts, S., Dras, M.: Choosing the right translation: A syntactically informed classification approach. In: Proceedings of the 22nd International Conference on Computational Linguistics (COLING 2008), pp. 1153–1160 (2008)

Towards Empirical Evaluation of Affective Tactical NLG

Ielka van der Sluis[1] and Chris Mellish[2]

[1] Dept. Computer Science, Trinity College Dublin, Ireland
ielka.vandersluis@cs.tcd.ie
https://www.cs.tcd.ie/Ielka.vanderSluis/
[2] Dept. Computing Science, University of Aberdeen, Scotland, UK
c.mellish@abdn.ac.uk
http://www.abdn.ac.uk/~csc248/

Abstract. One major aim of research in affective natural language generation is to be able to use language intelligently to induce effects on the emotions of the reader/ hearer. Although varying the *content* of generated language ("strategic" choices) might be expected to change the effect on emotions, it is not obvious that varying the *form* of the language ("tactical" choices) can do this. This chapter discusses two experiments carried out to show emotional effects of tactical variations. With the first experiment we were not able to show clear, statistically significant differences between the effects of the different texts in readers. We discuss a number of possible reasons and, building on our discoveries, we present a second experiment which does demonstrate such effects. This represents an important step towards the empirical evaluation of affective NLG systems.

1 Introduction

This chapter is about developing techniques for the empirical evaluation of affective natural language generation (NLG). Affective NLG has been defined as "NLG that relates to, arises from or deliberately influences emotions or other non-strictly rational aspects of the Hearer" [19]. It currently covers two main strands of work, the portrayal of non-rational aspects in an artificial speaker/writer (e.g. the work of [16] on projecting personality) and the use of NLG in ways sensitive to the non-rational aspects of the hearer/reader and calculated to achieve effects on these aspects (e.g. the work of [20] on generating instructions in an emotionally charged situation and that of [17] on producing appropriate tutorial feedback). Although there has been success in evaluating work of the first kind, it remains more problematic to evaluate whether work of the second type directly affects emotion or mood, or whether it succeeds or fails for other reasons. Existing work of this kind tends to be evaluated either by the reader's performance on a task not directly related to emotions (e.g. [5]) or by an assessment of system outputs by experts (e.g. [7]). Neither of these tells us in a direct way whether emotions have been affected.

E. Krahmer, M. Theune (Eds.): Empirical Methods in NLG, LNAI 5790, pp. 242–263, 2010.

Since the work of [23], NLG tasks have been considered to divide mainly into those involving *strategy* ("deciding what to say") and *tactics* ("deciding how to say it"). It seems clear that one can affect a reader's emotion differently by making different strategic decisions about content (e.g. telling someone that they have passed an exam will make them happier than telling them that they have failed), but it is less clear that tactical alternations (e.g. involving ordering of material, choice of words or syntactic constructions) can have these kinds of effects. Unfortunately, the exact dividing line between strategy and tactics remains a matter of debate. For the purpose of this chapter, we take "strategic" to cover matters of basic propositional content (the basic information to be communicated) and "tactical" to include most linguistic issues, including matters of emphasis and focus, inasmuch as they can be influenced by linguistic formulation. It is important to know whether tactical choices can influence emotions because to a large extent NLG research concentrates on tactical issues (partly because strategic NLG remains a rather domain-specific activity).

Some light on the effects of tactical variations in text is shed by work in psychology, where there has been a great deal of work on the effects of the "framing" of a text [24,18,22]. Other work in this area has been industrially funded, as there are considerable applications, for instance, in advertising. The alternative texts considered differ in ways that NLG researchers would call tactical. For instance, a piece of meat could be described as "75% lean" or "25% fat", and arguably these are alternative truthful descriptions of the same situation. However, evaluation of this work has been primarily in terms of whether it affects people's *choices* or *evaluations* of options available [14], or other aspects of task performance like motivation and beliefs [9,2,4]. As far as we know, nobody has detected *emotions* being affected in this way. There are therefore open questions about whether there can be non-rational effects of different tactical decisions on readers and whether it is possible to detect them. We believe that anwering these is important for the further scientific development of affective NLG.

In the rest of this chapter, we first discuss our choice of a method for measuring emotions (Section 2). This is followed by a section on the types of linguistic choices we consider. Then two attempts to measure emotional effects of tactical decisions in texts are described, in terms of the example texts we used, the text validation experiments we did to check our intuitions and the experimental set up and the results of the studies. Finally we reflect on the results in a concluding section.

2   Measuring Emotions

There are three broad ways of measuring the emotions of people – task performance methods, physiological methods and self-reporting. Task performance methods measure emotional effects via performance on a task that is known to be facilitated by particular emotions [27]. For instance, [5] measure persuasiveness of different textual realisations that may induce emotions. As indicated above, the trouble with such methods from our point of view is their indirectness – although performance could indeed be influenced by emotions, it could also be affected by other factors. Physiological methods, on the other hand, whilst measuring something objective, unfortunately tend to have the problems of complex setup and calibration, which mean that it is hard to transport them between tasks or individuals. In addition, although emotional states are undoubtedly connected to physiological variables, it is not always clear what is being measured by these methods (cf. [13,3]).

Because of the problems with the other two types of methods, we have turned to self-reporting methods as a way of measuring emotions. Although sometimes participants are not able objectively to report their own emotions (see our later discussion in Section 6), such methods as the Russell Affect Grid [21], the Positive and Negative Affect Scale (PANAS) [26], and the Self Assessment Manikin (SAM) [12] are widely used in Psychology. The PANAS test is a scale consisting of 20 words and phrases (10 for positive affect and 10 for negative affect) that describe feelings and emotions. Participants read each term and indicate to what extent they experience(d) the emotion indicated by that term using a five point scale ranging from (1) very slightly/not at all, (2) a little, (3) moderately, (4) quite a bit to (5) extremely. A total score for positive affect is calculated by simply adding the scores for the positive terms, and similarly for negative affect. The Russell Affect Grid and the SAM test both assess valence and arousal separately on a nine-point scale.

3   Linguistic Choice

3.1   Polarity and Magnitude

We decided that a safe way to start would be to choose primitive positive versus negative emotions (such as sadness, joy, disappointment, surprise, anger), as opposed to more complex emotions related to trust, persuasion, advice, reassurance. Therefore we focus here on alternatives that give a text a positive or negative "slant". These could be applied by an NLG system whose message has "positive" and "negative" aspects, where "positive" information conjures up scenarios that are pleasant and acceptable to the reader, makes them feel happy and cooperative etc. and "negative" information conjures up unpleasant or threatening situations and so makes them feel more unhappy, confused etc. For instance, [20] discuss generating instructions on how to take medication which have to both address positive aspects ('this will make you feel better if you do the following') and also negative ones ('this may produce side-effects, which I have to tell you about by law'). An NLG system in such a domain could make itself popular by only mentioning the positive information, but then it could leave itself open to later criticism (or litigation) if by doing so it clearly misrepresented the true situation. Although it may be inappropriate grossly to misrepresent the provided message, there are more subtle (tactical) ways to "colour" or "slant" the presentation of the message in order to emphasise either the positive or the negative aspects.

We assume that the message to be conveyed is a simple set of propositions, each classified in an application-dependent way as having positive or negative *polarity* according to whether the reader is likely to welcome it or be unhappy about it in the context of the current message.[1] In general, this classification could, for instance, be derived from the information that a planning system has about which propositions support which goals (e.g. to stay healthy one needs to eat healthy food). We also assume that a possible phrasing for a proposition has a *magnitude*, which indicates the degree of impact it has. This is independent of the polarity. We will not need to actually measure magnitudes, but when we make claims that one wording of a proposition has a smaller magnitude than another we indicate this with $<$. For instance, we would claim that usually:

"a few rats died" $<$ *"many rats died"*

Thus we claim that "a few rats died" has less impact than "many rats died", whether or not rats dying is considered a good thing (i.e. whether the polarity is positive or negative). In general, an NLG system can manipulate the magnitude of wordings of the propositions it expresses, to indicate its own (subjective) view of their importance. In order to slant a text positively, it can express positive polarity propositions in ways that have high magnitudes and negative polarity propositions in ways that have low magnitudes. The opposite applies for negative slanting. Thus, for instance, in an application where it is bad for rats to die, expressing a given proposition,"eight of twenty rats died" by "a few rats died" would be giving more of a positive slant, whereas saying "many rats died" would be slanting it more negatively.

3.2 Tactical Methods

For our experiments, we have considered a number of different types of tactical methods that could be implemented straightforwardly in an NLG system,[2] as follows. Here, the word "positive polarity" is used to refer to propositions giving good news to the reader or attributes which give good news to the reader if they have high values (such as thereader's intelligence). Similarly "negative polarity" refers to items that represent bad news, e.g. failing a test. In general, a further relevant concept is the "goal polarity" of a text in a context – which of positive and negative the text seeks to emphasise. A text which has a positive goal polarity will thus be "positively slanted" and one with a negative goal polarity will be "negatively slanted".

A. Sentence emphasis - include explicit emphasis in sentences with the same polarity as the goal polarity (e.g. exclamation marks and phrases such as "on top of this").

B. Choice of vague evaluative adjectives - when evaluating attributes with the same polarity as the goal, choose vague evaluative adjectives that are more extreme over ones that are less extreme (e.g. "excellent", rather than "ok" for positive polarity).

---

[1] Note that this sense of "polarity" is not the same as the one used to describe "negative polarity items" in Linguistics.

[2] Though the choice about *when* to apply them might not be so straightforward.

C. Choice of vague adverbs - provide explicit emphasis to propositions with the same polarity as the goal by including vague adverbs expressing great extent (e.g. "significantly", rather than "to some extent" or no adverb).

D. Choice of verbs - for a proposition with the same polarity as the goal, choose a verb that emphasises the great extent of the proposition (e.g. "outperformed", rather than "did better than").

E. Choice of realisation of rhetorical relations - when realising a concession/contrast relation between a positive polarity proposition and one that is negative or neutral, word it so that the proposition with the goal polarity is in the nucleus (more emphasised) position (e.g. say "although you did badly on X, you did well on Y" instead of "although you did well on Y, you did badly on X" for a positive goal polarity).

Most of these methods have corresponding methods for when some polarity is opposite to the goal polarity. E.g. a sentence or adjective might be de-emphasised by the inclusion of a "hedge" in such cases.

The idea is that an NLG system would employ methods of this kind in order to "slant" a message in a particular direction, rather that to present a message in a more neutral way. This might be done, for instance, to induce positive emotions in a reader who needs encouragement or negative emotions in a reader who is over-confident.

We claim that these choices can be viewed as tactical, i.e. that they are "allowable" alternative realisations of the same underlying content. For instance, we believe a teacher could use such methods in giving feedback to a student needing encouragement without fear of prosecution for misrepresenting the same truth that would be expressed without the use of these methods.

Whenever one words a proposition in different ways, it can be claimed that a (perhaps subtle) change of meaning is involved. However, in these cases we claim that it is the *writer's attitudes* that are being manipulated (and reflected in the text). We can therefore choose between these alternatives by varying the writer, not the underlying message. Our view is supported by a number of current accounts of the semantics of vague adjectives (though this is not an area without controversy). Many accounts of vagueness appeal to the idea that there is a norm which an adjective like "tall" implicitly refers to, and some of these argue both that the norm itself can be contextually determined and also that the amount by which the norm has to be exceeded has to be "significant" to a degree which is "relativized to some agent" [11]. For instance, with the phrase "John is tall"

> "the property [...] attributed to John is not an intrinsic property, but rather a relational one. Moreover, it is not a property the possession of which depends only on the difference between Johns height and some norm, but also on whether that difference is a significant one. I take it that whether or not a difference is a significant difference does not depend only on its magnitude, but also on what our interests are" [8]

It is compatible with these accounts that different agents, with different interests and notions of what is noteworthy, can use vague adjectives in different ways.[3]

---

[3] Though there are certainly *some* limits on the situations where a word like "tall" can be truthfully used to describe a height.

Probably the best way to check that we are using tactical alternations (according to our definition) is via some kind of text validation experiment in which human participants are asked to judge this. A good approximation seems to be to ask participants whether particular alternative sentences could be used to describe the same situation. Below we describe two studies in which we employ such text validation experiments, which provide strong support for our position.

4   Study I

4.1   Background for the Study

The goal of this first study was to experiment with emotion self-reporting methods as described in Section 2 and to investigate whether we could reliably measure differences in emotions arising from tactical variations of texts.

4.2   Test Texts

We started by composing by hand two messages containing mainly negative and positive polarity propositions respectively. The negative message tells the reader that a cancer-causing colouring substance is found in some foods available in the supermarkets. The positive message tells the reader that foods that contain Scottish water contain a mineral which helps to fight cancer. The first paragraph of both texts states that there is a substance found in consumer products that has an effect on people's health and it addresses the way in which this fact is handled by the relevant authorities. The second paragraph of the text elaborates on the products that contain the substance and the third paragraph explains in what way the substance can affect people's health.

To study the effects of different wordings, for each text a positive and a negative version (i.e. a version with positive or negative goal polarity) was produced by slanting propositions in either a positive or a negative way. This resulted in four texts in total, two texts with a negative message one positively and one negatively phrased (NP and NN), and two texts with a positive message one positively and one negatively verbalised (PP and PN). To maximise the impact aimed for, various slanting techniques were used by hand as often as possible without loss of believability (this was assessed by the intuition of the researchers). The positive and negative texts were slanted in parallel as far as possible, that is in both texts similar sentences were adapted so that they emphasised the positive or the negative aspects of the message. The linguistic variation used in the texts was algorithmically reproducible and the techniques are illustrated below. A number of these were suggested by work on "framing" in Psychology [18,22]. Indeed, that work also suggests further variations that could be manipulated, for instance, the choice between using numerical and non-numerical values for expressing quantities.

SLANTING EXAMPLES FOR THE NEGATIVE MESSAGE

Here it is assumed that recalls of products, risks of danger etc. involve negative polarity propositions. Therefore negative slanting will amongst other things choose high magnitude realisations for these.

Techniques involving adjectives and adverbs:
- "*A recall*" < "*A large-scale recall*" of infected merchandise was triggered

Techniques involving quantification:
- Sausages, tomato sauce and lentil soup are "*some*" < "*only some*" of the affected items

Techniques involving a change in polarity
Proposition expressed with positive polarity:
- Tests on monkeys revealed that as many as "*40 percent*" of the animals infected with this substance "*did not develop any tumors*"

Proposition expressed with negative polarity:
- Tests on monkeys revealed that as many as "*60 percent*" of the animals infected with this substance "*developed tumors*".

Techniques manipulating rhetorical prominence
Positive slant:
- "So your health is at risk, but every possible thing is being done to tackle this problem"

Negative slant:
- "So although every possible thing is being done to tackle this problem, your health is at risk"

SLANTING EXAMPLES FOR THE POSITIVE MESSAGE

Here it is assumed that killing cancer, promoting Scottish water etc. involve positive polarity propositions. Therefore positive slanting will amongst other things choose high magnitude realisations for these.

Techniques involving adjectives and adverbs:
- Neolite is a "*detoxifier*" < "*powerful detoxifier*" preventing cancer cells

Techniques involving quantification:
- "*Cancer-killing Neolite*" < "*Substantial amounts of cancer-killing Neolite*" was found in Scottish drinking water

Techniques involving a change in polarity
Proposition expressed with negative polarity:
- A study on people with mostly stage 4 cancer revealed that as many as "*40 percent*" of the patients that were given Neolite "*still had cancer*" at the end of the study.

Proposition expressed with positive polarity:
- A study on people with mostly stage 4 cancer revealed that as many as "*60 percent*" of the patients that were given Neolite "*were cancer free*" at the end of the study.

Techniques manipulating rhetorical prominence
Negative slant:
- "Neolite is certainly advantageous for your health, but it is not a guaranteed cure for, or defence against cancer"

Positive slant:
- "So Although Neolite is not a guaranteed cure for, or defence against cancer, it is certainly advantageous for your health"

4.3   Text Validation

To check our intuitions on the effects of the textual variation between the four texts described above, a text validation experiment was conducted in which 24 colleagues participated. The participants were randomly assigned to one of two groups (i.e. P and N), group P was asked to validate 23 sentence pairs from the positive message (PN versus PP) and group N was asked to validate 17 sentence pairs from the negative message (NN versus NP). Each pair consisted of two sentences intended to be alternative realisations of the same underlying content (as in the examples in the last section). Both the N and the P group sentence pairs included four filler pairs which were meant to keep participants alert. The participants in group P were asked which of the two sentences in each pair they thought most positive in the context of the message about the positive effects of Scottish water. The participants in group N were asked which of the two sentences in each pair they found most alarming in the context of the message about the contamination of food available for consumption. All participants were asked to indicate if they thought the sentences in each pair could be used to report on the same event (i.e. represented purely tactical variations).

Results in the N group indicated that in 89.75% of the cases participants agreed with our intuitions about which one of the two sentences was most alarming. On average, per sentence pair 1.08 of the 12 participants judged the sentences differently than what we expected. In 7 of the 13 sentence pairs (17 - 4 fillers) participants unanimously agreed with our intuitions. In the other sentence pairs 1 to, maximally, 4 participants did not share our point of view. In the two cases in which four participants did not agree with or were unsure about the difference we expected, we adapted our texts. One of these cases was the pair:

"*just 359*" infected products have been withdrawn < "*as many as 359*" infected products have been withdrawn "*already*"

We thought that the latter of the two would be more alarming (and correspond to negative slanting) because it is a bad thing if products have to be withdrawn (negative polarity). However, some participants felt that products being withdrawn was a good thing (positive polarity), because it meant that something was being done to tackle the problem, in which case the latter would be imposing a positive slant. As a consequence of the validation results, it was decided to 'neutralise' this sentence in both the NP and NN versions of the text to "359 infected products have been withdrawn". Overall, in 78.85% of the cases the participants thought that both sentences in a pair could report on the same event.

Results in the P group were similar. In 82.46% of the cases participants agreed with our intuitions about which one of the two sentences was most positive. In two cases, minor changes were made to make the texts clearer. Overall, in 86.84 % of the cases the participants thought that both sentences in a pair could report on the same event.

4.4   Experiment

Participants. Because a pilot experiment using 24 of our sceptical colleagues failed to produce clear patterns of behaviour, we decided to increase the likelihood of finding measurable emotional effects of text by targeting a group of participants likely to be especially affected by the subject material. It has been shown that young women are highly interested in health issues and especially health risks [6] and so we decided on young female students as our participants. In total 60 female students took part in the experiment and were paid a small fee for their efforts. The average age of the participants was about 20.57 (std. 2.41) years old.

Materials. For this experiment, we used versions of the NN, NP, NN and PP texts described above, with the modifications made after the text validation experiment. The texts were tailored in superficial ways to the participant group, by for example mentioning food products that are typically consumed by students as examples in the texts and by specifically mentioning young females as targets of the consequences of the message. On a more general level, the texts were adapted to a Scottish audience by, for instance, mentioning Scottish products and a Scottish newspaper as the source of the article. We thought that the presentation of the texts could be improved by making them look more like newspaper articles, with a date and a source indication. Before and after the participants read a test text, they were asked to fill out a questionnaire with a PANAS test and some questions for collecting demographical information. All materials, the test texts and questionnaires, as well as the experiment introduction, consent form and debriefing were presented to the participants printed on A4 pages.

Procedure. Participants were asked to fill in questionnaires before and after reading a text about a general topic that would have particular consequences for them. For ethical reasons, both in this experiment and the following one, the main experimental procedure was followed by a debriefing session in which the participants were informed that they had been deceived by the texts presented and during which it was possible to provide support for participants if their emotional reactions had been especially strong.

The participants were evenly and randomly distributed over the four texts (i.e. NN, NP, PN, PP) tested in this study, that is 15 participants per group. As the SAM test and the Russell Grid had caused confusion for participants in our previous work, we elected to use a version of the PANAS test. However, in the pilot study some participants had showed signs of boredom or disinterest while rating the PANAS terms some just marked all the terms as 'slightly/not at all' by circling them all in one go instead of looking at the terms separately. Also, some participants indicated that they found it difficult to distinguish particular terms. For example the PANAS test includes both 'scared' and 'afraid'. For these two reasons, a reduced (but still validated) version of the PANAS test [15] was used, in which the number of emotion terms that participants had to rate for themselves was decreased from 20 to 10. This PANAS set, consisting of five positive (i.e. alert, determined, enthusiastic, excited, inspired) and five negative terms (i.e. afraid, scared, nervous, upset, distressed), was used both before and after participants read the test text. Before the participants read the test text, they were asked to indicate how they felt at that point in time using the PANAS terms. After the participants read the test text, they were asked to rate the affect terms with respect to their feelings about the text. Note that this is different from asking them about their current feelings, because we wanted to emphasise that we wanted to know about their emotions related to the content of the text they just read and not about their feelings in general. We expected that the reduced PANAS test would produce reliable results because of its previous successful use. In the questionnaires, the PANAS terms were interleaved with other questions about recall and opinions to further avoid boredom.

Hypotheses. Four texts were tested on four different groups of participants. Two groups read the positive message (PP-group and PN-group) two groups read the negative message (NN-group and NP-group). Of the two groups that read the positive message, we expected the positive emotions of the participants that read the positive version of this message (PP-group) to be stronger than the positive emotions of the participants that read the negative version of this message (PN-group). Of the two groups that read the negative message, we expected the participants that read the negative version of this message (NN-group) to have stronger negative emotions than the participants that read the positive version of the message (NP-group).

Results. The bar chart presented in Figure 1 illustrates the results of the PANAS questionnaire before and after reading the texts for each condition. The data resulting from the study did not confirm our hypotheses. The hoped-for results for the positive/negative slanting are also not forthcoming - t-tests show no significant differences, the PP-group did not report stronger positive emotions than the PN-group and the NN-group did not report stronger negative emotions than the NP-group. Cronbach Alpha scores show that participants consistently rate the positive and negative terms before (α Positive $= .775$; α Negative $= .823$) and after (α Positive $= .791$; α Negative $= .911$) reading the text.

Post-hoc we could analyse the data as follows: In general, subjects in all conditions reported similar positive and negative emotions before reading the texts; there were no outliers. After reading the text, in terms of the differences in message content (P* vs N*), there is a difference between the ratings of the negative terms; as one would expect the NN-group and the NP-group report stronger negative emotions than the PP-group and the PN group. However, there is no such difference for the positive terms, which were rated fairly similarly for all groups. Also, contrary to what was expected, the rating of the negative PANAS terms by both N* groups is lower than their rating of the positive terms. When looking at these results in more detail, it appears that, of the positive PANAS terms, only 'excited' and 'inspired' had a higher mean for PP than for PN. When comparing the positive and the negative version of the negative message (NP vs

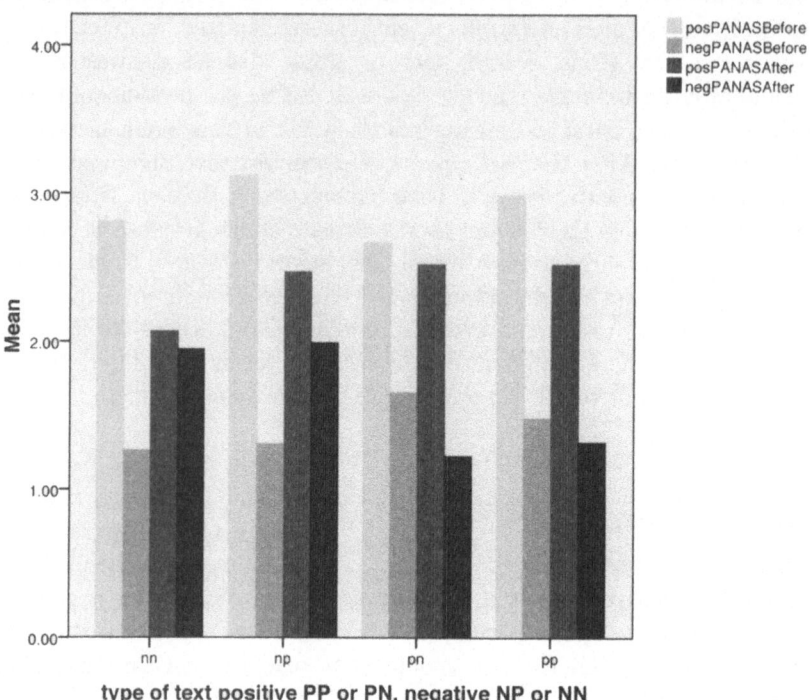

Fig. 1. Positive and negative PANAS means before and after the Participants read the test text

NN), as expected, the NN-group had lower means for all 5 positive terms than the NP group. But none of these results reached statistical significance.

Overall, participants in this study were highly interested in the experiment and in the text they were asked to read. Participants that read the positive message, about the benefits of Scottish water, appeared very enthusiastic and expressed disappointment when they read the debriefing from which they learned that the story contained no truth. Similarly, participants that read the negative message expressed anger and fear in their comments on the experiment and showed relief when the debriefing told them that the story on food poisoning was completely made up for the purposes of the experiment. Only a few participants that read a version of the negative message commented that they had got used to the fact that there was often something wrong with food and were therefore less scared. Table 1 shows some descriptives that underline these impressions. For instance, on a 5-point scale the participants rated the texts they read more than moderately interesting (average of $po\text{-}i = 3.74$). They also found the text informative (average of $inform = 3.82$) and noted that it contained new information (average of $new = 4.05$). These are surprisingly positive figures when we consider that the participants indicated only an average interest in food (average of $pr\text{-}i = 2.89$) before they read the test text. The participants

Table 1. Means and Standard deviations (between brackets) for the PN, PP, NP and NN texts for various variables: *pr-i* interest in food before reading the text, the *inf*ormativeness of the message, the *positive* or *negative* polarity of the message, *new* information and the *po-i* post interest in the message. All measured on a 5-point scale: 1 = not at all, ..., 5 = extremely.

	PN	PP	NN	NP
pr-i	2.47(1.13)	3.07(1.03)	3.00(.85)	3.00(1.25)
inf	3.87(.83)	3.80(.94)	3.67(1.05)	3.93(.70)
pos	3.93(.96)	4.27(1.03)	1.67(.98)	1.67(.97)
neg	1.53(.64)	1.27(.59)	4.07(1.22)	3.53(1.19)
new	4.13(1.18)	4.53(.64)	3.87(1.30)	3.67(1.59)
po-i	3.67(.82)	3.80(.78)	3.67(.72)	3.80(1.01)

that read the negative messages (NN and NP) recognised that the message was negative (cf. *pos* and *neg* in Table 1). Moreover, the NN-group rated the text more negatively than the NP-group (4.07 vs 3.53). The participants that read the positive message found that they had read a positive message. The PP-group rated their text slightly more positive than the PN-group rated theirs.

4.5   Discussion

From this study various conclusions can be drawn. First of all, from the fact that only the lower half of the 5-point PANAS scale was used it can be concluded that the participants in this study seem to have difficulties with reporting on their emotions. This was the case both before and after the test text was read. Furthermore, participants seem to have a preference for reporting their positive emotions and focused less on their negative emotions. The inference that self-reporting of emotions is troublesome is also indicated by the fact that the participants of this full study seemed highly interested and involved in the experiment and in what they read in the experiment texts. The participants generally believed the story they read and they expressed disappointment or relief when they were told the truth after the experiment. In addition, the descriptives in Table 1 show that participants generally correctly identified the text they read as either positive or negative. Note that in this respect the more fine-grained differences between the PP-group and the PN-group as well as the differences between the NN-group and the NP-group also confirm our expectations.

In this first attempt, we were not able to measure significant differences between the emotions evoked in readers dependent on the way their texts were phrased There are several reasons that may have played a role in this. It may be that the emotion measuring methods we tried were not fine-grained enough to measure the emotions that were invoked by the texts. As mentioned above, participants only used part of the PANAS scale and seemed to be reluctant to record their emotions (especially negative ones). Other ways of recording levels of emotional response that are more fine-grained than a 5-point scale, such as magnitude estimation [1], might be called for here. Carrying out experiments with even more participants might reveal patterns that are obscured by noise in the current study, but this would be expensive.

Alternatively, it could be that the differences between the versions of the messages were just too subtle and/or that there was not enough text for these subtle differences to produce measurable effects. Indeed, we are not aware of PANAS being used to assess purely textual effects before. It may not have helped that we asked for feelings about the text rather than simply current feelings. Perhaps it is necessary to immerse participants more fully in slanted text in order to really affect them differently. Or perhaps more extreme versions of slanting could be found. Perhaps indeed the main way in which NLG can achieve effects on emotions is through appropriate content determination (strategy), rather than through lexical or presentation differences (tactics).

Another reason could still be a lack of involvement of the participants of the study. Although the participants of the full study indicated their enthusiasm for the study as well as their interest in the topic and the message, they may have felt that the news did not affect them too much, because they considered themselves as responsible people when it comes to health and food issues.

5   Study II

5.1   Background for the Study

In Study 1 we investigated different methods of measuring the effects of texts on emotions to demonstrate that tactical differences would lead to differences in effects. However, we were unable to show statistically significant results of tactical variations and suggested various possible explanations. One involved the reliability of the reduced PANAS test, but this had been validated and used in multiple previous studies in Psychology, and so there is no reason to suggest that they would have fundamentally failed in this new context. Another was a problem with the limited granularity of the measurement method and partial use of the scale by participants. We addressed this in a simple way in Study II by having participants respond to the PANAS questions using a slider, rather than a five point scale. This means that only two terms were put at the extreme ends of the slider (i.e. 'very slightly/not at all' and 'extremely' were presented but not 'a little', 'moderately' or 'quite a bit').

Of the possible explanations for the Study I result, we believe that the possibility that the participants were not involved enough in the task to get strong emotions is the most compelling. It is very believable that the participants would fail to be really concerned by the texts in the experiments reported since the source was unclear, the message a general one not addressed to them individually and the topic (healthy and unhealthy food) one that occurs often enough in newspapers to fail to overcome natural boredom. Therefore the main innovation of Study II was in our method of seeking the emotional involvement of the participants. The texts that the participants read took the form of "feedback" on a (fake) IQ test that they undertook as part of the experiment. We selected university students as the participants, as they would likely be concerned about their intelligence, especially as compared to their peers. The texts appeared to be written individually for the participants and so sought to engage them directly.

5.2   Test Texts

For the experiment, we produced two feedback texts describing the same set of intelligence test results, one relatively neutral and one "positively slanted" using the above methods. For ethical reasons we did not use negative feedback. In the experiment, the feedback texts were given to participants in two groups, named "0" and "+" respectively. Each text consisted of 7 sentences, with a direct correspondance between the sentences of the two texts. Figure 2 presents the variations used in the feedback used in the experiment for group + (i.e. positively slanted) and group 0 (i.e. neutrally slanted). Note that the actual numbers are the same in both texts.

5.3   Text Validation

A text validation study was conducted in which 15 colleagues participated. The participants were asked to comment on 12 sentence pairs, the 7 shown in Figure 2 and 5 additional filler pairs. The following analysis reports on our findings on the 7 sentence pairs shown in Figure 2 only.

In order that we could test our intuitions about the tactical nature of the linguistic alternations (discussed in Section 3.2 above), the participants were presented with a scenario where there were two different teachers, Mary Jones and Gordon Smith, both completely honest but with very different ideas about teaching (Mary believing that any pupil can succeed, given encouragement, but Gordon believing that most pupils are lazy and have overinflated ideas about their abilities). Given a positively slanted sentence (e.g. +7) from Mary and a corresponding more neutrally slanted one (e.g. 07) from Gordon, addressed to one or more pupils, participants were asked to indicate:

1. "Is it possible that Mary and Gordon might actually be (honestly) giving different feedback to the *same* pupil on the same task?"
2. "If the two pieces of feedback were given to the same pupil (for the same task) and the pupil's parents found out, do you think they would have grounds to make a complaint that one of the teachers is lying?"

The hypothesis was that (for the 7 pairs of sentences from Figure 2) in general participants would answer "yes" to question 1 and "no" to question 2. Indeed, for 6 pairs at least 14 out of the 15 participants answered as we had predicted. For the other pair (+4/04), 12 out of 15 agreed with both predictions. We see this as very strong evidence for our position (the participants gave different answers for the filler pairs, and so were not just producing these answers blindly).

No alterations were made to the two feedback texts on the basis of the text validation results.

+1: Your Baumgartner score of 7.38 is excellent!

01: Your Baumgartner score of 7.38 is ok.

+2: You did distinctively better than the average score obtained by other people in your age group.

02: You did somewhat better than the average score obtained by other people in your age group.

+3: Especially your scores on Imagination/Creativity and on Clarity of Thought were great and considerably higher than average.

03: Your scores on Imagination/Creativity and on Clarity of Thought were good and a little higher than average.

+4: A factor analyses of your Baumgartner score results in an overall excellent performance.

04: A factor analyses of your Baumgartner score results in an overall reasonable performance.

+5: Although, compared to your peers, you have only slightly higher Spatial Intelligence (7.5 vs 7.0) and Visual Intelligence (7.2 vs 6.8) scores, your Clarity of Thought Score is very much better (7.2 vs 6.3).

05: Compared to your peers, you have a somewhat better Clarity of Thought Score (7.2 vs 6.3), but you have only slightly higher Spatial Intelligence (7.5 vs 7.0) and Visual Intelligence (7.2 vs 6.8) scores.

+6: On top of this you also outperformed most people in your age group with your exceptional scores for Imagination and Creativity (7.9 vs 7.2) and Logical-Mathematical Intelligence (7.1 vs. 6.5).

06: You did better than most people in your age group with your scores for Imagination and Creativity (7.9 vs 7.2) and Logical-Mathematical Intelligence (7.1 vs. 6.5).

+7: There is a lot of variation in your age group, but your score is significantly higher than average.

07: Your score is higher than average, but there is a lot of variation in your age group.

Fig. 2. Linguistic variation used in the IQ test feedback

5.4   Experiment

Participants. 30 participants, all female university students, took part in the experiment. All participants except two were in age band 18-24. The exceptions were in age band 25-29 (group +) and 30-34 (group 0).

Materials. As stated above, the texts that we presented to our participants were portrayed as giving feedback on an IQ test that the participants had just taken. This feedback first explained the test and its type of scoring:

> The Baumgartner test which you have just undertaken tests various kinds of intelligence, for instance, your visual intelligence, your logical-mathematical intelligence and your spatial intelligence. These various aspects of your intelligence contribute to an overall Baumgartner Score. The Baumgartner Score rates your intelligence on a 10-point scale with 10 as the highest possible score. Note that your Baumgartner Score can change over time dependent on experience and practice. Below your test score is presented in comparison with the average score in your age group.

The introduction to the test was followed by either the positively $(+1\ldots+7$, Figure 2) or the relatively neutrally $(01\ldots07$, Figure 2) phrased test results.

Before and after the participants took the IQ test, they were asked to fill out a questionnaire with a PANAS test and some questions for collecting demographical information. The materials, the test texts and questionnaires, as well as the experiment introduction and consent form were presented to the participants as a web experiment. For ethical reasons, participants received a debriefing about the aims of the study on paper from the experimenter in person.

Procedure. The participants could linearly traverse through the various phases of the experiment. An outline of the set up is given in Figure 3. In the general introduction to the experiment, participants were told that the experiment was 'an assessment of a new kind of intelligence test which combines a number of well-established methods that are used as indicators of human brain power'. To make it more difficult for the participant to keep track of how well/poorly she performed over the course of the test, it also said that the test consisted of open and multiple choice questions that had different weight factors in the calculation of the overall score and that would assess various aspects of their intelligence. Subsequently, the participant was asked to tick a consent form to participate in the study. Then a questionnaire followed in which the participant was asked about her age, gender and the quality of her English. She was also asked if she had any experience with IQ tests and how she expected to score on this one. These questions were interleaved with an emotion assessment test (reduced PANAS) in which the participant was asked 'how do you feel right now?'.

After filling out the questionnaire, the participant could start the "IQ test" whenever she was ready. The "IQ test" consisted of 30 questions which she had to answer one at a time. The participant could not skip a question and also had to indicate for each of the questions how confident she was about her answer. The questions that were used for the test were carefully collected from the internet and included items from various tests and games. Different types of questions were used: questions about logical truths, mathematical questions that required some calculations, questions about words and letter sequences, questions including pictures and questions about the participant's personality. They were ordered randomly (but with the same order for each participant).

When the participant had finished the test, she was asked to wait patiently while the system calculated the test scores. When enough calculation time had passed the participant was presented with the test feedback (one of the two texts, regardless of their actual performance). After the participant had processed the feedback, she was asked to fill out one more questionnaire to assess her emotions (i.e. 'How do you feel right now knowing your scores on the test'). This time the simplified PANAS test was interleaved with questions about the participant's results, (e.g. were they as expected and how did she value them), the test (e.g. was it difficult, doable or easy?) and space for comments on the test and the experiment. Finally the participant was debriefed about the experiment and about the goal of the study.

Although our particular experiment focussed on positive affect, we included the negative affect terms partly so that we could detect outliers in our participant set – people who were perhaps extremely nervous about the test or sensitive about their IQ. In fact, we did not find any such outliers.

The participants were randomly distributed over group + and group 0 and (for ethical reasons) did the test one by one in a one-person experiment room while the experimenter was waiting outside the room. As soon as the participant indicated that she had finished the task (i.e. stepped out of the experiment room), she was debriefed about the study by the experimenter and was paid with a voucher worth 5 pounds.

1. General introduction to the experiment;
2. Consent form;
3. Questionnaire on participant's background and familiarity with IQ-test interleaved with a PANAS test to assess the participant's current emotional state;
4. Message: 'Please press the next button at the bottom of this page whenever you are ready to start the intelligence test';
5. IQ test questions;
6. Message: Please be patient while your answers are being processed and your test score is computed. After the result page, you will be asked another set of questions about the test, your performance and the way you feel about it. This information is very important for this study, so please answer the questions as honestly as possible.';
7. Feedback + or 0;
8. Questionnaire: PANAS test to assess how the participants felt after reading the test feedback interleaved with questions about the test, their expectations and space for comments;
9. Debriefing which informed participants about the study's purpose and stated that the IQ test was not real and that their test results did not contain any truth.

Fig. 3. Phases in the experiment set up

Hypotheses. Since the message of the feedback texts was relatively positive and there is no necessary correlation between positive and negative PANAS scores [25], we expected the main effects of the texts to be on the average evaluation of the positive PANAS terms. The hypothesis for this study was that participants who received the positively phrased feedback would show a larger change in their positive emotions than the participants who received the neutrally phrased feedback.

Results. The bar chart presented in Figure 4 illustrates the results of the PANAS questionnaire after reading the feedback texts. Table 2 indicates that on average after they had received their test results, participants in the +-group were more positively tuned than participants in the 0-group. Participants in the +-group also rated the positive emotion terms higher than they had done before they undertook the IQ test. No such results were found for the 0-group. In contrast, compared to their responses before the IQ test, participants in the

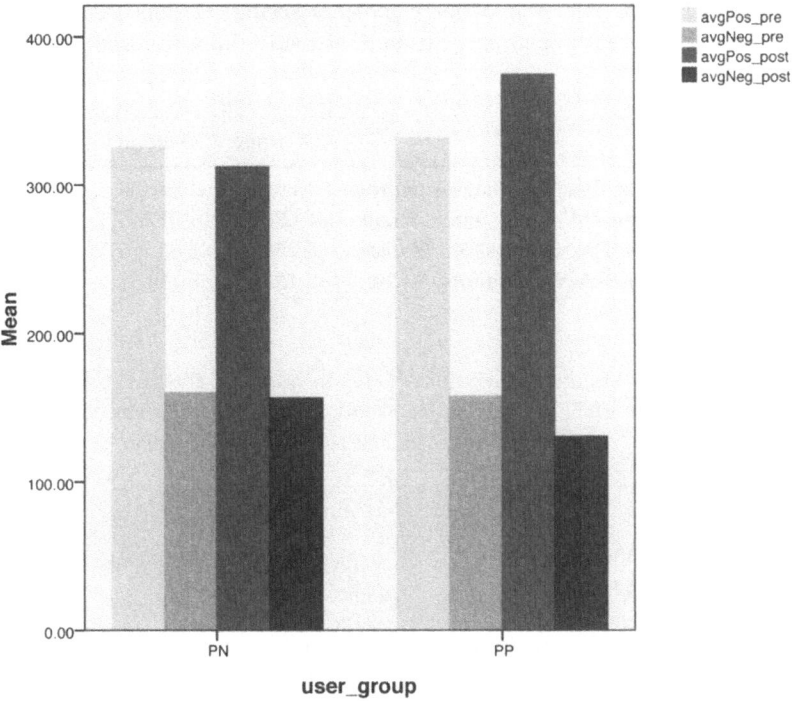

Fig. 4. Positive and negative PANAS means before and after the Participants read the feedback about their test results

0-group rated the positive terms slightly lower after they had processed their neutrally phrased feedback. With respect to the negative PANAS terms, participants in the +-group report slightly less negative emotions after they read their test scores, but none of the differences found in the negative PANAS scores were significant. Cronbach Alpha scores show that participants consistently rate the positive and negative terms before (α Positive $= .797$; α Negative $= .883$) and after (α Positive $= .758$; α Negative $= .879$) reading the feedback text.

A 2 (feedback type) * 2 (before/after) * 2 (positive/negative mean) repeated measures ANOVA was carried out on the average PANAS scores. This showed no main effect of feedback type (+ vs 0) and no main effect of before/after on average PANAS scores. However, there was a highly significant interaction between feedback type and before/after, which indicates that the change in PANAS mean before and after the text was strongly dependent on feedback type[4] ($F(1, 28) = 10.246$, $p < .003$; power $= .871$). We interpret this to mean that the (after minus before) value is significantly greater for the +-group. A two-tailed, two sample t-test verifies this ($t = 3.2$, $p < 0.004$).

---

[4] An ANOVA test on the positive means only produces a similar result.

Table 2. Means and Standard deviations (between brackets) for the negative and positive PANAS terms as indicated before and after the IQ test undertaken by participants that received neutral and participants that received positive feedback on their performance.

	0-group	+-group
Negative PANAS terms Before	1.60(.76)	1.58(.68)
Negative PANAS terms After	1.57(.68)	1.31(.45)
Positive PANAS terms Before	3.25(.78)	3.32(.55)
Positive PANAS terms After	3.13(.58)	3.75(.55)

Table 3. Means and Standard deviations (between brackets) for the positive PANAS terms as indicated after the IQ test undertaken by participants that received positive and participants that received neutral feedback on their performance.

	0-group	+-group
Alert Before	3.96(.80)	3.17(.99)
Alert After	3.45(.76)	3.65(.75)
Determined Before	3.49(1.02)	3.60(.50)
Determined After	3.50(1.13)	3.74(.61)
Enthusiastic Before	3.52(1.05)	3.49(.72)
Enthusiastic After	2.97(.81)	3.84(.66)
Excited Before	2.74(.97)	3.28(.61)
Excited After	2.64(.75)	3.69(.83)
Inspired Before	2.56(1.21)	3.06(.77)
Inspired After	3.06(1.05)	3.81(.78)

We did some post-hoc investigation in an attempt to understand the main result more fully. When looking at the positive PANAS scores in more detail (see Table 3), it turns out that only three of the five positive PANAS terms included in the simplified PANAS test render promising results. Interactions were found for the terms 'alert' ($F(1, 28) = 10.291$, $p < .003$) and 'enthusiastic' ($F(1, 28) = 5.651$, $p < .025$). No interactions were found for the terms 'determined' and 'inspired'. For 'inspired' however, we found a main effect of feedback type : ($F(1, 28) = 8.755$, $p < .006$), which indicates that participants in the +-group could have been more inspired because of their test scores than participants in the 0-group. Not all of these results would be significant if Bonferroni corrections were made.

5.5   Discussion

Compared with Study I, we expected participants to indicate stronger emotional effects, because the text participants were asked to read was about their own capabilities instead of about something in the world around them which they could think would not affect them. Indeed, this seems to have been the case. In Study I, all responses used the lower half of the scale, whereas with the slider our participants indicated values up to both extremes of the range available. Unfortunately, the fact that one set of values is discrete and the other continuous means that it is hard to carry out a simple statistical comparison.

6   Conclusions and Future Directions

This chapter presented our efforts to measure differences in emotional effects invoked in readers. These efforts were based on our assumption that the wording used to present a particular proposition matters in how the message is received.

In Study I, participants' judgements of the negative or positive nature of a text (in both the text validation and in the full study) are in accord with our predictions. In terms of *reflective analysis* of the text, therefore, participants behave as we expected. Although we strongly emphasised that we were interested in emotions with respect to the test text, our attempts to measure the *emotional effects* invoked in readers caused by tactical text differences did, however, not produce any significant results.

In Study II, we aimed to increase the reader's involvement. We used the technique of using a feedback task, where participants play a game or answer some questions after which they receive feedback on their performance. The study aimed to measure the emotional effects of slanting this feedback text in a positive or a neutral way. As in such a feedback situation the test text is directly related to the participants' own performance, we expected an increased involvement and stronger emotions.

The fact that we have been able to show a significant difference in the emotions induced by the two texts in Study II is very encouraging. It suggests that there *is* a possible methodology for directly evaluating affective NLG and that the tactical concerns with which much of NLG research is occupied are relevant to affective NLG. A similar methodology could perhaps now be used to determine the effectiveness of specific NLG methods and mechanisms in terms of inducing emotions. This is an important first step, but there is still a lot more to do. Although we have now shown that NLG tactical decisions can affect emotions, it remains to be seen what kind of changes in strategy, learning, motivation, etc., can be induced by positive or negative affect and thus how these framing decisions can best be made by an NLG system.

In Study II, a number of different techniques (e.g. emphasis, vague adjectives and adverbs) were used to phrase the various propositions in the feedback. In future work we aim to identify the relative importance of the individual techniques. We also aim to investigate more fully the algorithms that might be used by an NLG system wishing to employ these slanting techniques. In particular, techniques such as "monitoring" [10] might be useful to ensure that they are used enough, but not excessively, for a given purpose.

As argued above, the results of our study seem to indicate that self-reporting of emotions is difficult. This could be because participants do not like to show their emotions, because the emotions invoked by what they read were just not very strong or because they do not have good conscious access to their emotions. Although self-reporting is widely used in Psychology, it could be that participants are not (entirely) reporting their true emotions, and that maybe this matters more when effects are likely to be subtle. In all of these situations, the solution could be to use additional measuring methods (e.g. physiological methods), and to check if the results of such methods can strengthen the results of the questionnaires. Another option is to use an objective observer during the experiment (e.g. videotaping the participants and observing the duration of smiles or frowns) to judge whether the participant is affected.

Acknowledgments. This work was supported by the EPSRC platfrom grant 'Affecting people with natural language' (EP/E011764/1) and also in part by Science Foundation Ireland under a CSET grant (CNGL/CSET). We would like to thank the people who contributed to this study, most notably Louise Phillips, Emiel Krahmer, Linda Moxey, Graeme Ritchie, Judith Masthoff, Albert Gatt, Kees van Deemter and Nikiforos Karamanis.

References

1. Bard, E.G., Robertson, D., Sorace, A.: Magnitude estimation of linguistic acceptability. Language 72(1), 32–68 (1996)
2. Brown, R., Pinel, E.: Stigma on my mind: Individual differences in the experience of stereotype threat. Journal of Experimental Social Psychology 39, 626–633 (2003)
3. Cacioppo, J., Bernston, G., Larson, J., Poehlmann, K., Ito, T.: The psychophysiology of emotion. In: Lewis, M., Haviland-Jones, J. (eds.) Handbook of Emotions, pp. 173–191. Guilford Press, New York (2000)
4. Cadinu, M., Maass, A., Rosabianca, A., Kiesner, J.: Why do women underperform under stereotype threat? Psychological Science 16(7), 572–578 (2005)
5. Carenini, G., Moore, J.D.: An empirical study of the influence of argument conciseness on argument effectiveness. In: Proceedings of the 38th annual meeting of the Association for Computational Linguistics, pp. 150–157 (2000)
6. Finucane, M., Slovic, P., Mertz, C., Flynn, J., Satterfield, T.: Gender, race, and perceived risk: the 'white male' effect. Health, Risk & Society 2(2), 159–172 (2000)
7. Fleischman, M., Hovy, E.: Towards emotional variation in speech-based natural language generation. In: Proceedings of the Second International Natural Language Generation Conference, INLG 2002 (2002)
8. Graff, D.: Shifting sands: An interest-relative theory of vagueness. Philosophical Topics 20, 45–81 (2000)
9. O'Hara, L., Sternberg, R.: It doesn't hurt to ask: Effects of instructions to be creative, practical, or analytical on essay-writing performance and their interaction with students' thinking styles. Creativity Research Journal 13(2), 197–210 (2001)
10. Hovy, E.: Pragmatics and natural language generation. Artificial Intelligence 43(2), 153–198 (1990)
11. Kennedy, C.: Vagueness and grammar: the semantics of relative and absolute gradable adjectives. Linguistics and Philosophy 30, 1–45 (2007)
12. Lang, P.: Behavioral treatment and bio-behavioral assessment: Computer applications. In: Sidowske, J., Johnson, J., Williams, T. (eds.) Technology in Mental Health Care Delivery Systems, pp. 119–137. Ablex, Norwood (1980)

13. Lazarus, R., Kanner, A., Folkman, S.: Emotions: A cognitive-phenomenological analysis. In: Plutchik, R., Kellerman, H. (eds.) Emotion, theory, research, and experience, pp. 189–217. Academic Press, New York (1980)
14. Levin, I., Schneider, S., Gaeth, G.: All frames are not created equal: A typology and critical analysis of framing effects. Organizational behaviour and human decision processes 76(2), 149–188 (1998)
15. Mackinnon, A., Jorm, A., Christensen, H., Korten, A., Jacomb, P., Rodgers, B.: A short form of the positive and negative affect schedule: evaluation of factorial validity and invariance across demographic variables in a community sample. Personality and Individual Differences 27(3), 405–416 (1999)
16. Mairesse, F., Walker, M.: Trainable generation of big-five personality styles through data-driven parameter estimation. In: Proceedings of the 46th Annual Meeting of the ACL: HLT, pp. 165–173 (2008)
17. Moore, J., Porayska-Pomsta, K., Varges, S., Zinn, C.: Generating tutorial feedback with affect. In: Proceedings of the 7th International Florida Artificial Intelligence Research Symposium Conference (FLAIRS), pp. 123–130 (2004)
18. Moxey, L., Sanford, A.: Communicating quantities: A review of psycholinguistic evidence of how expressions determine perspectives. Applied Cognitive Psychology 14(3), 237–255 (2000)
19. De Rosis, F., Grasso, F.: Affective natural language generation. In: Paiva, A.C.R. (ed.) IWAI 1999. LNCS, vol. 1814, pp. 204–218. Springer, Heidelberg (2000)
20. De Rosis, F., Grasso, F., Berry, D.: Refining instructional text generation after evaluation. Artificial Intelligence in Medicine 17(1), 1–36 (1999)
21. Russell, J., Weiss, A., Mendelsohn, G.: Affect grid: A single-item scale of pleasure and arousal. Journal of Personality and Social Psychology 57, 493–502 (1989)
22. Teigen, K., Brun, W.: Verbal probabilities: A question of frame. Journal of Behavioral Decision Making 16, 53–72 (2003)
23. Thompson, H.: Strategy and tactics: A model for language production. In: Proceedings of the Chicago Linguistics Society, Chicago (1977)
24. Tversky, A., Kahneman, D.: The framing of decisions and the psychology of choice. Science 211, 453–458 (1981)
25. Watson, D., Clark, L.: Manual for the Positive and Negative Affect Schedule - Expanded Form. The University of Iowa (1999)
26. Watson, D., Clark, L., Tellegen, A.: Development and validation of brief measures of positive and negative affect: The PANAS scales. Journal of Personality and Social Psychology 54, 1063–1070 (1988)
27. Wilson, E., MacLeod, C., Campbell, L.: The information processing-approach to emotion research. In: Coan, J., Allen, J. (eds.) Handbook of emotion elicitation and assessment. Oxford University Press, New York (2007)

Introducing Shared Tasks to NLG: The TUNA Shared Task Evaluation Challenges

Albert Gatt[1,2] and Anja Belz[3]

[1] Institute of Linguistics, Centre for Communication Technology, University of Malta
`albert.gatt@um.edu.mt`
[2] Communication and Cognition, Faculty of Arts, Tilburg University
[3] NLTG, School of Computing, Mathematical and Information Sciences,
University of Brighton
`asb@bton.ac.uk`

Abstract. Shared Task Evaluation Challenges (STECs) have only recently begun in the field of NLG. The TUNA STECs, which focused on Referring Expression Generation (REG), have been part of this development since its inception. This chapter looks back on the experience of organising the three TUNA Challenges, which came to an end in 2009. While we discuss the role of the STECs in yielding a substantial body of research on the REG problem, which has opened new avenues for future research, our main focus is on the role of different evaluation methods in assessing the output quality of REG algorithms, and on the relationship between such methods.

1 Introduction

Evaluation has long been an important topic for Natural Language Processing. Among the developments that have helped to bring it to the forefront of the research agenda of many areas of NLP, two are particularly important. Starting with the work of Spärck Jones and Galliers [71], there has been an attempt to identify the main methodological issues involved, coupled with a more recent drive towards the development of evaluation methods, especially metrics that can be automatically computed [26,54,58]. A second development has been the organisation of a number of Shared Task Evaluation Challenges (STECs) in areas such as Summarisation[1], Information Retrieval[2] and Machine Translation,[3] which bring together a number of researchers with solutions to a problem based on shared datasets. In addition, several contributions exploring the significance and validity of different evaluation methods, as well as the relationship between them, have appeared in recent years [7,20,64]. Some of these have discussed the relationship between *intrinsic* evaluation, which evaluates the output of an algorithm in its own right, either against a reference corpus or by eliciting human judgements of quality, and

---

[1] See, for example, `http://duc.nist.gov/pubs.html`
[2] See `http://trec.nist.gov/`
[3] See `http://www.itl.nist.gov/iad/mig/tests/mt`

E. Krahmer, M. Theune (Eds.): Empirical Methods in NLG, LNAI 5790, pp. 264–293, 2010.

extrinsic evaluation, where the system's output is assessed in the context of an externally defined task [7,11,27]. However, the relationship between results from these two kinds of evaluation remains an open question.

Within NLG, evaluation has typically relied on methods which involve 'humans in the loop'. For example, many intrinsic evaluations have taken the form of elicitation of ratings or responses to questionnaires about various aspects of generated text [19,31,53]. To take a specific example, the STORYBOOK system [19] was evaluated using questionnaires designed to elicit judgements of quality of different versions of an automatically generated story, obtained by suppressing or including the contribution of different modules to the generation process. Similarly, the SUMTIME weather forecasting system [67] was evaluated by asking participants various questions to assess the extent to which they liked a forecast. The evaluations discussed in the present chapter (see especially Section 4) and elsewhere in this volume [9] also make use of judgements of this kind, among other methods.

Starting with the work of Langkilde [51], automatic intrinsic methods exploiting corpora have mostly been used in evaluations of realisers [5,17,18,78], though some corpus-based experiments involving Referring Expressions Generation (REG) algorithms have been reported recently [37,39,41,76], including in the TUNA evaluations discussed in the present chapter, and in the GREC shared task competitions [9].

Another evaluation methodology in NLG involves the use of extrinsic, task-based methods to assess the fitness for purpose of an end-to-end generation system within its application domain. Examples include the STOP system [65], which generated smoking cessation letters; BT-45 [62], which produced textual summaries of neonatal intensive care data, and MPIRO [44] and PEACH [72], both of which involved the generation of descriptions of museum artifacts. While such studies tend to be more expensive and labour-intensive, it is often taken for granted within the NLG community that extrinsic methods give a more reliable assessment of a system's utility in doing what it was designed to do.

As the above examples suggest, there is a fairly strong evaluation tradition in NLG. However, *comparative* evaluation is a relatively recent development. This is in part due to the fact that in many cases, researchers develop systems and methods which are not directly comparable, answering to different needs depending on the domain of application. Thus, it is not surprising that comparative studies have either focused on specific subtasks of NLG on which there is significant consensus regarding problem definitions and input/output pairings, or on end-to-end generation tasks. To date, the only example of an evaluation experiment involving directly comparable, independently developed end-to-end generation systems comes from weather forecast generation [8,10]. As for evaluation of subtasks, the best example is perhaps realisation, for which there is a readily available dataset in the form of the Penn Treebank. Indeed, several groups of researchers have reported results for regenerating the Wall Street Journal corpus [17,51,78], though the systems are not directly comparable as they use inputs of different levels of specificity.

There are a number of arguments in favour of comparative evaluation which are worth rehearsing briefly here (they are discussed in greater detail by Belz and Kilgarriff [12]). Perhaps the most obvious immediate benefit is the creation of shared, publicly available data on the basis of which systems can be directly compared. A second argument is that a focus on a common problem speeds up technological progress while conferring additional benefits on the community, such as a growth in size through increased participation. Third, a body of research with results on a single dataset provides baselines against which to compare novel approaches. Finally, given several sets of comparable results from such an evaluation exercise, there is the possibility of follow-up studies, whether these are concerned with identifying properties of systems or algorithms that enhance performance [33], or with methodological issues such as the significance and validity of different evaluation methods [7]. It is with the latter question that the second part of this chapter will be primarily concerned.

The growing interest in comparative evaluation in NLG gave rise to some expressions of interest in the organisation of Shared Tasks, which began to be sounded during discussions at the UCNLG and ENLG workshops in 2005, and continued during a special session on the topic at INLG'06, culminating in an NSF-funded workshop on Shared Tasks and Comparative Evaluation in NLG in April 2007 [22]. At that meeting, various work groups were formed to discuss different aspects of the topic, including the identification of tasks that could potentially form the basis of a STEC. Two of the task types proposed at that meeting – Referring Expression Generation (REG) and Giving Instructions in Virtual Environments (GIVE) – have since been used in a STEC. Two series of REG Shared Tasks have been organised since 2007, one focusing on attribute selection and realisation of references to objects in visual domains (the TUNA STECs, which are the focus of this chapter); the other on generating referring expressions to discourse entities in the context of running text (the Generation of Referring Expressions in Context, or GREC, Tasks; see the chapter by Belz *et al.* [9]). As for instruction giving, the first GIVE Challenge was presented in 2009 (see the chapter by Koller *et al.* [47]). Since 2008, all of these STECs have been presented as part of Generation Challenges, an umbrella event which is now expanding further, with a new Shared Task on Question Generation planned for 2010.[4]

This chapter discusses the first of these series of STECs, those using the TUNA Corpus, which were organised over three years. Our aim is twofold. First, we give an overview of the structure and scope of the Shared Tasks, describing the evaluation procedures as well as the degree of participation and the nature of the contributions (Section 3). Second, we discuss the variety of evaluation methods used and the relationship between them (Section 4). Our findings in this regard have consistently pointed to a problematic relationship between intrinsic and extrinsic methods, echoing findings in other areas of NLP such as Summarisation [27]. Before turning to these topics, we first give some background on the REG task and the TUNA corpus (Section 2).

---

[4] See http://www.nltg.brighton.ac.uk/research/genchal10/ for details of the tasks forming part of Generation Challenges 2010.

2   Background to the TUNA Tasks

Referring Expression Generation (REG) is one of the most intensively studied subtasks of NLG. Like realisation, it has tended to be studied independently of its role within larger NLG systems. Since the seminal work by Appelt, Kronfeld, Dale and Reiter starting in the late 1980s [1,2,21,23,50] widespread consensus has developed regarding the problem that REG algorithms seek to solve, as well as the nature of their inputs and outputs. The problem definition, subscribed to by most traditional approaches to REG (see [16] for a related formulation), is as follows:

Definition 1. *Given a domain of discourse U, consisting of entities and their attributes, and a target referent $r \in U$, find a set of attributes D (the description), such that $[\![D]\!] = \{r\}$.*

This is not to imply that unique identification is *all* that there is to REG; indeed, the foundations laid in the work cited above have been extended in several directions, for example to deal with discourse-anaphoric NPs, issues related to expressiveness and logical completeness (such as dealing with n-place relations, negation and plurality), as well as vagueness. Nevertheless, attribute selection and identification were two central issues in REG and this consensus made it a good candidate for the organisation of a STEC. This was compounded by a growing interest in empirically evaluating REG algorithms, which had hitherto often been justified on the basis of theoretical and psycholinguistic principles, but lacked a sound empirical grounding. Initial forays into empirical evaluation of REG algorithms either used existing corpora such as COCONUT [39,43], or constructed new datasets based on human experiments [37,68,76]. These evaluation studies focused on classic approaches to the REG problem, such as Dale's Full Brevity and Greedy algorithms [23] and Dale and Reiter's Incremental Algorithm [21]. In the studies by Gupta and Stent [39] and by Jordan and Walker [43], these algorithms were evaluated in a dialogue context and were extended with novel features to handle dialogue and/or compared to new frameworks such as Jordan's *Intentional·Influences* model [42].

2.1   The TUNA Corpus

One of the datasets constructed as part of the REG evaluations mentioned in the previous section was the TUNA Corpus,[5] a collection of descriptions of objects elicited in an experiment involving human participants. The corpus is made up of several files, each of which pairs a domain (a representation of entities and their attributes) with a human-authored description for one or two entities (the target referent(s)). In each case, there are six further entities in addition to the target(s). In line with the familiar REG terminology, these are referred to as *distractors*.

---

[5] http://www.csd.abdn.ac.uk/research/tuna/

Table 1. Attributes and values in the two TUNA domain types

Attribute	Possible values Furniture	Possible values People
TYPE	*chair, sofa, desk, fan*	*person*
ORIENTATION	*front, back, left, right*	*front, left, right*
X-DIMENSION (column number)	$1, 2, 3, 4, 5$	$1, 2, 3, 4, 5$
Y-DIMENSION (row number)	$1, 2, 3$	$1, 2, 3$
SIZE	*large, small*	-
COLOUR	*blue, red, green, grey*	-
AGE	-	*young, old*
BEARD	-	0 (**false**), 1 (**true**)
HAIR-COLOUR	-	*dark, light, other*
HASHAIR	-	0 (**false**), 1 (**true**)
HASGLASSES	-	0 (**false**), 1 (**true**)
HASSHIRT	-	0 (**false**), 1 (**true**)
HASTIE	-	0 (**false**), 1 (**true**)
HASSUIT	-	0 (**false**), 1 (**true**)

The descriptions were collected in an online elicitation experiment which was advertised mainly on the University of Zurich Web Experimentation List[6] [63] and in which participation was unrestricted, though data from persons who did not complete the experiment, or who reported a low level of fluency in English, was later removed.

During the experiment, participants were shown visual domains of objects of two types: furniture or people. Domains of the former type consisted of pictures obtained from the Object Databank[7], a set of realistic, digitally created images of familiar objects. These were digitally manipulated to create versions of the same object in different colours and orientations. In the people domain type, the pictures consisted of real black and white photographs of males, which had been used in a previous study by van der Sluis and Krahmer [69]. The attributes of the objects in each domain type are summarised in Table 1. As the table indicates, the two types of images differ in their complexity, in that the photographs of people have a broader range of attributes from which to select. This difference was one of the motivations for eliciting descriptions of two different classes of objects. Apart from attributes such as COLOUR or TYPE, objects were also defined by two numeric attributes (X-DIMENSION and Y-DIMENSION), corresponding to their coordinates in a sparse 3 (row) × 5 (column) matrix in which the objects were visually presented during the experiment.

Design. A full description of the design of the experiment can be found in Gatt *et al.* [37]. Here, we focus primarily on those aspects of the design which are

---

[6] http://genpsylab-wexlist.unizh.ch

[7] The Object Databank was constructed by Michael Tarr and colleagues and is available on http://alpha.cog.brown.edu:8200/stimuli/objects/ objectdatabank.zip/view

relevant to the STECs. The main within-participants condition manipulated in the experiment was ±LOC: in the +LOC condition, participants were told that they could refer to entities using any of their properties (including their location on the screen). In the −LOC condition, they were discouraged from doing so, though a small number of participants nevertheless included this information. The corpus annotation indicates which of the two conditions a description was elicited in (see 2 and the description of the annotation below). The other within-participants variable manipulated in the experimental design concerned the number of entities referred to. One third of the trials were singular, that is, there was one target referent and six distractor objects. The rest were plural, with two target referents, which were SIMILAR one-third of the time (that is, had identical values on their attributes[8]), and DISSIMILAR in the rest of the trials. Since plurals did not feature in any of the TUNA STECs, we shall not discuss the distinction any further here.

The experimental design also contained a further between-groups condition, which manipulated whether the communicative situation in which descriptions were produced was perceived as 'fault-critical' or not (±FC). The way this was done is explained below. However, for the TUNA STECs, data from the ±FC conditions was merged. There were two reasons for this. First, distinguishing (±FC) descriptions would have complicated the design of the shared tasks, taking them beyond the conventional task definition given above. Second, preliminary analysis of the corpus suggested that this variable did not significantly affect the content of descriptions. However, it is worth noting that other researchers have shown an impact of communicative intent and 'importance' on descriptions produced by humans [56,73].

The experiment consisted of a total of 38 trials, each consisting of a visual domain for which a participant typed a description for the target referent. The trials were balanced in the following sense. For each possible subset of the attributes defined for a given domain type (shown in Table 1), there was an equal number of trials in which an identifying description of the target(s) required the use of all those attributes (i.e. leaving one of the attributes out would have resulted in an unsuccessful identifying description) *unless* it included locative information. For example, there was a furniture trial in which a target could be distinguished by using COLOUR and SIZE. TYPE was never included in this calculation because previous research has suggested that this attribute is included by human speakers irrespective of whether it has any discriminatory value in an identification task [60]. The locative attributes (X-DIMENSION and Y-DIMENSION) were not balanced in the same manner, because the position of objects in the grid in which they were presented was determined randomly at runtime for each trial and each participant.

---

[8] The exception was TYPE in the furniture domain: SIMILAR plurals in this domain were identical on all their visual attributes, including COLOUR and ORIENTATION, but had different TYPE values. For example, one object could be a chair and the other a sofa. This does not apply in the people domain, since all entities were of the same type.

Procedure. All participants were exposed to all 38 trials in randomised order. Examples of trials with both people and furniture are shown in Figure 1. A trial consisted of a visual domain consisting of one or two target referents and six distractor entities in a sparse 3×5 grid. The target(s) were indicated by a red border. Participants typed a description of the referent in a text box in answer to the question: *Which object is surrounded by a red border?*

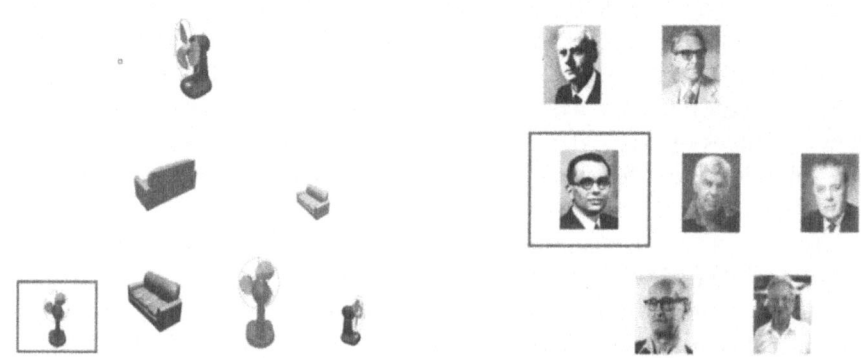

(a) Furniture trial (images by Michael Tarr and colleagues, distributed with the Object Databank).

(b) People trial (images from a previous study by van der Sluis and Krahmer [69]).

Fig. 1. Trials in the TUNA elicitation experiment

Participants were told that their descriptions were being used to test a language understanding system in real time. Following each trial, some objects were removed from the screen and participants were given feedback as to whether the 'system' had managed to interpret their description correctly. In actual fact, the objects to be removed were determined in advance, so that the 'system' was successful 75% of the time. As noted above, one of the conditions manipulated in the experiment was ±FC: in the −FC condition, participants were told that, should the system make an error, they could correct it (by clicking on the right target objects). In the −FC condition, this was not possible.

Participation and corpus size. Sixty participants completed the experiment on a voluntary basis. All were self-reported native or fluent speakers of English. This gave rise to $(60 \times 38 =)$ 2280 descriptions paired with their domain representation, of which 780 were singular (420 furniture and 360 people descriptions), and 1500 were plural (780 furniture descriptions and 720 people descriptions). Only the singular sub-corpus was used for the TUNA STECs.

Annotation. The XML format used in the TUNA STECs, shown in Figure 2, is a variant of the original format of the TUNA corpus.[9] The root TRIAL node

---

[9] The annotation manual for the original TUNA Corpus is available from the corpus website.

```
<TRIAL CONDITION="+/-LOC" ID="...">
  <DOMAIN>
    <ENTITY ID="..." TYPE="target" IMAGE="...">
      <ATTRIBUTE NAME="..." VALUE="..." />
      ...
    </ENTITY>
    <ENTITY ID="..." TYPE="distractor" IMAGE="...">
      <ATTRIBUTE NAME="..." VALUE="..." />
      ...
    </ENTITY>
    ...
  </DOMAIN>
  <WORD-STRING>
    string describing the target referent
  </WORD-STRING>
  <ANNOTATED-WORD-STRING>
    string in WORD-STRING annotated
    with attributes in ATTRIBUTE-SET
  </ANNOTATED-WORD-STRING>
  <ATTRIBUTE-SET>
    set of domain attributes in the description
  </ATTRIBUTE-SET>
</TRIAL>
```

Fig. 2. XML format of corpus items

has a unique ID and an indication of the ±LOC experimental condition. The DOMAIN node contains 7 ENTITY nodes (1 target and 6 distractors), which themselves contain a number of ATTRIBUTE nodes defining the possible properties of an entity in attribute-value notation. The attributes cover all properties defined in Table 1, such as an object's colour or a person's clothing, and the location of the image in the visual display which the DOMAIN represents. Each ENTITY node indicates whether it is the target referent or one of the six distractors, and also has a pointer to the corresponding image that was used in the experiment. The WORD-STRING is the actual description typed by one of the human authors, the ANNOTATED-WORD-STRING is the description with substrings annotated with the attributes they realise, while the ATTRIBUTE-SET contains this set of attributes only.

3   The TUNA Challenges 2007-2009

Starting in 2007, three Shared Task Evaluation Challenges (STECs) were organised using the TUNA corpus.[10] In what follows, we refer to these as TUNA'07, TUNA'08, and TUNA'09.[11] TUNA'07 [6] was presented five months after the NSF

---

[10] The Participants' Packs from all three years, including evaluation tools and documentation, are available from http://www.nltg.brighton.ac.uk/home/Anja.Belz
[11] Though these were not the original titles of the STECs, we have adopted this naming convention for ease of reference throughout this chapter.

workshop, in September 2007 at the UCNLG+MT workshop at MT Summit XI in Copenhagen. It was organised in the spirit of a pilot event to gauge community interest in STECs, with encouraging results. This was followed by TUNA'08 [34], held as part of Generation Challenges 2008, which also included the first GREC task. The results were presented during a special session at INLG'08 in Salt Fork, Ohio. The third and final edition, TUNA'09 [35], was presented at ENLG'09 in Athens as part of Generation Challenges 2009, which also included two GREC tasks and the first GIVE challenge.

In addition to the shared tasks proper, all of these events included two special tracks: (i) an open submission track in which participants could submit any work involving the data from any of the shared tasks, while opting out of the competitive element, and (ii) an evaluation track, in which proposals for new evaluation methods for the shared task could be submitted. Generation Challenges 2009 additionally offered a task proposal track inviting proposals for new shared tasks.

The idea behind the Open Track was partly motivated by discussions at the INLG'06 Special Session and the NSF Workshop, in which one of the reservations expressed in relation to STECs was that they might result in a narrowing of the scope of research on a topic by emphasising a small set of tasks at the risk of stifling novel ideas. The Evaluation Methods track was intended to complement the growing interest within the NLP community in evaluation methods mentioned in Section 1, especially in view of the cautionary notes sounded from various quarters concerning the validity of certain metrics and the relationship between them. In the event, there was only one submission to the Evaluation Methods track in ASGRE'07 [25], and no submissions in subsequent editions. Similarly, the Open Track attracted one submission during TUNA-REG'08, proposing a method for exploiting TUNA ATTRIBUTE-SETs to generate descriptions in Brazilian Portugese [61].

In the remainder of this section, we give an overview of the three TUNA STECs, focusing on their structure and the evaluation methods used. We also briefly discuss the level of participation in the various sub-tasks of each STEC. An exhaustive description of all the different submissions is beyond the scope of this chapter; the relevant publications by participants can be found in the proceedings of the special sessions of the events at which they were presented.

3.1   Structure and Scope of the TUNA Shared Tasks

The structure of the Shared Tasks is summarised in Table 2. As the Table indicates, over the three editions of the TUNA STECs, there was a gradual broadening of the scope of the REG tasks involved.

TUNA'07 consisted of a attribute selection single task, TUNA-AS. Although the task definition was close to the one traditionally used in the REG literature and given above in Definition 1, no emphasis was made on unique identification in order to avoid narrowing its scope. (As it turned out, all peer systems submitted to the STEC performed unique identification.) In the context of the TUNA Corpus, the input to peer systems was a DOMAIN node, and the output was

Table 2. Summary of tasks in the three editions of the TUNA Challenges

Edition	Tasks	Attribute Selection	Realisation
TUNA'07	TUNA-AS	y	n
TUNA'08	TUNA-AS	y	n
	TUNA-R	n	y
	TUNA-REG	y	y
TUNA'09	TUNA-REG	y	y

an `ATTRIBUTE-SET` which represented the selected attributes for the designated target referent in the `DOMAIN`. The scope of the Shared Tasks was extended in TUNA'08, which kept the TUNA-AS task from the 2007 edition, but included two others, TUNA-R and TUNA-REG, both of which included realisation. Thus, the output of systems in the latter two tasks was no longer an `ATTRIBUTE-SET`, but a `WORD-STRING`. In TUNA-R, which focused exclusively on realisation, the task was defined as mapping from an input `ATTRIBUTE-SET` to an actual Natural Language description, while TUNA-REG was defined as an end-to-end referring expression generation task, which went from an input `DOMAIN` to an output `WORD-STRING`, combining content determination (the TUNA-AS Task) and realisation (the TUNA-R Task). However, no constraints were imposed on the way this was to be achieved (for example, content determination and realisation could be implemented as separate modules in a pipeline or interleaved). This broadening of scope resulted in several submissions which presented novel approaches to realisation in addition to content determination. Finally, TUNA'09 was organised as a follow-up to TUNA'08 and consisted of the single end-to-end TUNA-REG task. The aim was to provide participants of TUNA'08 with the possibility of improving on the results obtained in the 2008 edition, while also including new participants.

In all editions, there was a three-stage participation procedure. Potential participants first registered to express their interest in participating. Shortly after, a participants' pack consisting of training and development data (see Section 3.2 below), software to compute some of the relevant evaluation measures (see Section 3.4), and detailed documentation for both data and software, was made available. Participating teams were then required to submit a writeup describing their system(s), including evaluation scores computed on the development set with evaluation software provided by us. Once this was submitted, they could download the test data inputs and submit their system outputsfor the test data, after which the organisers computed the evaluation scores on the test data outputs.

3.2   Datasets

In each edition of the TUNA tasks, participants were provided with both training and development datasets. Unlike the test data, which was provided to participants in the form of inputs only, both the training and development sets consisted of paired inputs and outputs. Although no restriction was placed on the way training and development data were to be used in the process of designing a system, the purpose of the division was twofold. First, participants could conduct an iterative training and testing process in the course of building their systems, using the development dataset for the latter purpose. Second, the development data was also used by participants to report scores on a subset of the evaluation methods used for a given task. These scores, computed on 'seen' data, were then compared by the organisers to the final evaluation scores obtained on the test data outputs.

For the TUNA'07 edition, test data as well as training and development data came from the original TUNA corpus, which had not been publicly released at that point. The corpus was divided into 60% training, 20% development and 20% test data. For TUNA'08, the original corpus data was divided into 80% training and 20% development data, and two new test sets were constructed by partially replicating the original TUNA experiment described in Section 2.1. One of the new test sets was used for the TUNA-AS and TUNA-REG tasks in 2008, and later re-used in the 2009 replication of TUNA-REG; the other was used exclusively for TUNA-R in 2008.

The experiment to construct the new test sets was designed to ensure that each DOMAIN had two reference outputs. Thus, the corpus-based evaluations for the 2008 and 2009 STECs were conducted against multiple descriptions for each input DOMAIN. Like the original TUNA experiment, the new one was conducted over the web, and advertised mainly among staff and students of the Universities of Aberdeen and Brighton. The task given to experimental participants was identical to the original, except for the following differences. First, no plurals were included, since the TUNA tasks only focused on singular descriptions. Second, the new experiment did not include the 'interactive' component of the original, whereby participants were told that they were interacting with a language understanding system. This is because we decided not to replicate the ±FC manipulation, given that data from the +FC and −FC conditions had been collapsed for the purposes of the TUNA tasks. The experiment was completed by 218 participants, of whom 148 were native speakers of English. The test sets were constructed by randomly sampling from the data gathered in the experiment from native speakers only: both sets consisted of 112 items, divided equally into furniture and people descriptions and sampled evenly from both experimental conditions (±LOC).

3.3   Participation

Over the three editions, the various tasks in the TUNA STECs attracted a total of 61 different systems from 10 different teams, based in 10 countries and 5 continents. This turnout suggests that exercises in comparative evaluation are positively viewed on the whole within the NLG community. On the other hand, participation was no doubt enhanced by the fact that the REG task is very familiar to many members of the community (this was the reason why it appeared to be a good candidate for a first try at an NLG STEC).[12] While TUNA'07 had 22 systems from 6 teams, and the three tasks in TUNA'08 had a total of 33 systems from 8 teams, TUNA'09 had 6 systems from 4 teams. These participation levels reflect the different numbers of tasks on offer, but may also be an indication that the TUNA tasks had reached a saturation point after three years.

On the whole, these events yielded a substantial body of new research on the REG problem. Many of the proposed solutions to the attribute selection problem built on classic approaches such as Dale and Reiter's Incremental Algorithm [21] or Dale's Full Brevity Algorithm [23]; or on well-understood formalisms such as graphs [49]. However, there were also a number of new algorithmic approaches, perhaps the most notable of which were those using evolutionary techniques [40,46].

In general, the various approaches to attribute selection in the TUNA-AS task had a markedly empirical orientation, relying on corpus-based frequencies to drive attribute selection (e.g. [55,70]). This was the case even when the algorithms were extensions of existing frameworks. For instance, Theune et al. [75] used the graph-based approach to REG proposed by Krahmer et al. [49], in which a REG domain is represented as a graph whose nodes are entities and whose edges represent properties. Theune et al. used cost functions derived from corpus data to weight graph edges, thus making some attributes more likely to be selected than others.

There was also a notable interest in using individual profiles, that is, speaker-based corpus-derived statistics to guide selection. Thus, Bohnet [14,15] proposed a content determination algorithm that combined well-known heuristics (such as Full Brevity) with individual author identities to prioritise attributes during selection. A similar rationale was followed by di Fabbrizio et al. [29,30], who showed that using frequencies from individual speakers to guide attribute selection results in improvements on intrinsic evaluation scores. These approaches exploited the fact that the corpus contains several descriptions collected from different authors in different experimental trials.

TUNA'08 was also the point where realisation was introduced, and this resulted in an increased focus on realisation methods, which are often ignored in the REG literature. This too yielded a broad spectrum of proposals, again with a decidedly empirical flavour, for example relying on corpus-derived templates or language models [29], as well as case-based reasoning [40,45].

The breadth of approaches to both attribute selection and realisation may serve to open up new avenues of research in future work on REG. Moreover, the various TUNA tasks have now yielded a sizeable body of evaluation data which can act as a baseline against which to compare new systems, but also as the underlying dataset for 'meta-analyses' of various kinds, for example comparing different algorithmic approaches (see [33] for an example based on the TUNA'08 data) or different evaluation methods (see Section 4).

---

[12] Tasks which began to be organised more recently, namely GREC and GIVE, have so far attracted fewer submissions, and this may be due to their relative novelty.

Table 3. Intrinsic evaluation methods used in the three editions of the TUNA tasks

Criterion	Method	Type	Tasks
Humanlikeness	DICE	automatic	TUNA-AS'07-8
	MASI	automatic	TUNA-AS'08
	ACCURACY	automatic	all tasks
	BLEU	automatic	TUNA-R, TUNA-REG
	NIST	automatic	TUNA-R, TUNA-REG
	EDIT	automatic	TUNA-R, TUNA-REG
Theoretical	MINIMALITY	automatic	TUNA-AS
	UNIQUENESS	automatic	TUNA-AS
Judged appropriateness	ADEQUACY	human	TUNA-REG'09
	FLUENCY	human	TUNA-REG'09

Table 4. Extrinsic evaluation methods used in the three editions of the TUNA tasks

Criterion	Method	Type	Tasks
Task effectiveness	Identification Time	human	TUNA-AS'07, TUNA-REG
	Identification Accuracy	human	TUNA-AS'07, TUNA-REG
Ease of comprehension	Reading Time	human	TUNA-AS'07, TUNA-REG'08

3.4   Evaluation Methods

Perhaps the most important aspect of the TUNA STECs was their use of a broad variety of evaluation methods. These included both intrinsic methods, which ranged from automatically computed metrics to methods involving human participants (Table 3), and extrinsic methods involving humans (Table 4).

The decision to use a variety of methods was motivated by the experience of other shared task evaluation challenges, which has shown that the use of a single method of evaluating participating systems can cause loss of trust and/or loss of interest, in particular if the method is seen as biased in favour of a particular type of system (a criticism that has been levelled at BLEU in Machine Translation research [20]) or if it severely restricts the definition of quality (as in the case, for example, of the PARSEVAL metric in syntactic parsing). An additional motivation, already hinted at in Section 1, came from recent work suggesting that different evaluation methods can yield very different results and that the relationship between them is often not straightforward, making comparisons based on multiple methods even more desirable. The particular choice of methods was informed by the following criteria.

Humanlikeness: It is more or less taken as read in many areas of NLP evaluation that the greater the similarity between a system's output and a particular corpus of human-produced reference outputs, the better the system. This is the rationale behind metrics such as BLEU, ROUGE and NIST. The TUNA Challenges used a range of methods that assessed humanlikeness. For the TUNA-AS task, this was defined in terms of set-theoretic metrics comparing attribute sets produced by systems to those produced by human authors on the same DOMAIN. We used the Dice coefficient in the 2007 version of the task, adding the MASI score [59] in 2008. Both metrics measure the degree of overlap between sets. Let A and B be two sets of attributes, corresponding to two descriptions. Dice, which is computed as shown in (1), ranges between 0 and 1, where 1 indicates identity.

$$Dice(A, B) = \frac{2 \times |A \cap B|}{|A| + |B|} \tag{1}$$

The MASI score, which is based on the Jacard coefficient, is also a set comparison metric that ranges between 0 and 1:

$$MASI(A, B) = \delta \times \frac{|A \cap B|}{|A \cup B|} \tag{2}$$

where δ is a *monotonicity coefficient* defined as follows:

$$\delta = \begin{cases} 0 \text{ if } A \cap B = \emptyset \\ 1 \text{ if } A = B \\ \frac{2}{3} \text{ if } A \subset B \text{ or } B \subset A \\ \frac{1}{3} \text{ otherwise} \end{cases} \tag{3}$$

Dice and MASI differ in two respects. First, the Jacard coefficient in MASI tends to penalise a low degree of overlap between sets more than Dice. Second, the monotonicity coefficient (3) biases the score in favour of those cases where one set is a subset of the other.

In the TUNA-R and TUNA-REG tasks, where system outputs were WORD-STRINGs, we used Levenshtein (String Edit) Distance, as well as BLEU and NIST. Edit Distance computes the cost of transforming one string into another based on the number of insertions, deletions and substitutions. In the TUNA STECs, the cost of insertions and deletions was set to 1, while substitutions had a cost of 2. Levenshtein distance has a value of 0 when the strings are identical, and its upper bound depends on the length of the longest string in the comparison pair. BLEU and NIST are n-gram based aggregate scores. For BLEU, we set the parameter that controls the maximum length of n-grams taken into account to the standard value of 4 in TUNA'08. However, this gave odd results due to the brevity of the strings being compared. Hence, we used a maximum n-gram length of 3 in TUNA'09. The primary difference between BLEU and NIST is that the latter gives more importance to less frequent n-grams.

Finally, all tasks used a measure of ACCURACY, which computed the proportion of times a system output was identical to a corpus instance. The precise definition of this depends on the task: in the case of TUNA-AS, the item of comparison was the ATTRIBUTE-SET, while for the TUNA-R and TUNA-REG tasks, it was the WORD-STRING.

Theoretically motivated criteria: In the case of REG, the Shared Tasks also provided an opportunity to evaluate systems in terms of some assumptions that have dominated the field since the work of Appelt [1] and Dale and Reiter [21]. The first measure, UNIQUENESS, was motivated by the emphasis in the REG literature on *identification* (see Definition 1). It was defined as the proportion of ATTRIBUTE-SETs produced by a system which uniquely identified the intended referent. All systems scored 100% on UNIQUENESS; therefore, we shall not discuss it further. In addition to identification, a lot of research on REG has also revolved around a (theoretical) interpretation of the Gricean maxims [38], particularly the Maxim of Quantity, which was interpreted as a constraint on REG algorithms to include no more information in a description than absolutely required for identification (see Section 4.1 for further discussion). This criterion was straightforwardly applicable in the TUNA-AS: the MINIMALITY measure computed the proportion of attribute sets produced by a system that (a) uniquely identified their target referent and (b) were minimal, in the sense that any alternative attribute set that distinguished the target referent within the domain would be at least as long as the one returned by the system.

Judged appropriateness: While automatically computed metrics compare system outputs against a gold standard corpus, an alternative way to assess intrinsic quality is to obtain human judgements, a method which is frequently employed in NLG. Human judgements were included in the final edition of the TUNA STECs, in an experiment that included both system outputs and the human-authored reference descriptions from the corpus. The experiment involved 8 native speakers of English who were doing a Masters degree in a linguistics-related subject, recruited from among post-graduate students at UCL and the Universities of Sussex and Brighton. Participants were presented with a description of a target referent, together with a visual representation of the input DOMAIN (see Figure 1). They were asked to rate the description based on two criteria, ADEQUACY and FLUENCY. The following are the relevant excerpts from the instructions given to them:

1. (ADEQUACY) How clear is this description? Try to imagine someone who could see the same grid with the same pictures, but didn't know which of the pictures was the target. How easily would they be able to find it, based on the phrase given?
2. (FLUENCY) How fluent is this description? Here your task is to judge how well the phrase reads. Is it good, clear English?

In addition, the instructions to participants emphasised that the two measures are distinct, that is, it is perfectly possible for a description to be highly adequate but non-fluent. It appears that participants were able to assess the two criteria independently to some extent: subsequent analysis revealed a correlation of just 0.6 between the two measures which was not statistically significant [35].

For both kinds of judgements, ratings were obtained using a slider, whose position returned an integer between 1 (worst) and 100 (best). Participants were not shown the actual integer values, but only manipulated the slider's position. For each trial, the slider was initially placed in the middle. The motivation for using a slider rather than the more standard rating scale (e.g. a scale from 1 to 5) was that means obtained from such ordinal scales tend to be hard to interpret. This is partly because participants may attach different meanings to the values on the scale and the distances between them (for example, interpreting the distance between 1 and 2 and between 4 and 5 differently); moreover, participants who provide the ratings in the first place have no opportunity to rate an item as falling within an interval. The use of parametric methods to analyse this type of ordinal data is not generally considered appropriate. The method used here, by contrast, relied exclusively on spatial distance in a visual modality: it was the degree of displacement of the slider that reflected participants' judgements since, although the slider position mapped to a numeric value, the value itself was not known to participants. Our assumption was that the physical position of a slider would not be as susceptible to differences in 'meaning' as would a value on an ordinal scale.

Our use of sliders was partially inspired by studies using *Magnitude Estimation* (ME) [3], which requires participants to make judgements by comparing an item to a modulus (i.e. an item that serves as a comparison point throughout the experiment) using a numeric scale of their choice. In one version of this task, participants carry out judgements in a visual modality rather than using numbers, e.g. by indicating the degree of acceptability of an item relative to the modulus using the length of a line. Bard *et al.* [3] have shown that participants are remarkably self-consistent, with high correlations obtained between grammaticality judgements in the numeric and visual modalities. Similarly, Gatt and van Deemter [36] asked participants to rate the plausibility of plural referring expressions by comparing them to a modulus using both numeric scales and sliders. This experiment, which focused on plausibility rather than grammaticality, also found very high degrees of self-consistency in the ratings given by participants.

Task effectiveness and comprehension: All the TUNA STECs involved a task-based (extrinsic) evaluation which was based on the task definition typically associated with REG (Definition 1), that is, the idea that the output of REG algorithms aims to help a reader or hearer identify the intended referent in the relevant domain of discourse. The ability of algorithms to do this was tested in a series of experiments in which participants were exposed to a description (produced either by a system or by a human author), and were shown the visual domain. The task was to identify the intended referent based on the description, by clicking on the corresponding picture. The relevant dependent variables were the speed with which a referent was identified (identification time), and the proportion of correct identifications per system (identification accuracy). In addition, reading time was a relevant variable in one of the experiments. Over the course of the three editions, we experimented with different methods, which differed primarily in whether or not the presentation of the description was made at the same time as the presentation of the visual domain, and in whether the description was read or heard:

1. In the experiment carried out as part of TUNA'07, a trial consisted of a visual domain and a written description presented on screen in tandem. Thus, in this experiment, reading/comprehension time and identification time could not be distinguished. As a result, a follow-up experiment was performed after the Shared Task [7], which separated reading/comprehension and identification. This method, which was also used for TUNA'08, is described below. Since in this task submissions consisted of ATTRIBUTE-SETs, all submitted outputs were realised using a very simple template-based realiser (written by Irene Langkilde-Geary, Brighton University, for this purpose) which always realises each attribute in the same way and in the same position regardless of context, except that it disambiguates negated attributes by always realising them at the end of the description.[13]

2. In TUNA'08, as well as in the follow-up identification experiment for TUNA'07 referred to in (1) above, reading/comprehension and identification were separated, by splitting each trial so that a description was first presented and read by a participant, and the visual domain was subsequently presented once a button was pressed. The first phase of a trial was therefore a self-paced reading task, while the second was the identification task proper. This is why *ease of comprehension*, operationally defined in terms of reading time, emerged as a separate evaluation criterion in this experiment (see Table 4). In this edition, only systems submitted to the end-to-end task (that is, TUNA-REG) were included in the evaluation experiment, obviating the need to use a realiser to render attribute sets as strings.

3. In TUNA'09, reading was eliminated in favour of a paradigm in which a description was heard over a headset *while* a domain was visually presented. This method was adopted because it seemed to better approximate realistic comprehension settings; there are psycholinguistic findings to the effect that comprehension of spoken language descriptions and search/identification are tightly coupled, with identification occurring incrementally as a description is understood [74]. In the present case, we used a speech synthesiser to read system-produced and human-authored descriptions.[14]

4   Comparing Different Evaluation Methods

Having described the evaluation methods used and the criteria motivating them, we now turn to a comparison of the different classes of methods. In what follows, we identify points of correspondence and divergence between the various methods

---

[13] For example, the realiser generates *the person with glasses and no beard* rather than *the person with no beard and glasses*, which is potentially ambiguous.

[14] We used the University of Edinburgh's Festival speech generation system [13] in combination with the nitech_us_slt_arctic_hts voice, a high-quality female American voice.

Table 5. Pearson's r correlations between MINIMALITY and humanlikeness as measured by the Dice and MASI coefficients. $**$: significant at $p \leq .001$; $*$ $p \leq .05$.

Task	DICE	MASI
TUNA-AS 2007	-.90$**$	-.790$**$
TUNA-AS 2008	-.96$*$	-.90$*$

using Pearson's r product-moment correlations. All analyses reported below were carried out by comparing means of different systems on the relevant measures. All significance tests are two-tailed. In the following subsections, we first report the results and then discuss some of their implications in Section 4.4.

4.1 Minimality versus Humanlikeness

In the REG literature, the role of minimality (or 'brevity'; [23]) received prominence following the work of Dale and Reiter [21]. In that paper, various computational interpretations of the Gricean Maxim of Quantity were discussed. Echoing an earlier observation by Appelt [1], the authors showed that a literal interpretation of this maxim (along the lines of the MINIMALITY criterion used in TUNA-AS) led to an exponential running time for REG algorithms, since exhaustive search needs to be performed in order to identify the smallest possible set of attributes that identifies an entity. The authors proposed the Incremental Algorithm as a compromise, since it does not perform backtracking, performing a hillclimbing search along a predefined order of attributes, and halting once a referent has been distinguished. This algorithm does not guarantee that the solution returned is the briefest available, but the authors argued that this is in fact a psycholinguistically plausible outcome, since speakers appear to produce overspecified descriptions in referential tasks. In fact, psycholinguistic results to this effect have been consistently reported for some time [4,28,60]. Nevertheless, brevity has remained a central concern of many recent approaches to REG. For example, Gardent [32] has proposed a constraint-based approach to generating minimal plural descriptions, in order to avoid the excessively complex outputs that are produced by a generalisation of the Incremental Algorithm proposed by van Deemter to handle negation and disjunction [24].

Using the results from the 2007 and 2008 editions of the TUNA-AS task, we computed the correlation between the MINIMALITY scores (the proportion of minimal descriptions produced by a system) and the humanlikeness of the ATTRIBUTE-SETs, as reflected by the DICE and MASI coefficients.[15] For both tasks, there was a strong negative correlation between MINIMALITY and both humanlikeness scores, as shown in Table 5, indicating that systems which produced more minimal outputs were overall less likely to match corpus descriptions. Given the robust psycholinguistic findings showing that people overspecify descriptions, these results are relatively unsurprising.

---

[15] For the TUNA-AS'07 task, MASI was calculated after the Shared Task, as it was not originally included as an evaluation measure; see Table 3.

We also looked at the correlation between minimality and identification time in the TUNA-AS'07 data, using the results from the follow-up identification experiment reported in Gatt and Belz [33] and described in the previous section (see also Section 4.3 below). Minimality did not correlate significantly with Reading Time ($r = .18; p > .05$) or with Identification Accuracy, the proportion of identification errors made by experimental participants ($r = .51; p > .05$). However, we found a significant *positive* correlation between Identification Time and Minimality ($r = .56; p < .05$), suggesting that minimal descriptions slowed down identification. However, a caveat needs to be raised. Minimal descriptions in the present study typically consisted of a set of attributes such as COLOUR and SIZE, with no TYPE information, because TYPE was never required to distinguish an object in the TUNA domains. For the purposes of the experiment, such descriptions were rendered by the realiser using a default head noun to realise the TYPE attribute. Furthermore some attributes were realised using several words, others as single words. As a result, minimal attribute sets could conceivably have been realised as longer word strings than non-minimal ones.

4.2   Measures of Humanlikeness versus Task Effectiveness

The comparison of humanlikeness and task effectiveness in this section is based on results from all three editions of the TUNA STECs. We make the following comparisons:

1. For the TUNA-AS'07 task, we use the results of the *follow-up* identification experiment. Unlike the original experiment reported for that STEC, the follow-up utilised the same experimental paradigm as for TUNA'08, separating reading time and identification time, as well as computing identification accuracy (see Section 3). We report correlations for these measures with DICE and MASI scores on attribute sets. In addition, we also report correlations with the *string* metrics used in subsequent STECs, namely EDIT, BLEU and NIST. These were computed as part of our follow-up evaluation, using the strings generated by the template-based realiser described in Section 3.4.
2. For the 2008 and 2009 tasks, we report correlations between the extrinsic measures and string-based metrics only (i.e. not DICE or MASI). This is because only the TUNA-REG submissions were included in the extrinsic evaluation for the 2008 edition (to the exclusion of the TUNA-AS and TUNA-R systems), while the 2009 edition consisted only of the TUNA-REG task. In both cases, participants were not required to submit ATTRIBUTE-SETs (only WORD-STRINGs); hence, computation of DICE and MASI is not possible. Note that reading time is not available for TUNA'09, since descriptions were presented as auditory stimuli.

All correlations are displayed in Table 6.

The most striking feature of the results in Table 6 is that none of the automatically computed metrics that assess humanlikeness have a significant correlation with any of the task-performance measures. The one exception to this trend is

Table 6. Pearson's r correlations between intrinsic and extrinsic measures. RT: Reading Time; IT: Identification Time; IA: Identification Accuracy; *: significant at $p \leq 0.05$.

Measure	Task	RT	IT	IA
DICE	TUNA-AS (2007)	0.12	-0.28	-0.39
MASI	TUNA-AS (2007)	0.23	-0.17	-0.29
EDIT	TUNA-AS (2007)	-0.30	0.09	0.22
	TUNA-REG (2008)	0.1	0.09	0.22
	TUNA-REG (2009)	-	0.68	-0.01
BLEU	TUNA-AS (2007)	0.39	0.04	-0.08
	TUNA-REG (2008)	0.22	0.04	-0.37
	TUNA-REG (2009)	-	-0.51	0.49
NIST	TUNA-AS (2007)	0.54*	0.22	0.03
	TUNA-REG (2008)	0.54	0.27	-0.45
	TUNA-REG (2009)	-	0.06	0.60

the significant (though not very strong) positive correlation between NIST and reading time in the TUNA-AS'07 results. In general, there is no evidence of a relationship between humanlikeness as measured by the metrics used here, and task performance defined in terms of reading speed, identification speed and identification accuracy. On the other hand, separate analyses showed strong correlations *within* the two classes of measures, that is, systematic covariation is observed between different extrinsic measures on the one hand, and intrinsic ones on the other (for details, see the TUNA Shared Task reports [6,34,35]).

There have been some other publications reporting a similar lack of correlation between intrinsic and extrinsic evaluation results [27,57], as well as experimental studies showing divergent trends between human judgements and/or preferences and actual task performance [28,52]. However, our automatic intrinsic scores also do not correlate at all with the *human* intrinsic scores, and this finding is not supported by a set of similar results in the literature (quite the contrary; see next section). At the same time, the human intrinsic scores do correlate with extrinsic results very straightforwardly.

So we have a situation where (surprisingly) human and automatic intrinsic results do not correlate, and automatic intrinsic and extrinsic results do not correlate either, whereas human intrinsic and extrinsic results do correlate. If human and automatic intrinsic results did correlate (as they should), then (because human intrinsic and extrinsic correlate) one might also find a correlation between automatic intrinsic and extrinsic results.

We return to this set of interrelated issues in the next section, where we discuss a possible explanation.

4.3 Human Intrinsic Measures versus Automatic Intrinsic and Human Extrinsic Measures

Finally, we turn to the relationship between judged appropriateness, on the one hand, and extrinsic and automatically computed intrinsic measures, on the other. In this case, we focus exclusively on the results from TUNA-REG'09, since this is the only task for which judged appropriateness ratings (FLUENCY and ADEQUACY) were collected. The correlations between the judgements and the extrinsic and automatic intrinsic measures are displayed in Table 7.

Table 7. Correlations between judged appropriateness and extrinsic measures. IT: Identification Time; IA: Identification Accuracy; ** : $p \leq .001$; * : $p \leq .05$.

Measure	IT	IA	EDIT	BLEU	NIST
FLUENCY	-0.89*	0.50	-0.57	0.66	0.30
ADEQUACY	-0.65	0.95**	-0.29	0.60	0.48

The table shows no significant correlation between any of the automatically computed intrinsic measures and the two kinds of ratings obtained from participants However, some of the correlations between the ratings and the extrinsic measures are significant. In particular, there is a strong negative correlation between FLUENCY and identification time, suggesting that descriptions judged as highly fluent led to faster identification. On the other hand, there was no significant correlation between FLUENCY and identification accuracy. The relationship between ADEQUACY and identification time goes in the same direction as that between FLUENCY and identification time, but fails to reach significance. In contrast, the correlation of ADEQUACY with identification accuracy is very strong and highly significant. These correlations are exactly as one would expect.

In other NLP subfields – MT and summarisation in particular – judged appropriateness ratings of fluency and adequacy are used as a kind of gold standard against which to measure automatic metrics. In other words, the higher its correlation with the human ratings, the better an automatic metric is deemed to be. In those fields, very strong correlations with human judged appropriateness ratings are typically reported for BLEU (and its NIST variant). In NLG too, correlations above .8 have been reported for weather forecast texts [10]. Closest to the present task, the GREC evaluations have also revealed strong and highly significant correlations between automatic intrinsic metrics and fluency ratings in particular, for both tasks involved [9].

So how are we to interpret the complete lack of significant correlations between our intrinsic automatic and intrinsic human-assessed scores? Is the TUNA Task so different not just from MT, SUMMARISATION and data-to-text generation, but even from the GREC Tasks (which are also REG tasks), that the same metrics that usually achieve high correlations with human judgements for these tasks simply do not work for referring expression generation from attribute-based representations of sets of entities?

As this seems an unconvincing proposition, we need to look for further explanation. A methodological issue that might have affected the results reported in this section concerns the nature of the reference data against which automatic scores are computed. The TUNA Corpus was constructed through an experiment which, being web-based, was relatively unrestricted. The corpus contains a high degree of variation among individual authors, with some adopting a highly telegraphic style, while others produced full NPs of the sort that are typically exemplified in the REG literature.[16] Some variation may also be due to the fact that some of the people who wrote the REs in the corpus were not native speakers.

We have some insight into the perceived quality of these descriptions, as some of them were included in our TUNA'09 judged appropriateness experiments. In these experiments we asked participants to evaluate the competing systems' outputs for the test data inputs, so the human-produced descriptions we included in the experiments were the corresponding descriptions in the corpus (the test data reference 'outputs'). Because we had two descriptions for each input in the corpus, we randomly split them into two groups of single outputs, HUMAN-1 and HUMAN-2. Despite the fact that the TUNA'08/09 test set was obtained in a more controlled setting than the original TUNA data (see Section 3.2 above), HUMAN-1 and HUMAN-2 were never rated top (and always beaten by the two top ranking systems), neither for the furniture nor for the people subdomain, neither for Fluency nor Adequacy. In fact, overall, HUMAN-1 and HUMAN-2 ranked fourth and sixth (out of 8) for Adequacy, and third and fourth for Fluency.

From this it is clear that systems were able to produce better quality outputs, at least as perceived by our assessors, than the data they were trained on. This was possible, because most systems used information other than frequency of occurrence in the training data; some systems hardwired defaults such as always including the type (e.g. *chair*) of an entity in a description; and many of the low-quality aspects of descriptions in the training data, provided they were of sufficiently low frequency, would have simply 'come out in the statistical wash'. However, in computing the automatic intrinsic scores, system outputs are assessed in terms of their similarity with the original corpus data (which our human judges considered less than ideal, certainly worse than some systems).

If the test data reference descriptions had been such that the human assessors rated them highly (higher than the system outputs at least), then it is reasonable to expect that scores computing string similarity with those reference descriptions would have also had some correlation with the human assessors' scores. Given that there is good correlation between the assessors' scores and extrinsic scores, one might then also find some correlation between the extrinsic scores and the automatic intrinsic ones.

---

[16] Some of the more idiosyncratic examples from the test data include the following: "a red chair, if you sit on it, your feet would show the south east", "male dark hair grey beard and black rimmed glasses", "left hand photo on 2nd row", "the dark haired, younger looking aged gentleman with spectacles", "the picture of the person on the far right in the top row is surrounded by a red border", "top right".

While the quality of the TUNA data may offer a *plausible* explanation, we have at present no way of confirming that this is the *right* explanation. The way to test it would be to redo the experiments with a test set that the human assessors judge to be of high quality. Here, however, there are two issues to be considered. First, even if such an experiment revealed a better correlation between automatic and human intrinsic scores, it may of course have no effect on the lack of correlation between automatic intrinsic and extrinsic scores. Second, it is an open question whether such a test set could be produced for a task such as REG, without introducing artificially stringent instructions to participants aiming to minimise certain types of variation (for example, instructions to the effect that only full definite NPs can be produced). Individual variation is a well-known problem in the use of corpora texts as reference outputs for NLG. This has been discussed in the context of REG by Viethen and Dale [77], and also by Reiter and Sripada [66] in the context of a different task. As we suggested above, some of the variation observed in our test data may be due to the fact that the authors represented in the corpus were self-reported native *or* fluent speakers of English, at least when the original corpus was constructed. However, it seems unlikely that this is the sole explanation, since the native speakers were in the majority, and the test sets constructed for TUNA'08 and TUNA'09 consisted exclusively of data from native speakers. Moreover, variation also seems to arise in more controlled studies of reference. For example, a recent study which replicated the TUNA experiment with Dutch speakers, under controlled laboratory conditions, found substantial variation among authors as in the original TUNA corpus [48]. It is quite possible that the kind of telegraphic style we observed among some of the authors in the TUNA corpus is simply due to the nature of the task which, unlike tasks involving the production of longer sentences or texts, can be carried out (and indeed is often carried out in normal everyday speech) using relatively 'degraded' forms without necessarily compromising the success of the main communicative goal, namely, to identify an object.

4.4 Summary and Implications

The three sets of results reported above can be summarised as follows.

The role of minimality: While minimality or brevity emerged as the main theoretically motivated measure of quality of a referring expression in the traditional REG literature, our findings suggest that it negatively impacts the humanlikeness of a referring expression, as reflected in significant negative correlations. In spite of a long tradition of psycholinguistic research showing that human speakers tend to overspecify when referring, our results remain topical for REG, where appeals to brevity have continued to be made until recently (compare, for example, the psycholinguistically-motivated arguments given by Dale and Reiter in favour of overspecification in the Incremental Algorithm [21] with the argument for Brevity offered by Gardent [32] for the generation of plurals, where minimality is adopted in order to avoid excessive complexity in the logical forms generated by a content determination algorithm).

While brevity no doubt has *some* role to play (one would not argue, for example, that *all* possible attributes should be included in a description by default), looking at human-produced descriptions shows that they do not tend to opt for the absolute minimal number of attributes required to identify a referent. Our results furthermore give some preliminary indication that descriptions that are minimal in this sense may actually delay identification (Section 4.1).

Automatic measures of humanlikeness: Concerns with the validity of automatically computed metrics have been voiced in other fields, notably Summarisation [27] and Machine Translation [20]. Alongside several existing reports of discrepancies between intrinsic and extrinsic evaluation measures, our results raise the possibility that automatic, corpus-based metrics of humanlikeness focus on very different aspects of the quality of human referring expressions than those tapped into by extrinsic, task-based measures. Nevertheless, we have argued that the results reported here need to be interpreted with caution and require further follow-up research to avoid some potential methodological pitfalls.

Human intrinsic measures of quality: The relationship between human intrinsic measures of quality (in the form of judged appropriateness) and extrinsic measures in our 2009 results appears to be more straightforwardly interpretable. There therefore appears to be an interesting asymmetry between the two classes of intrinsic measures considered in this chapter, namely corpus-based/automatic and human. We have discussed some possible reasons for this, foremost among which is the fact that our reference material (the human-authored referring expressions in the test data part of the TUNA corpus) are not regarded as high quality by human judges. Another possible explanation is that there is an underlying asymmetry in the standards of evaluation adopted in the two classes of intrinsic measures. Corpus-based, automatic measures compare outputs to what people do (i.e. to human production), whereas in a judgement experiment, participants are asked to tap into their own individual subjective views of what makes a referring expression good or bad.

Because of the methodological caveats we have noted with some aspects of our study, a replication of the studies reported here involving a larger sample and a new dataset produced in a more controlled setting seems particularly worthwhile, given that the question of the relationship between different classes of methods is crucial to a better understanding of evaluation methodology in NLG.

5   Conclusion

This chapter has given an overview of the structure and scope of the three editions of the TUNA Shared Task Evaluation Challenges. We have argued that such exercises in comparative evaluation can be highly beneficial, not least because a concentrated effort on the part of several research teams on a single problem can enhance progress and the range of approaches to the problem can be broadened as a result. In the case of Referring Expression Generation, we believe that these events have helped to broaden the field by including realisation in addition to the more familiar attribute selection task and also through the variety of algorithmic approaches adopted by the various participants, including many trainable systems (new to REG).

These events have yielded a rich source of comparative evaluation data based on a large number of different evaluation measures. Such data can function both as a baseline against which to compare future systems, and as a repository to be exploited for further follow-up investigation. We have presented results from some of our follow-up investigations [7,33] in this chapter, where we focused on the relationship between different kinds of evaluation criteria. Our results suggest that different classes of methods give different results whose relationship is not always predictable or transparent. This is especially true of the relationship between metrics computed against corpora and measures obtained through task-based experiments. Unlike corpus-based automatic metrics, human judgements constitute a class of intrinsic measures which do evince a systematic relationship with extrinsic ones. In future work, we are planning to follow up these studies further, using a new corpus and experimental data.

Acknowledgments. We are grateful to all the participants in the three editions of the TUNA Challenges, whose enthusiasm and input made the work reported here possible. Thanks to Ehud Reiter and Jette Viethen, who co-organised the first event, and to Irene Langkilde-Geary for her work on the template-based realiser used in TUNA-AS'07. The participants of our extrinsic evaluation experiments were drawn from the staff and students of UCL, Sussex and Brighton universities. We are also grateful to Emiel Krahmer for helpful comments on earlier drafts of this paper. Referring Expression Generation Challenges 2008 and Generation Challenges 2009 were organised with the financial support of the UK Engineering and Physical Sciences Research Council (EPSRC).

References

1. Appelt, D.: Planning English referring expressions. Artificial Intelligence 26(1), 1–33 (1985)
2. Appelt, D., Kronfeld, A.: A computational model of referring. In: Proceedings of the 10th International Joint Conference on Artificial Intelligence (IJCAI 1987), pp. 640–647 (1987)
3. Bard, E.G., Robertson, D., Sorace, A.: Magnitude estimation of linguistic acceptability. Language 72(1), 32–68 (1996)
4. Belke, E., Meyer, A.: Tracking the time course of multidimensional stimulus discrimination: Analysis of viewing patterns and processing times during same-different decisions. European Journal of Cognitive Psychology 14(2), 237–266 (2002)
5. Belz, A.: Statistical generation: Three methods compared and evaluated. In: Proceedings of the 10th European Workshop on Natural Language Generation (ENLG 2005), pp. 15–23 (2005)
6. Belz, A., Gatt, A.: The attribute selection for GRE challenge: Overview and evaluation results. In: Proceedings of UCNLG+MT: Language Generation and Machine Translation, pp. 75–83 (2007)

7. Belz, A., Gatt, A.: Intrinsic vs. extrinsic evaluation measures for referring expression generation. In: Proceedings of the 46th Annual Meeting of the Association for Computational Linguistics (ACL 2008), pp. 197–200 (2008)
8. Belz, A., Kow, E.: System-building cost vs. output quality in data-to-text generation. In: Proceedings of the 12th European Workshop on Natural Language Generation (ENLG 2009), pp. 16–24 (2009)
9. Belz, A., Kow, E., Viethen, J., Gatt, A.: Generating referring expressions in context: The GREC task evaluation challenges. In: Krahmer, E., Theune, M. (eds.) Empirical Methods in NLG. LNCS (LNAI), vol. 5790, pp. 294–328. Springer, Heidelberg (2010)
10. Belz, A., Reiter, E.: Comparing automatic and human evaluation of NLG systems. In: Proceedings of the 11th Conference of the European Chapter of the Association for Computational Linguistics (EACL 2006), pp. 313–320 (2006)
11. Belz, A.: That's nice.. what can you do with it? Computational Linguistics 35(1), 111–118 (2009)
12. Belz, A., Kilgarriff, A.: Shared task evaluations in HLT: Lessons for NLG. In: Proceedings of INLG 2006, pp. 133–135 (2006)
13. Black, A., Taylor, P., Caley, R.: The Festival speech synthesis system: System documentation. Tech. Rep. 1.4 edition., University of Edinburgh (1999)
14. Bohnet, B.: IS-FBN, IS-FBS, IS-IAC: The adaptation of two classic algorithms for the generation of referring expressions in order to produce expressions like humans do. In: Proceedings of UCNLG+MT: Language Generation and Machine Translation, pp. 84–86 (2007)
15. Bohnet, B.: The fingerprint of human referring expressions and their surface realization with graph transducers. In: Proceedings of the 5th International Conference on Natural Language Generation (INLG 2008), pp. 207–210 (2008)
16. Bohnet, B., Dale, R.: Viewing referring expression generation as search. In: Proceedings of the 19th International Joint Conference on Artificial Intelligence (IJCAI 2005), pp. 1004–1009 (2005)
17. Cahill, A., van Genabith, J.: Robust pcfg-based generation using automatically acquired lfg approximations. In: Proceedings of the 21st International Conference on Computational Linguistics and 44th Annual Meeting of the Association for Computational Linguistics (COLING/ACL 2006), pp. 1033–1040 (2006)
18. Callaway, C.B.: Evaluating coverage for large symbolic NLG grammars. In: Proceedings of the 18th International Joint Conference on Artificial Intelligence (IJCAI 2003), pp. 811–817 (2003)
19. Callaway, C.B., Lester, J.C.: Narrative prose generation. Artificial Intelligence 139(2), 213–252 (2002)
20. Calliston-Burch, C., Osborne, M., Koehn, P.: Re-evaluating the role of BLEU in machine translation research. In: Proceedings of the 11th Conference of the European Chapter of the Association for Computational Linguistics (EACL 2006), pp. 249–256 (2006)
21. Dale, R., Reiter, E.: Computational interpretation of the Gricean maxims in the generation of referring expressions. Cognitive Science 19(8), 233–263 (1995)
22. Dale, R., White, M. (eds.): Shared Tasks and Comparative Evaluation in Natural Language Generation: Workshop Report (2007), http://www.ling.ohio-state.edu/nlgeval07/NLGEval07-Report.pdf
23. Dale, R.: Cooking up referring expressions. In: Proceedings of the 27th Annual Meeting of the Association for Computational Linguistics, ACL 1989, pp. 68–75 (1989)

24. van Deemter, K.: Generating referring expressions: Boolean extensions of the incremental algorithm. Computational Linguistics 28(1), 37–52 (2002)
25. van Deemter, K., Gatt, A.: Content determination in GRE: Evaluating the evaluator. In: Proceedings of the 2nd UCNLG Workshop: Language Generation and Machine Translation (2007)
26. Doddington, G.: Automatic evaluation of machine translation quality using n-gram co-occurrence statistics. In: Proceedings of the 2nd International Conference on Human Language Technology Research (HLT 2002), pp. 138–145 (2002)
27. Dorr, B.J., Monz, C., President, S., Schwartz, R., Zajic, D.: A methodology for extrinsic evaluation of text summarization: Does ROUGE correlate? In: Proceedings of the ACL Workshop on Intrinsic and Extrinsic Evaluation Measures for Machine Translation and/or Summarisation, pp. 1–8 (2005)
28. Engelhardt, P.E., Bailey, K., Ferreira, F.: Do speakers and listeners observe the Gricean Maxim of Quantity? Journal of Memory and Language 54, 554–573 (2006)
29. Fabbrizio, G.D., Stent, A.J., Bangalore, S.: Referring expression generation using speaker-based attribute selection and trainable realization (ATT-REG). In: Proceedings of the 5th International Conference on Natural Language Generation (INLG 2008), pp. 211–214 (2008)
30. Fabbrizio, G.D., Stent, A.J., Bangalore, S.: Trainable speaker-based referring expression generation. In: Proceedings of the 12th Conference on Computational Natural Language Learning (CONLL 2008), pp. 151–158 (2008)
31. Foster, M.: Automated metrics that agree with human judgements on generated output for an embodied conversational agent. In: Proceedings of the 5th International Conference on Natural Language Generation (INLG 2008), pp. 95–103 (2008)
32. Gardent, C.: Generating minimal definite descriptions. In: Proceedings of the 40th Annual Meeting of the Association for Computational Linguistics (ACL 2002), pp. 96–103 (2002)
33. Gatt, A., Belz, A.: Attribute selection for referring expression generation: New algorithms and evaluation methods. In: Proceedings of the 5th International Conference on Natural Language Generation (INLG 2008), pp. 50–58 (2008)
34. Gatt, A., Belz, A., Kow, E.: The TUNA Challenge 2008: Overview and evaluation results. In: Proceedings of the 5th International Conference on Natural Language Generation (INLG 2008), pp. 198–206 (2008)
35. Gatt, A., Belz, A., Kow, E.: The TUNA-REG Challenge 2009: Overview and evaluation results. In: Proceedings of the 12th European Workshop on Natural Language Generation (ENLG 2009), pp. 198–206 (2009)
36. Gatt, A., van Deemter, K.: Lexical choice and conceptual perspective in the generation of plural referring expressions. Journal of Logic, Language and Information 16(4), 423–443 (2007)
37. Gatt, A., van der Sluis, I., van Deemter, K.: Evaluating algorithms for the generation of referring expressions using a balanced corpus. In: Proceedings of the 11th European Workshop on Natural Language Generation (ENLG 2007), pp. 45–56 (2007)
38. Grice, H.: Logic and conversation. In: Cole, P., Morgan, J. (eds.) Syntax and Semantics: Speech Acts, vol. III. Academic Press, London (1975)
39. Gupta, S., Stent, A.J.: Automatic evaluation of referring expression generation using corpora. In: Proceedings of the 1st Workshop on Using Corpora in NLG (UCNLG 2005), pp. 1–6 (2005)

40. Hervás, R., Gervás, P.: Evolutionary and case-based approaches to REG. In: Proceedings of the 12th European Workshop on Natural Language Generation (ENLG 2009), pp. 187–188 (2009)
41. Jordan, P., Walker, M.: Learning attribute selections for non-pronominal expressions. In: Proceedings of the 38th Annual Meeting of the Association for Computational Linguistics (2000)
42. Jordan, P.W.: Can nominal expressions achieve multiple goals? In: Proceedings of the 38th Annual Meeting of the Association for Computational Linguistics (ACL 2000), pp. 142–149 (2000)
43. Jordan, P.W., Walker, M.: Learning content selection rules for generating object descriptions in dialogue. Journal of Artificial Intelligence Research 24, 157–194 (2005)
44. Karasimos, A., Isard, A.: Multi-lingual evaluation of a natural language generation system. In: Proceedings of the 4th International Conference on Language Resources and Evaluation, LREC 2004 (2004)
45. Kelleher, J., Namee, B.M.: Referring expression generation challenge 2008 DIT system descriptions. In: Proceedings of the 5th International Conference on Natural Langauge Generation (INLG 2008), pp. 221–224 (2008)
46. King, J.: OSU-GP: Attribute selection using genetic programming. In: Proceedings of the 5th International Conference on Natural Language Generation (INLG 2008), pp. 225–226 (2008)
47. Koller, A., Striegnitz, K., Byron, D., Cassell, J., Dale, R., Moore, J., Oberlander, J.: The first challenge on generating instructions in virtual environments. In: Krahmer, E., Theune, M. (eds.) Empirical Methods in NLG. LNCS (LNAI), vol. 5790, pp. 329–353. Springer, Heidelberg (2010)
48. Koolen, R., Gatt, A., Goudbeek, M., Krahmer, E.: Need I say more? On factors causing referential overspecification. In: Proceedings of the Workshop on Production of Referring Expressions: Bridging Computational and Psycholinguistic Approaches (PRE-COGSCI 2009) (2009)
49. Krahmer, E., van Erk, S., Verleg, A.: Graph-based generation of referring expressions. Computational Linguistics 29(1), 53–72 (2003)
50. Kronfeld, A.: Conversationally relevant descriptions. In: Proceedings of the 27th Annual Meeting of the Association for Computational Linguistics (ACL 1989), pp. 60–67 (1989)
51. Langkilde-Geary, I.: An empirical verification of coverage and correctness for a general-purpose sentence generator. In: Proceedings of the 2nd International Conference on Natural Language Generation, INLG 2002 (2002)
52. Law, A.S., Freer, Y., Hunter, J., Logie, R., McIntosh, N., Quinn, J.: A comparison of graphical and textual presentations of time series data to support medical decision making in the neonatal intensive care unit. Journal of Clinical Monitoring and Computing 19, 183–194 (2005)
53. Lester, J., Porter, B.: Developing and empirically evaluating robust explanation generators: The KNIGHT experiments. Computational Linguistics 23(1), 65–101 (1997)
54. Lin, C.Y., Hovy, E.: Automatic evaluation of summaries using n-gram co-occurrence statistics. In: Proceedings of HLT-NAACL 2003, pp. 71–78 (2003)
55. de Lucena, D., Paraboni, I.: USP-EACH frequency-based greedy attribute selection for referring expressions generation. In: Proceedings of the 5th International Conference on Natural Language Generation (INLG 2008), pp. 219–220 (2008)
56. Maes, A., Arts, A., Noordman, L.: Reference management in instructive discourse. Discourse Processes 37(2), 117–144 (2004)

57. Miyao, Y., Saetre, R., Sagae, K., Matsuzaki, T., Tsujii, J.: Task-oriented evaluations of syntactic parsers and their representations. In: Proceedings of the 46th Annual Meeting of the Association for Computational Linguistics (ACL 2008), pp. 46–54 (2008)

58. Papineni, S., Roukos, T., Ward, W., Zhu., W.: BLEU: A method for automatic evaluation of machine translation. In: Proceedings of the 40th Annual Meeting of the Association for Computational Linguistics (ACL 2002), pp. 311–318 (2002)

59. Passonneau, R.: Measuring agreement on set-valued items (MASI) for semantic and pragmatic annotation. In: Proceedings of the 5th International Conference on Language Resources and Evaluation, LREC 2006 (2006)

60. Pechmann, T.: Incremental speech production and referential overspecification. Linguistics 27, 89–110 (1989)

61. Pereira, D.B., Paraboni, I.: From TUNA attribute sets to Portuguese text: A first report. In: Proceedings of the 5th International Conference on Natural Language Generation (INLG 2008), pp. 232–234 (2008)

62. Portet, F., Reiter, E., Gatt, A., Hunter, J., Sripada, S., Freer, Y., Sykes, C.: Automatic generation of textual summaries from neonatal intensive care data. Artificial Intelligence 173(7–8), 789–816 (2009)

63. Reips, U.D.: The Web Experimental Psychology Lab: Five years of data collection on the Internet. Behavioral Research Methods and Computers 33(2), 201–211 (2001)

64. Reiter, E., Belz, A.: An investigation into the validity of some metrics for automatically evaluating Natural Language Generation systems. Computational Linguistics 35(4), 529–558 (2009)

65. Reiter, E., Robertson, R., Osman, L.: Lessons from a failure: Generating tailored smoking cessation letters. Artificial Intelligence 144, 41–58 (2003)

66. Reiter, E., Sripada, S.: Should corpora texts be gold standards for NLG? In: Proceedings of the 2nd International Conference on Natural Language Generation, INLG 2002 (2002)

67. Reiter, E., Sripada, S., Hunter, J., Yu, J., Davy, I.: Choosing words in computer-generated weather forecasts. Artificial Intelligence 167, 137–169 (2005)

68. van der Sluis, I., Gatt, A., van Deemter, K.: Evaluating algorithms for the generation of referring expressions: Going beyond toy domains. In: Proceedings of the Conference on Recent Advances in Natural Language Processing, RANLP 2007 (2007)

69. van der Sluis, I., Krahmer, E.: Generating multimodal referring expressions. Discourse Processes 44(3), 145–174 (2007)

70. Spanger, P., Kurosawa, T., Tokunaga, T.: TITCH: Attribute selection based on discrimination power and frequency. In: Proceedings of UCNLG+MT: Language Generation and Machine Translation, pp. 98–100 (2008)

71. Spärck Jones, K., Galliers, J.R.: Evaluating natural language processing systems: An analysis and review. Springer, Berlin (1996)

72. Stock, O., Zancanaro, M., Busetta, P., Callaway, C., Krueger, A., Kruppa, M., Kuflik, T., Not, E., Rocchi, C.: Adaptive, intelligent presentation of information for the museum visitor in PEACH. User Modeling and User-Adapted Interaction 17(3), 257–304 (2007)

73. von Stutterheim, C., Mangold-Allwinn, R., Barattelli, S., Kohlmann, U., Kölbing, H.G.: Reference to objects in text production. Belgian Journal of Linguistics 8, 99–125 (1993)

74. Tanenhaus, M.K., Spivey-Knowlton, M.J., Eberhard, K.M., Sedivy, J.G.: Integration of visual and linguistic information in spoken language comprehension. Science 268, 1632–1634 (1995)
75. Theune, M., Touset, P., Viethen, J., Krahmer, E.: Cost-based attribute selection for generating referring expressions (GRAPH-FP and GRAPH-SC). In: Proceedings of UCNLG+MT: Language Generation and Machine Translation, pp. 95–97 (2007)
76. Viethen, J., Dale, R.: Algorithms for generating referring expressions: Do they do what people do? In: Proceedings of the 4th International Conference on Natural Language Generation (INLG 2006), pp. 63–70 (2006)
77. Viethen, J., Dale, R.: Evaluation in natural language generation: Lessons from referring expression generation. Traitement Automatique des Langues 48(1), 141–160 (2007)
78. White, M., Rajkumar, R., Martin, S.: Towards broad coverage surface realization with CCG. In: Proceedings of the Workshop on Using Corpora for NLG: Language Generation and Machine Translation, UCNLG+MT (2007)

Generating Referring Expressions in Context: The GREC Task Evaluation Challenges

Anja Belz[1], Eric Kow[1], Jette Viethen[2], and Albert Gatt[3,4]

[1] NLTG, School of Computing, Mathematical and Information Sciences,
University of Brighton, Brighton BN2 4GJ, UK
{a.s.belz,e.y.kow}@brighton.ac.uk
[2] Macquarie University, Sydney NSW 2109, Australia
jviethen@ics.mq.edu.au
[3] Institute of Linguistics, Centre for Communication Technology, University of Malta
albert.gatt@um.edu.mt
[4] Communication and Cognition, Faculty of Arts, Tilburg University,
The Netherlands

Abstract. Until recently, referring expression generation (REG) research focused on the task of selecting the semantic content of definite mentions of listener-familiar discourse entities. In the GREC research programme we have been interested in a version of the REG problem definition that is (i) grounded within discourse context, (ii) embedded within an application context, and (iii) informed by naturally occurring data. This paper provides an overview of our aims and motivations in this research programme, the data resources we have built, and the first three shared-task challenges, GREC-MSR'08, GREC-MSR'09 and GREC-NEG'09, we have run based on the data.

1 Background

Referring Expression Generation (REG) is one of the most lively and thriving subfields of Natural Language Generation (NLG). Traditionally, it has addressed the following question:

> [G]iven a symbol corresponding to an intended referent, how do we work out the semantic content of a referring expression that uniquely identifies the entity in question? [5, p. 1004]

Realisation, i.e. turning the resulting semantic content representation into a string of words, is not part of this problem specification, and REG has moreover predominantly considered the task in isolation—taking into account neither the discourse context (it was assumed that attributes were being selected for definite mentions of listener-familiar entities), nor the context of the language generation process (it was assumed that a REG module is called at some point in the generation process and that referent, potential distractors, and their possible attributes will be provided as parameters to the REG module).

In the 1990s, REG research looked at two main factors in selecting attributes (semantic content): unique identification (of the intended referent from a set

E. Krahmer, M. Theune (Eds.): Empirical Methods in NLG, LNAI 5790, pp. 294–327, 2010.

including possible distractors), and brevity [9,31]. The most influential of these algorithms, the Incremental Algorithm (IA) [10], originally just selected attributes for a single entity from a given set, but a range of extensions have been reported, including van Deemter's SET algorithm which can generate REs to sets of entities [11], and Siddharthan and Copestake's algorithm [32] which is able to identify attributes that are particularly discriminating given the entities in the contrast set of distractor entities.

Work in the 2000s increasingly took into account that there is more to REG than attribute selection, identification and brevity. Krahmer and Theune [21] moved away from the simplifying assumption, made by Dale and Reiter among others, that the contextually specified set of salient destractors would be provided to the REG algorithm. Their context-sensitive version of the IA took context into account, replacing the requirement that the intended referent be the *only* entity that matches the RE, to the requirement that it be the *most salient* in a given context. Jordan [18] showed that REs used by people do not always follow the brevity principle: she found a large proportion of over-specified redescriptions in the Coconut corpus of dialogues and showed that some dialogue states and communicative goals make over-specified REs more likely. Viethen & Dale [37] pointed out that the question why people choose different REs in different contexts has not really been addressed:

> Not only do different people use different referring expressions for the same object, but the same person may use different expressions for the same object on different occasions. Although this may seem like a rather unsurprising observation, it has never, as far as we are aware, been taken into account in the development of any algorithm for generation of referring expressions. [37, p. 119]

Researchers also started looking at real REG data, e.g. Viethen & Dale [37] and Gatt et al. [12] collected corpora of referring expressions elicited by asking participants to describe entities in scenarios of furniture items and faces. Others have looked at REs in discourse context. Nenkova's thesis [26] looked at rewriting mentions of people in extractive summaries with the aim of improving them within their new context, focusing on first mentions. Belz & Varges [3] collected a corpus of Wikipedia articles in order to investigate the question of how writers select mentions of named entities (cities, countries, rivers, people, mountains) in discourse context.

Other resources exist within which REs have been annotated in some way. In the GNOME Corpus [28,29] different types of discourse and semantic information are annotated, including reference and semantic attributes. The corpus annotation was, for example, used to train a decision tree learner for NP modifier generation [7]. The RE annotations in the Coconut corpus represent information at the discourse level (reference and attributes used) and at the utterance level (information about dialogue state); the 400 REs with their annotations in the corpus were used to train a REG module [19]. Gupta and Stent [16] annotated both the Maptask and Coconut corpora with POS-tags, NP boundaries, referents and knowledge representations for each speaker which included values for different attributes for potential referents.

Choosing Main Subject Reference

Please read the text below carefully and fill in all gaps by clicking on the arrows and selecting a reference from the drop-down menu that will appear. There are no right or wrong answers.

In order to select an empty reference, please select the underscore character (_). You can select each option as many times as you wish (including not at all).

Isaac Newton

[⬦], FRS (4 January 1643 – 31 March 1727) [OS: 25 December 1642 – 20 March 1727] was an English physicist, mathematician, astronomer, alchemist, and natural philosopher, regarded by many as the greatest figure in the history of science. [⬦] treatise Philosophiae Naturalis Principia Mathematica, published in 1687, described universal gravitation and the three laws of motion, laying the groundwork f [Isaac Newton's Newton's] cs. By deriving Kepler's laws of planetary motion from this system, [⬦] was the first to show that the motion of objects on Earth an [Sir Isaac Newton's] are governed by the same set of natural laws. The unifying and deterministic power of [⬦] laws was integral to the scientific revolution and the ad [His Whose] ntrism.

In mechanics, [⬦] also notably enunciated the principles of con———————————tum and angular momentum. In optics, [⬦] invented the reflecting telescope and [⬦] discovered that the spectrum of colours observed when white light passes through a prism is inherent in the white light and not added by the prism (as Roger Bacon had claimed in the thirteenth century). [⬦] notably argued that light is composed of particles. [⬦] also formulated an empirical law of cooling, [⬦] studied the speed of sound, and [⬦] proposed a theory of the origin of stars. In mathematics, [⬦] shares the credit with Gottfried Leibniz for the development of calculus. [⬦] also demonstrated the generalized binomial theorem, [⬦] developed the so-called "Newton's method" for approximating the zeroes of a function, and [⬦] contributed to the study of power series.

French mathematician Joseph-Louis Lagrange often said that [⬦] was the greatest genius who ever lived, and once added that [⬦] was also "the most fortunate, for we cannot find more than once a system of the world to establish." English poet Alexander Pope was moved by [⬦] accomplishments to write the famous epitaph:

Nature and nature's laws lay hid in night; God said "Let Newton be" and all was light.

(Save text)

Fig. 1. Screenshot of experiment in which participants performed the GREC tasks manually

2 Overview of GREC Research Programme

Extending the existing body of REG work that has started taking discourse context and real data into account, and building on earlier work [3], in the GREC research programme (Generating Referring Expressions in Context) we have been interested in a version of the REG problem that is (i) grounded within discourse context, (ii) embedded within an application context, and (iii) informed by naturally occurring data. This paper provides an overview of our aims and motivations in this research programme, the data resources we have built (the GREC-2.0 corpus and the GREC-People corpus), and the first three shared-task challenges we have run based on these data resources (GREC-MSR'08, GREC-MSR'09 and GREC-NEG'09.

In the next two sections (Sections 3 and 4), we describe the two data resources and annotation schemes we have created for GREC. In Section 5 we outline the two GREC task definitions (GREC-MSR and GREC-NEG), and in Section 6 the evaluation procedures we have applied to the two tasks. In Section 7 we provide a brief overview of systems and results in the two GREC-MSR challenges we ran in 2008 and 2009, and in Section 8 of the GREC-NEG challenge which ran for the first time in 2009 and for the second time in 2010. In Section 9 we discuss some of the issues and outcomes of the GREC evaluations.

In the remainder of this section, we briefly introduce the common features of the two GREC tasks and summarise the differences between the annotation schemes we have developed for them.

2.1   The GREC Task in General Terms

In general terms, the GREC tasks are about how to generate appropriate references to an entity in the context of a piece of discourse longer than a sentence. Rather than requiring participants to generate referring expressions from scratch, the GREC-MSR and GREC-NEG tasks provide sets of possible referring expressions for selection. Figure 1 shows a human-readable version of the GREC task: all references to *Isaac Newton* have been deleted and lists of possible referring expressions are provided instead. The task is to select a sequence of referring expressions to insert into the gaps such that the resulting text is fluent and coherent.

The immediate motivating application context for the GREC Tasks is the improvement of referential clarity and coherence in extractive summaries and multiply edited texts (such as Wikipedia articles) by regenerating referring expressions contained in them. The motivating theoretical interest for the GREC Tasks is to discover what kind of information is useful for making choices between different kinds of referring expressions in context.

In both the GREC-2.0 and the GREC-People annotation schemes, a distinction is made between *reference* and *referential expression*. A reference is 'an instance of referring' which is unique, whereas a referential expression is a word string and each reference can be realised by many different referential expressions. In the GREC corpora, each time an entity is referred to, there is a single reference, but there may be several referring expressions corresponding to it: in the training/development data, there is a single RE for each reference (the one found in the corpus), and a set of possible alternative REs is also provided; in the test set, there are four REs for each reference (the one from the corpus and three additional ones selected by subjects in a manual selection experiment), as well as the list of alternative REs.

2.2   Summary of Differences between the Two GREC Datasets (GREC-People and GREC-2.0)

The GREC-MSR and GREC-NEG tasks used different datasets, namely the GREC-2.0 corpus and the GREC-People corpus, respectively. The main difference between these is that in GREC-2.0 only references to the main subject of the text (MSRs) have been annotated, whereas in GREC-People, references to more than one discourse entity have been annotated. Other differences are that in GREC-People, (i) we have corrected spelling errors, (ii) the annotations have been extended to plural references, attributive complements, appositive supplements and subjects of gerund-participials, (iii) integrated dependents are included within the annotation of an RE, and (iv) the SYNCAT attribute has been split into SYNCAT and SYNFUNC attributes, indicating (just) syntactic category of the RE and syntactic function, respectively. See below for explanations of all these terms.

3   The GREC-2.0 Corpus

The GREC-2.0 corpus consists of 1,941 introduction sections from Wikipedia articles in five different domains (cities, countries, rivers, people and mountains). The corpus texts have been annotated for three broad categories of references to the main subject[1] of each text, called Main Subject References (MSRs). These are categories which were relatively simple to identify and achieve high inter-annotator agreement on (complete agreement among four annotators in 86% of MSRs).

The GREC-2.0 corpus has been divided into training, development and test data for the purposes of the GREC-MSR Task. The number of texts in the three data sets and the five subdomains is as follows:

	All	Mountains	People	Countries	Cities	Rivers
Total	1941	932	442	251	243	73
Training	1658	791	373	216	213	65
Development	97	46	24	12	11	4
Test	183	92	45	23	19	4

3.1   Types of Referential Expressions Annotated

In terminology and view of grammar our approach to annotating REs relies heavily on Huddleston and Pullum's Cambridge Grammar of the English Language [17]. We have annotated three broad categories of referential expression (RE) in the GREC-2.0 corpus: (i) subject NPs, (ii) object NPs and (iii) genitive NPs including pronouns which function as subject-determiners within their matrix NP.

I **Subject NPs**: referring subject NPs, including pronouns and special cases of VP coordination where the same RE functions as the subject of the coordinated VPs (see Section 3.2), e.g:

   1. _He was proclaimed dictator for life._
   2. _Alexander Graham Bell (March 3, 1847 - August 2, 1922) was a Scottish scientist and inventor who emigrated to Canada._
   3. _Most Indian and Bangladeshi rivers bear female names, but this one has a rare male name._
   4. _"The Eagle" was born in Carman, Manitoba and __ grew up playing hockey._

II **Object NPs:** referring NPs that function as direct or indirect objects of VPs and prepositional phrases; e.g.:

   1. _These sediments later deposit in the slower lower reaches of the river._
   2. _People from the city of São Paulo are called paulistanos._
   3. _His biological finds led him to study the transmutation of species._

---

[1] An example of a main subject of a text in the cities domain is e.g. _London._

III Subject-determiner genitives: genitive NPs that function as subject-determiners[2] including genitive forms of pronouns. Note that this excludes genitives that are the subject of a gerund-participial[3]:

1. _Its_ estimated length is 4,909 km.
2. _The country's_ culture, heavily influenced by neighbours, is based on a unique form of Buddhism intertwined with local elements.
3. _Vatican City_ is a landlocked sovereign city-state _whose_ territory consists of a walled enclave within the city of Rome.

3.2 Comments on Some Aspects of Annotation

Some types of relative pronoun, those in supplementary relative clauses (as opposed to integrated relative clauses, see Huddleston and Pullum, 2002, p. 1058), are interpreted as anaphorically referential (I(2) and III(3) above). These differ from integrated relative clauses in that in supplementary relative clauses, the relative clause can be dropped without affecting the meaning of the clause containing it. From the point of view of generation, the meaning could be equally expressed in two independent sentences or in two clauses of which one is a supplementary relative clause. An example of the single-sentence construction is shown in (1) below, with the semantically equivalent two-sentence alternative shown in (2):

(1) _Hristo Stoichkov is a football manager and former striker who was a member of the Bulgaria national team that finished fourth at the 1994 FIFA World Cup._
(2) _Hristo Stoichkov is a football manager and former striker. He was a member of the Bulgaria national team that finished fourth at the 1994 FIFA World Cup._

The GREC-2.0 annotation scheme also includes 'non-realised' subject REs in a restricted set of cases of VP coordination where an RE is the subject of the coordinated VPs. Consider the following example, where the subclausal coordination in (3) is semantically equivalent to the clausal coordination in (4):

(3) _He stated the first version of the Law of conservation of mass, __ introduced the Metric system, and __ helped to reform chemical nomenclature._
(4) _He stated the first version of the Law of conservation of mass, he introduced the Metric system, and he helped to reform chemical nomenclature._

According to Huddleston and Pullum, utterances as in (3) can be thought of as a reduction of longer forms as in (4), even though the former are not syntactically derived by ellipsis from the latter p. 1280, and from the point of view of language analysis there is no need for an analysis involving a null anaphoric reference. The motivation for annotating the approximate place where the subject NP would be if it were realised (the gap-like underscores above) is that from a generation

---

[2] I.e. "they combine the function of determiner, marking the NP as definite, with that of complement (more specifically subject)." (Huddleston and Pullum (2002), p. 56)

[3] E.g. _His early career was marred by *his being involved in a variety of social and revolutionary causes._

perspective there is a choice to be made about whether to realise the subject NP in the second (and subsequent) coordinate(s) or not. Note that only cases of subclausal coordination at the level of VPs have been annotated in this way. Therefore these are all cases where *only* the subject NP is 'missing'.[4]

There are some items that could be construed as main subject reference which we decided not to include in teh GREC-2.0 annotation scheme. These include those that are, according to Huddleston and Pullum, true gaps and ellipses, adjective and noun modifiers, and implicit or anaphorically derivable references (other than those mentioned above). Some of these we added to the annotation in the separate GREC-People corpus (see Section 4 below). Furthermore, we did not annotate the following types of text elements at all: the main title of the article, titles of books, films, etc. mentioned in the article; citations from articles, books, films, etc.; names and titles of organisations and persons (except where they are used in their entirety to refer to the main subject).

3.3   XML Format

The XML format described below was intended only for the purpose of the GREC-MSR Task. While attempts have been made to make it linguistically plausible and generic, certain aspects of it have been determined solely by the requirements of the GREC-MSR Task.

Figure 2 shows one of the texts from the GREC-2.0 corpus. Each item in the corpus is an XML annotated text file, and is of document type TEXT.

A TEXT is composed of one TITLE followed by any number of PARAGRAPHs. A TITLE is just a string of characters. A PARAGRAPH is any combination of character strings and REF elements. The REF element indicates a reference, in the sense of 'an instance of referring' (as discussed above). A REF is composed of one REFEX element (the 'selected' referential expression for the given reference; in the training data texts it is just the referential expression found in the corpus) and one ALT-REFEX element which in turn is a list of REFEXs (alternative referential expressions obtained by other means, as explained in Section 4.3).

The attributes of the REF element are ID, a unique reference identifier taking integer values; SEMCAT, indicating the semantic category of the referent and ranging over city, country, river, person, and mountain; and SYNCAT, the syntactic category required of referential expressions for the referent in this context (values np-obj, np-subj, subj-det[5]).

The SYNCAT attribute does not so much indicate a property of reference, as a constraint on the referential expressions that can realise it in a given context. Because in the GREC-MSR Task the context is fully realised text, the constraint is on syntactic category.

---

[4] E.g. we would not annotate a non-realised RE in *She wrote books for children and books for adults.*

[5] These stand for NP in object position, NP in subject position and NP that is both a determiner and a subject, respectively. See Section 3.1 for explanation of these terms.

```
<?xml version="1.0" encoding="utf-8"?>
<!DOCTYPE TEXT SYSTEM "reg08-grec.dtd">
<TEXT ID="36">

<TITLE>Jean Baudrillard</TITLE>

<PARAGRAPH>
 <REF ID="36.1" SEMCAT="person" SYNCAT="np-subj">
  <REFEX REG08-TYPE="name" EMPHATIC="no" HEAD="nominal" CASE="plain">Jean Baudrillard</REFEX>
  <ALT-REFEX>
   <REFEX REG08-TYPE="name" EMPHATIC="no" HEAD="nominal" CASE="plain">Jean Baudrillard</REFEX>
   <REFEX REG08-TYPE="name" EMPHATIC="yes" HEAD="nominal" CASE="plain">Jean Baudrillard
    himself</REFEX>
   <REFEX REG08-TYPE="empty">_</REFEX>
   <REFEX REG08-TYPE="pronoun" EMPHATIC="no" HEAD="pronoun" CASE="nominative">he</REFEX>
   <REFEX REG08-TYPE="pronoun" EMPHATIC="yes" HEAD="pronoun" CASE="nominative">he
    himself</REFEX>
   <REFEX REG08-TYPE="pronoun" EMPHATIC="no" HEAD="rel-pron" CASE="nominative">who</REFEX>
   <REFEX REG08-TYPE="pronoun" EMPHATIC="yes" HEAD="rel-pron" CASE="nominative">who
    himself</REFEX>
  </ALT-REFEX>
 </REF>
 (born June 20, 1929) is a cultural theorist, philosopher, political commentator,
 sociologist, and photographer.
 <REF ID="36.2" SEMCAT="person" SYNCAT="subj-det">
  <REFEX REG08-TYPE="pronoun" EMPHATIC="no" HEAD="pronoun" CASE="genitive">His</REFEX>
  <ALT-REFEX>
   <REFEX REG08-TYPE="name" EMPHATIC="no" HEAD="nominal" CASE="genitive">Jean
    Baudrillard's</REFEX>
   <REFEX REG08-TYPE="pronoun" EMPHATIC="no" HEAD="pronoun" CASE="genitive">his</REFEX>
   <REFEX REG08-TYPE="pronoun" EMPHATIC="no" HEAD="rel-pron" CASE="genitive">whose</REFEX>
  </ALT-REFEX>
 </REF>
 work is frequently associated with postmodernism and post-structuralism.
</PARAGRAPH>

</TEXT>
```

Fig. 2. Example text from GREC-2.0 corpus

The distinction between reference and referential expression is useful, because a single reference can have multiple possible realisations, but also because the two have distinct properties. For example, a reference does not have lexical and syntactic properties, whereas referential expressions do; a distinction between reference and referential expression in the annotation scheme allows such properties to be annotated only where appropriate.

A REFEX element indicates a referential expression (a word string that can be used to refer to an entity). It has four attributes. HEAD is the category of the head of the RE (values: nominal, pronoun, rel-pron). CASE indicates the case of the head (values for pronouns: nominative, accusative, genitive; for nominals: plain, genitive). EMPHATIC is a Boolean attribute and indicates whether the RE is emphatic. In the GREC-2.0 corpus, the only type of RE that has this attribute is one which incorporates a reflexive pronoun used emphatically (e.g. *India itself*).

The REG08-TYPE attribute (values name, common, pronoun, empty) indicates basic RE type as required for the GREC-MSR task definition. The choice of types is motivated by the hypothesis that one of the most basic decisions to be taken in RE selection for named entities is whether to use an RE that includes a name, such as *Modern India* (the corresponding REG08-TYPE value is `name`); whether to go for a common-noun RE, i.e. with a category noun like *country* as the head (`common`); whether to pronominalise the RE (`pronoun`); or whether it can be left unrealised (`empty`).

Finally, an ALT-REFEX element is a list of REFEX elements (corresponding to different possible realisations).

4   The GREC-People Corpus

The GREC-People corpus (a separate corpus that has no overlap with GREC-2.0) consists of 1,000 annotated introduction sections from Wikipedia articles in the category People. Each text therefore has a person as the main subject. There are three subcategories: inventors, chefs and early music composers. For the purposes of the GREC-NEG competitions, the GREC-People corpus was divided into training, development and test data. The number of texts in the three data sets and the three subdomains are as follows:

	All	Inventors	Chefs	Composers
Total	1,000	307	306	387
Training	809	249	248	312
Development	91	28	28	35
Test	100	31	30	39

As in GREC-2.0, we have annotated mentions of people by marking up the word strings that function as referential expressions (REs) and annotating them with coreference information as well as syntactic and semantic features. Since the subject of each text is a person, there is at least one coreference chain in each text. The numbers of coreference chains (entities) in the 900 texts in the training and development sets are as follows (e.g. there are 38 texts with 5 person discourse entities):

x coref chains	1	2	3	4	5	6	7	8	9	10	11	12	13	14	15	16	17	18	19	20	21	22	23
in y texts	437	192	80	63	38	31	16	18	4	7	9	1	1	0	0	0	0	0	0	1	1	0	1

The texts vary greatly in length, from 13 words to 935, the average being 128.98 words.

4.1   Annotation of Referring Expressions in GREC-People

This section describes the different types of referring expression (RE) that we annotated in the GREC-People corpus. As in GREC-2.0, we relied on Huddleston and Pullum's work for terminology and view of syntax. The manual annotations were automatically checked and converted to the XML format described in Section 4.3 (which encodes slightly less information, as explained below).

In the example sentences in the following sections, (unbroken) underlines are used for REs that are an example of the specific type of RE they are intended to illustrate, whereas dashed underlines are used for other REs that are also annotated in the corpus. Coreference between REs is indicated by subscripts $i, j, \ldots$ immediately to the right of an underline (the scope of the coindexing variables is one sentence, i.e. an i in one example sentence does not represent the same entity as an i in another example sentence). Square brackets indicate supplements. The syntactic component relativised by a relative pronoun is indicated by vertical bars. Supplements and their anchors (in the case of appositive supplements), and relative clauses and the component they relativise (in the case of relative-clause supplements) are co-indexed by superscript $x, y, \ldots$. Dependents integrated in an RE are indicated by curly brackets. Both supplements and dependents are highlighted in boldface font where they specifically are being discussed. All terms are explained below.

In the XML format of the annotations, the beginning and end of a reference is indicated by <REF><REFEX>... </REFEX></REF> tags, and other properties discussed in the following sections (such as syntactic category etc.) are encoded as attributes on these tags (for full details see Section 4.3 below). For the GREC-NEG'09 Task we decided not to transfer the annotations of integrated dependents and relative clauses to the XML format. Such dependents are included within <REFEX>...</REFEX> annotations where appropriate, but without being marked up as separate constituents.

We distinguish the following syntactic categories and functions:

I Subject NPs: referring subject NPs, including pronouns and special cases of VP coordination where the same referring expression functions as the subject of the coordinated VPs. For example:

1. $\underline{He}_i$ *was born in Ramsay township, near Almonte, Ontario, Canada, the eldest son of* $|Scottish\ immigrants,\ \{John\ Naismith\ and\ Margaret\ Young\}|^x_{j,k}$ $[\underline{who}_{j,k}$ *had arrived in the area in 1851 and* $\underline{\ \ }_{j,k}$ *worked in the mining industry*$]^x$.

2. $\underline{The\ Ban\bar{u}\ M\bar{u}s\bar{a}\ brothers}_{i,j,k}$ *were three 9th century Persian scholars, of Baghdad, active in the House of Wisdom.*

Ia Subjects of gerund-participials

1. $\underline{His}_i$ *research on hearing and speech eventually culminated in* $\underline{Bell}_i$ *being awarded the first U.S. patent for the invention of the telephone in 1876.*

2. $\underline{Fessenden}_i$ *used the alternator-transmitter to send out a short program from Brant Rock, which included* $\underline{his}_i$ *playing the song O Holy Night on the violin* ... .

3. *Many of* $\underline{his}_i$ *scientific contemporaries disliked* $\underline{him}_i$, *due in part to* $\underline{his}_i$ *using the title Professor which technically* $\underline{he}_i$ *wasn't entitled to do.*

II Object NPs: referring NPs including pronouns that function as direct or indirect objects of VPs and prepositional phrases; e.g.:

1. $\underline{He}_i$ *entrusted* $\underline{them}_{j,k,l}$ *to* $\underline{Ishaq\ bin\ Ibrahim\ al\text{-}Mus'abi}^x_m$, $[a\ former\ governor\ of\ Baghdad]^x_m$.

2. $\underline{He}_i$ *was the son of* $|Nasiruddin\ Humayun|^x_j$ $[\underline{whom}_j\ \underline{he}_i$ *succeeded as ruler of the Mughal Empire from 1556 to 1605*$]^x$.

IIa Reflexive pronouns:
1. *He_i committed $himself_i$ to design and development of rocket systems.*
2. *$Smith_i$ called $himself_i$ the "Komikal Konjurer".*

III Subject-determiner genitives:
1. *$They_{i,j,k}$ shared the 1956 Nobel Prize in Physics for $their_{i,j,k}$ invention.*
2. *He_i is best known as |a pioneer of human-computer interaction|^x [$whose_i$ team developed hypertext, networked computers, and precursors to GUIs]^x.*
3. *On the eve of his_i death in 1605, the Mughal empire spanned almost 500 million acres (doubling during $Akbar's_i$ reign).*

Note that this category excludes cases where the term has become lexicalised, such as *the so-called "Newton's method"; Koch's postulates*, which we take not to contain an embedded reference to a person.

IIIa REs in composite nominals: this is the only type of RE we have annotated that is not an NP, but a nominal. This type functions as integrated attributive complement, e.g.:
1. *The company was sold to Westinghouse in 1920, and the next year its assets, including numerous important $Fessenden_i$ patents, were sold to the Radio Corporation of America, which also inherited the $Fessenden_i$ legal proceedings.*
2. *The $Eichengrün_i$ version was ignored by historians and chemists until 1999.*
3. *These flights demonstrated the controllability of the $Montgomery_i$ design ... .*

Note that this category excludes cases where the term has become lexicalised: *the Nobel Prizes; the Gatling gun; the Moog synthesizer.*

In contrast to GREC-2.0 we also annotated supplements and RE-internal dependents, as described in detail in the GREC-NEG'09 participants' pack.[6]

4.2   Further Explanation of Some Aspects of the Annotations

Nested references: As can be seen from some of the previous examples, we annotated all embedded references, e.g.:

1. *He_i was named after his_i maternal grandfather^x_j [Shaikh Ali Akbar $Jami$]^x_j.*
2. *born in The Hague as the son of $Constantijn$ $Huygens^x_i$, [a friend of René $Descartes_j$]^x_i*
3. *after European pioneers such as George $Cayley's_i$ $coachman_j$*

The maximum depth of embedding in the GREC-People corpus is 3.

Plural REs: We annotated all plural REs that refer to groups of people where the number of group members is known.

---

[6] The complete Participants' Packs can be downloaded here: http://www.itri.brighton.ac.uk/home/Anja.Belz.

Unnamed references and indefinites: We annotated all mentions of individual person entities even if they are not actually named anywhere in the text, including cases of both definite and indefinite references.

1. *The resolution's sponsor_i described it as ...*
2. *On 25 December 1990 he _i implemented the first successful communication between an HTTP client and server via the Internet with the help of Robert Cailliau_j and a {young} student staff {at CERN}_k.*

4.3 XML Annotation

Each item in the corpus is an XML annotated text file (an example is shown in Figure 3), which is of type GREC-ITEM. A GREC-ITEM consists of a TEXT element followed by an ALT-REFEX element. A TEXT has one attribute (an ID unique within the corpus), and is composed of one TITLE followed by any number of PARAGRAPHs. A TITLE is just a string of characters. A PARAGRAPH is any combination of character strings and REF elements.

The REF element is composed of one REFEX element. The attributes of the REF element are shown in Figure 4. ENTITY and MENTION together constitute a unique identifier for a reference within a text; together with the TEXT ID, they constitute a unique identifier for a reference within the entire corpus.

A REFEX element indicates a referential expression (a word string that can be used to refer to an entity), and has two attributes: REG08-TYPE is as defined for GREC-MSR (see Section 3); CASE indicates the case of the head (values for pronouns: nominative, accusative, genitive; for nominals: plain, genitive; for any 'empty' reference: nocase).

We allow arbitrary-depth embedding of references. This means that a REFEX element may have REF element(s) embedded in it. See below on embedding in REFEX elements contained in ALT-REFEX lists.

An ALT-REFEX element is a list of REFEX elements. For the GREC-NEG Task, these are obtained by collecting the set of all REFEXs that are in the text, and adding the following defaults: for each REFEX that is a named reference in the genitive form, add the corresponding plain REFEX; conversely, for each REFEX that is a named reference not in the genitive form, add the corresponding genitive REFEX; for each REFEX that is a named reference add pronoun REFEXs of the appropriate number and gender, in the nominative, genitive and accusative forms, a relative pronoun REFEX in the nominative, and an empty REFEX (i.e. one with REG08-TYPE="empty").[7]

REF elements that are embedded in REFEX elements contained in an ALT-REFEX list have an unspecified MENTION id (the '?' value). Furthermore, such REF elements have had their enclosed REFEX removed, i.e. they are 'empty'. For example:

```
<ALT-REFEX>
    ...
    <REFEX ENTITY="2" REG08-TYPE="common" CASE="plain">a friend of <REF ENTITY="1" MENTION="?"
        SEMCAT="person" SYNCAT="np" SYNFUNC="obj"></REF></REFEX>
    ...
</ALT-REFEX>
```

---

[7] Any resulting duplicates are removed.

```
<?xml version="1.0" encoding="utf-8"?>
<!DOCTYPE GREC-ITEM SYSTEM "genchal09-grec.dtd">
<GREC-ITEM>
<TEXT ID="15">
<TITLE>Alexander Fleming</TITLE>

<PARAGRAPH>
<REF ENTITY="0" MENTION="1" SEMCAT="person" SYNCAT="np" SYNFUNC="subj">
  <REFEX ENTITY="0" REGO8-TYPE="name" CASE="plain">Sir Alexander Fleming</REFEX>
</REF>
(6 August 1881 - 11 March 1955) was a Scottish biologist and pharmacologist.
<REF ENTITY="0" MENTION="2" SEMCAT="person" SYNCAT="np" SYNFUNC="subj">
  <REFEX ENTITY="0" REGO8-TYPE="name" CASE="plain">Fleming</REFEX>
</REF>
published many articles on bacteriology, immunology, and chemotherapy.
<REF ENTITY="0" MENTION="3" SEMCAT="person" SYNCAT="np" SYNFUNC="subj-det">
  <REFEX ENTITY="0" REGO8-TYPE="pronoun" CASE="genitive">his</REFEX>
</REF>
best-known achievements are the discovery of the enzyme lysozyme in 1922 and the discovery
of the antibiotic substance penicillin from the fungus Penicillium notatum in 1928, for which
<REF ENTITY="0" MENTION="4" SEMCAT="person" SYNCAT="np" SYNFUNC="subj">
  <REFEX ENTITY="0" REGO8-TYPE="pronoun" CASE="nominative">he</REFEX>
</REF>
shared the Nobel Prize in Physiology or Medicine in 1945 with
<REF ENTITY="1" MENTION="1" SEMCAT="person" SYNCAT="np" SYNFUNC="obj">
  <REFEX ENTITY="1" REGO8-TYPE="name" CASE="plain">Florey</REFEX>
</REF>
and
<REF ENTITY="2" MENTION="1" SEMCAT="person" SYNCAT="np" SYNFUNC="obj">
  <REFEX ENTITY="2" REGO8-TYPE="name" CASE="plain">Chain</REFEX>
</REF>
.</PARAGRAPH>
</TEXT>

<ALT-REFEX>
  <REFEX ENTITY="0" REGO8-TYPE="empty" CASE="no_case">_</REFEX>
  <REFEX ENTITY="0" REGO8-TYPE="name" CASE="genitive">Fleming's</REFEX>
  <REFEX ENTITY="0" REGO8-TYPE="name" CASE="genitive">Sir Alexander Fleming's</REFEX>
  <REFEX ENTITY="0" REGO8-TYPE="name" CASE="plain">Fleming</REFEX>
  <REFEX ENTITY="0" REGO8-TYPE="name" CASE="plain">Sir Alexander Fleming</REFEX>
  <REFEX ENTITY="0" REGO8-TYPE="pronoun" CASE="accusative">him</REFEX>
  <REFEX ENTITY="0" REGO8-TYPE="pronoun" CASE="genitive">his</REFEX>
  <REFEX ENTITY="0" REGO8-TYPE="pronoun" CASE="nominative">he</REFEX>
  <REFEX ENTITY="0" REGO8-TYPE="pronoun" CASE="nominative">who</REFEX>
  <REFEX ENTITY="1" REGO8-TYPE="empty" CASE="no_case">_</REFEX>
  <REFEX ENTITY="1" REGO8-TYPE="name" CASE="genitive">Florey's</REFEX>
  <REFEX ENTITY="1" REGO8-TYPE="name" CASE="plain">Florey</REFEX>
  <REFEX ENTITY="1" REGO8-TYPE="pronoun" CASE="accusative">him</REFEX>
  <REFEX ENTITY="1" REGO8-TYPE="pronoun" CASE="genitive">his</REFEX>
  <REFEX ENTITY="1" REGO8-TYPE="pronoun" CASE="nominative">he</REFEX>
  <REFEX ENTITY="1" REGO8-TYPE="pronoun" CASE="nominative">who</REFEX>
  <REFEX ENTITY="2" REGO8-TYPE="empty" CASE="no_case">_</REFEX>
  <REFEX ENTITY="2" REGO8-TYPE="name" CASE="genitive">Chain's</REFEX>
  <REFEX ENTITY="2" REGO8-TYPE="name" CASE="plain">Chain</REFEX>
  <REFEX ENTITY="2" REGO8-TYPE="pronoun" CASE="accusative">him</REFEX>
  <REFEX ENTITY="2" REGO8-TYPE="pronoun" CASE="genitive">his</REFEX>
  <REFEX ENTITY="2" REGO8-TYPE="pronoun" CASE="nominative">he</REFEX>
  <REFEX ENTITY="2" REGO8-TYPE="pronoun" CASE="nominative">who</REFEX>
</ALT-REFEX>
</GREC-ITEM>
```

Fig. 3. Example text from GREC-People corpus

Name:	Values:	Explanation:
ENTITY	integer	identifier for the discourse entity that is being referred to, unique within the text
MENTION	integer, ?	identifier for references to a given entity, unique for the entity
SEMCAT	person	the semantic category of the referent
SYNCAT	np, nom	the syntactic category required of referential expressions for the referent in the given context
SYNFUNC	subj	subject of a clause other than below
	subj_ger-part	subject of a gerund-participial ([17], p. 1191–1193)
	subj_rel-clause	subject within a relative clause
	obj	(in)direct object or object of PP other than below
	obj_rel-clause	object within a relative clause
	obj_refl	object of verb referring to same entity as subject of same verb
	app-supp	appositive supplement (e.g.: *George Sarton, the father of the history of science*)
	attr_compl	attributive complement (e.g.: *the Fessenden patents*)
	subj-det	genitive subject-determiner (e.g.: *his parents*)
	subj-det_rel-clause	genitive subject-determiner within a relative clause

Fig. 4. REF attribute names and values in GREC-People

5 GREC-MSR and GREC-NEG Task Definitions

5.1 GREC-MSR

The training/development data in GRE-MSR is exactly as shown in Figure 2. The test data is the same, except that of course REF elements contain only an ALT-REFEX list, not the actual REFEX. The task for participating systems was to select one of the REFEXs in the ALT-REFEX list, for each REF in each TEXT in the test sets. The selected REFEX then had to be inserted into the REF in test set outputs submitted for the GREC-MSR Task.

In the first run of this task (part of REG'08[8]), the main task aim was to get the REG08-TYPE of selected referring expressions (REs) right, and REG08-Type Accuracy (for definition see Section 6) was the main evaluation metric. In the GREC-MSR'09 run of this task (part of GenChal'09[9]), the main task aim was to

---

[8] The Referring Expression Generation Challenge 2008, see
http://www.itri.brighton.ac.uk/research/reg08
[9] Generation Challenges 2009, see
http://www.itri.brighton.ac.uk/research/genchal09

get the actual RE (the word string) right, and the main evaluation criterion was therefore Word String Accuracy.

We created four test sets for the GREC-MSR Task:

1. GREC-MSR Test Set C-1: a randomly selected 10% subset (183 texts) of the GREC corpus (with the same proportions of texts in the 5 subdomains as in the training/testing data).
2. GREC-MSR Test Set C-2: the same subset of texts as in C-1; however, for C-2 we did not use the REs in the corpus, but replaced them with human-selected alternatives. These were obtained in an online experiment (with an interface designed as shown in Figure 1) where participants selected REs in a setting that duplicated the conditions in which the participating systems in the GREC-MSR Task make selections.[10] We obtained three versions of each text, where in each version all REs were selected by the same person. The motivation for this version of Test Set C was that having several human-produced chains of REs against which to compare the outputs of participating ('peer') systems is more reliable than having one only; and that Wikipedia texts are edited by multiple authors which sometimes adversely affects MSR chains; we wanted to have additional reference texts where all references are selected by a single author.
3. GREC-MSR Test Set L: 74 Wikipedia introductory texts from the subdomain of lakes (there were no lake texts in the training/development set).
4. GREC-MSR Test Set P: 31 short encyclopaedic texts in the same 5 subdomains as in the GREC-2.0 corpus, in approximately the same proportions as in the training/testing data, but of different origin. We transcribed these texts from printed encyclopaedias published in the 1980s which are not available in electronic form. The texts in this set are much shorter and more homogeneous than the Wikipedia texts, and the sequences of MSRs follow very similar patterns. It seems likely that it is these properties that have resulted in better scores overall for Test Set P than for the other test sets in both the 2008 and 2009 runs of the GREC-MSR task.

Each test set was designed to test peer systems for generalisation to different kinds of unseen data. Test Set C tests for generalisation to unseen material from the same corpus and the same subdomains as the training set; Test Set L tests for generalisation to unseen material from the same corpus but different subdomain; and Test Set P for generalisation to a different corpus but the same subdomains.

5.2   GREC-NEG

The training/development data in GREC-NEG is exactly as shown in Figure 3. The test data is identical to the training/development data, except that REF elements do not contain a REFEX element, i.e. they are 'empty'.

---

[10] The experiment can be tried out here:
   http://www.itri.brighton.ac.uk/home/Anja.Belz/TESTDRIVE/

Table 1. Overview of evaluation methods used in GREC Shared Task Evaluations

Evaluation criterion	Type of evaluation	Evaluation technique
Humanlikeness	intrinsic/automatic	REG08-Type Accuracy, String Accuracy, String-edit distance, BLEU-3, NIST
Referential clarity	intrinsic/human extrinsic/automatic	Native speakers' judgement of clarity Automatic coreference resolution experiment
Fluency	intrinsic/human	Native speakers' judgement of fluency
Ease of comprehension	extrinsic/human	Reading speed and comprehension accuracy measured in a reading experiment

The task is to select one REFEX from the ALT-REFEX list for each REF in each TEXT in the test sets. If the selected REFEX contains an embedded REF then participating systems also need to select a REFEX for this embedded REF and to set the value of the MENTION attribute (which has the '?' value in REFs that are embedded in REFEXs in ALT-REFEX lists). The same applies to all further embedded REFEXs, at any depth of embedding.

In the first run[11] of this task (part of GenChal'09, see Footnote 9), the main task aim was to get the REG08-TYPE of selected referring expressions (REs) right, and REG08-Type Accuracy (for definition see Section 6) was the main evaluation metric.

We created two versions of the test data for the GREC-NEG Task:

1. GREC-NEG Test Set 1a: randomly selected 10% subset (100 texts) of the GREC-People corpus (with the same proportion of texts in the 3 subdomains as in the training/development data).
2. GREC-NEG Test Set 1b: the same subset of texts as in (1a); for this set we did not use the REs in the corpus, but replaced each of them with human-selected alternatives obtained in an online experiment as for GREC-MSR.

6   GREC Evaluation Procedures

As in the TUNA evaluations (see Gatt & Belz elsewhere in this volume [15]), we developed a portfolio of intrinsic and extrinsic, human-assessed and automatically computed evaluation methods to assess the quality of the REs generated by GREC systems. Table 1 is an overview of the techniques we have used in the GREC evaluations. Each technique is explained in one of the subsections below.

---

[11] The second run was held in 2010 as part of GenChal'10, see
http://www.itri.brighton.ac.uk/research/genchal10

In all GREC shared tasks, the data is divided into training, development and test data. In each case, we, the organisers, performed evaluations on the test data, using a range of different evaluation methods. Participants computed evaluation scores on the development set, using the `geval-2.0.pl` code provided by us which (in its most recent version) computes Word String Accuracy, REG'08-Type Recall and Precision, string-edit distance and BLEU.

Some of the test sets have a single version of each text (the original corpus text, as in GREC-MSR test set C-1 and GREC-NEG test set 1a, see Sections 5.1 and 5.2), and the scoring metrics below that are based on counting matches (Word String Accuracy counts matching word strings, REG08-Type Accuracy, Recall and Precision count matching REG08-Type attribute values) simply count the number of matches a system achieves against that single text.

For each task we also created one test set which has three versions of each text with human-selected REs in them (C-2 in GREC-MSR and 1b in GREC-NEG). For these sets, the match-based metrics first calculate the number of matches for each of the three versions and then use (just) the highest number of matches in further calculations.

6.1   Automatic Intrinsic Evaluations of Humanlikeness

One set of humanlikeness measures we computed were REG08-Type Accuracy (GREC-MSR), and REG08-Type Recall and Precision (GREC-NEG). REG08-Type Precision is defined as the proportion of REFEXs selected by a participating system that mach the corresponding REFEXs in the evaluation corpus; REG08-Type Recall is defined as the proportion of REFEXs in the evaluation corpus for which a participating system has produced a match. For GREC-MSR Recall equals Precision, which is why we call it Accuracy there.

The reason why we use REG08-Type Recall and Precision for GREC-NEG rather than REG08-Type Accuracy as in GREC-MSR is that in GREC-NEG there may be a different number of REFEXs in system outputs and the reference texts in the test set (because there are embedded references in GREC-People, and systems may select REFEXs with or without embedded references for any given REF). In GREC-MSR, the number of REFEXs in a system output and the corresponding reference texts in the test set is the same, hence we compute just one score, REG08-Type Accuracy.

For both tasks, we computed String Accuracy, defined as the proportion of word strings selected by a participating system that match those in the reference texts. For GREC-NEG this was computed on the complete text within the outermost REFEX, including the text in embedded REFEX nodes.

We also computed BLEU-3, NIST, string-edit distance and length-normalised string-edit distance, all on word strings defined as for String Accuracy. As regards the tests sets with multiple REs, BLEU and NIST are designed for multiple output versions (so they could be applied as they are), whereas for the string-edit metrics we computed the mean of means over the three text-level scores (computed against the three versions of a text).

6.2 Automatic Extrinsic Evaluation of Clarity

In all three GREC shared-task evaluations, we used Coreference Resolver Accuracy (CRA), an automatic extrinsic evaluation method based on coreference resolution performance. The basic idea is that it seems likely that badly chosen reference chains affect the ability to resolve REs in automatic coreference resolution tools.

To counteract the possibility of results being a function of a specific coreference resolution algorithm or evaluation method, we used several resolution tools and several evaluation methods and averaged results. There does not appear to be a single standard evaluation metric in the coreference resolution community. We opted to use the following three: MUC-6 [38], CEAF [23], and B-CUBED [1], which seem to be the most widely accepted metrics. All three metrics compute Recall, Precision and F-Scores on aligned gold-standard and resolver-tool coreference chains. They differ in how the alignment is obtained and which components of coreference chains are counted for calculating scores.

In GREC-MSR'08, we used three different resolvers—those included in Ling-Pipe,[12] JavaRap [30] and OpenNLP [24]. However, for GREC'09 we overhauled the CRA tool; the current version no longer uses JavaRAP, and uses the most recent versions of the other resolvers; the GREC-MSR'08 and GREC-MSR'09 results for this method are not entirely comparable for this reason.

For each system, the CRA tool runs the coreference resolvers on each system output, then CRA computes the MUC-6, CEAF and B-CUBED F-Scores for each coreference resolver output, then their mean, and finally the mean over all system outputs.

6.3 Human-Assessed Intrinsic Evaluation of Clarity and Fluency

6.3.1 GREC-MSR'09

The intrinsic human evaluation in GREC-MSR'09 involved 24 randomly selected items from Test Set C and outputs for these produced by peer and baseline systems (described in Section 7.1) as well as those found in the original corpus texts (8 'systems' in total). We used a Repeated Latin Squares design which ensures that each participant sees the same number of outputs from each system and for each test set item. There were three 8×8 squares, and a total of 576 individual judgements in this evaluation (72 per system: 3 criteria $\times 3$ articles $\times 8$ evaluators).

We recruited 8 native speakers of English from among post-graduate students currently doing a linguistics-related degree at University College London (UCL) and Sussex University.

Following detailed instructions, participants did two practice examples, followed by the 24 texts to be evaluated, in random order. Subjects carried out the evaluation over the internet, at a time and place of their choosing. They were allowed to interrupt and resume the experiment (though discouraged from

---

[12] http://alias-i.com/lingpipe/

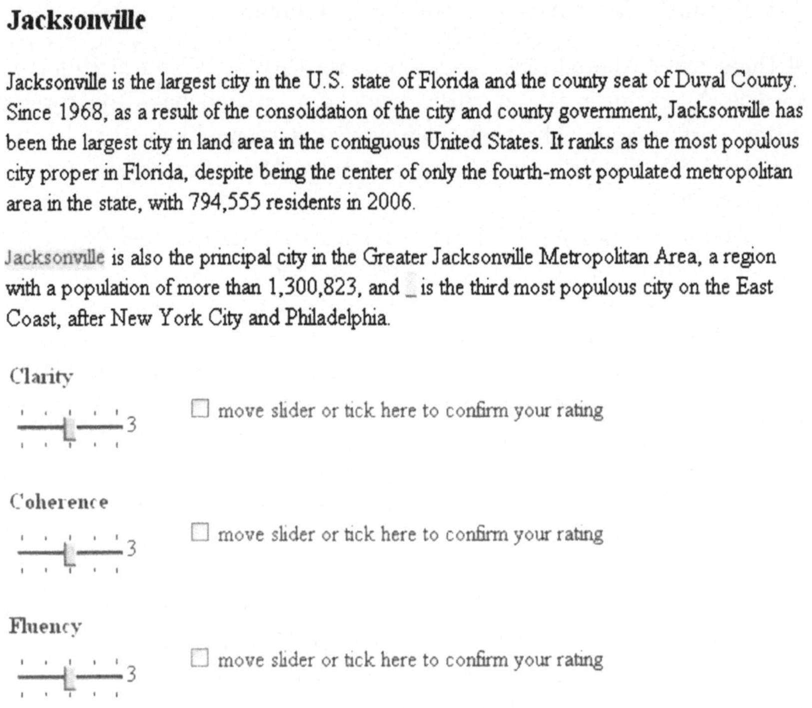

Jacksonville

Jacksonville is the largest city in the U.S. state of Florida and the county seat of Duval County. Since 1968, as a result of the consolidation of the city and county government, Jacksonville has been the largest city in land area in the contiguous United States. It ranks as the most populous city proper in Florida, despite being the center of only the fourth-most populated metropolitan area in the state, with 794,555 residents in 2006.

Jacksonville is also the principal city in the Greater Jacksonville Metropolitan Area, a region with a population of more than 1,300,823, and _ is the third most populous city on the East Coast, after New York City and Philadelphia.

Clarity

———┤——3   ☐ move slider or tick here to confirm your rating

Coherence

———┤——3   ☐ move slider or tick here to confirm your rating

Fluency

———┤——3   ☐ move slider or tick here to confirm your rating

Fig. 5. Example of text presented in human intrinsic evaluation of GREC-MSR systems

doing so). According to self-reported timings, participants took between 25 and 45 minutes to complete the evaluation (not counting breaks).

Figure 5 shows what participants saw during the evaluation of an individual text. All references to the MS were highlighted in yellow,[13] and the task is to evaluate the quality of the REs in terms of three criteria which were explained in the introduction as follows (the wording of the explanations of Criteria 1 and 3 were taken from the DUC evaluations):

1. **Referential Clarity:** It should be easy to identify who or what the referring expressions in the text are referring to. If a person or other entity is mentioned, it should be clear what their role in the story is. So, a reference would be unclear if an entity is referenced, but their identity or relation to the story remains unclear.
2. **Fluency:** A referring expression should 'read well', i.e. it should be written in good, clear English, and the use of titles and names etc. should seem natural. Note that the Fluency criterion is independent of the Referential Clarity criterion: a reference can be perfectly clear, yet not be fluent.
3. **Structure and Coherence:** The text should be well structured and well organised. The text should not just be a heap of related information, but should build from sentence to sentence to a coherent body of information about a topic. This criterion too is independent of the others.

---

[13] Showing up as pale shaded boxes around words in black and white versions of this document.

Ramon Pichot Gironès

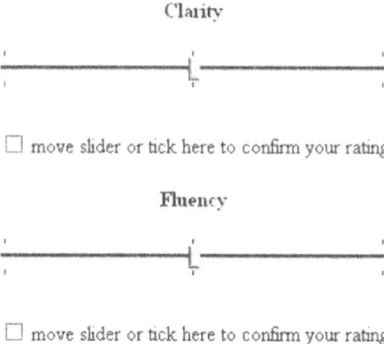

Ramon Pichot Gironès (1872 - 1 March 1925) was a Catalan and Spanish artist. He painted in an impressionist style.

He was a good friend of Pablo Picasso and _ acted as an early mentor to young Salvador Dalí. Salvador Dalí met Ramon Pichot Gironès in Cadaqués, Spain when Salvador was only 10 years old. Ramon also made many trips to France. Once in a while Salvador and his family would go on a trip with Ramon Pichot and his family.

Ramon Pichot Gironès (1872 - 1 March 1925) was a Catalan and Spanish artist. He painted in an impressionist style.

He was a good friend of Pablo Picasso and _ acted an early mentor to young Salvador Dalí. Salvador Dalí met him in Cadaqués, Spain when Salvador was only 10 years old. Ramon also made many trips to France. Once in a while Salvador Dalí and his family would go on a trip with Ramon Pichot and his family.

Clarity

☐ move slider or tick here to confirm your rating

Fluency

☐ move slider or tick here to confirm your rating

Fig. 6. Example of text pair presented in human intrinsic evaluation of GREC-NEG systems

Subjects selected evaluation scores by moving sliders (see Figure 5) along scales ranging from 1 to 5. Slider pointers started out in the middle of the scale (3). These were continuous scales and we recorded scores with one decimal place (e.g. 3.2). The meaning of the numbers was explained in terms of integer scores (1=very poor, 2=poor, 3=neither poor nor good, 4=good, 5=very good).

6.3.2 GREC-NEG'09

The motivating application context for the GREC-NEG Task is, as mentioned above (Section 2.1), improving referential clarity and coherence in multiply edited texts. We therefore designed the human-assessed intrinsic evaluation as a preference-judgement test where participants expressed their preference, in terms of two criteria, for either the original Wikipedia text or the version of it with system-generated referring expressions in it. The intrinsic human evaluation involved outputs for 30 randomly selected items from the test set from 5 of the 6 participating systems,[14] four baselines and the original corpus texts (10 'systems' in total, all described in Section 8.1). Again, we used a Repeated Latin Squares design. This time there were three 10×10 squares, and a total of 600 individual judgements in this evaluation (60 per system: 2 criteria $\times 3$ articles $\times 10$ evaluators). We recruited 10 native speakers of English from among students currently completing a linguistics-related degree at Kings College London and University College London.

As in the GREC-MSR and TUNA evaluation experiments, participants were given detailed instructions, two practice examples, and then the texts to be evaluated, in random order. Subjects did the evaluation over the internet, at a time and place of their choosing, and were allowed (though discouraged) to interrupt and resume.

Figure 6 shows what participants saw during the evaluation of an individual text pair. The place (left/right) of the original Wikipedia article was randomly determined for each individual evaluation. People references are highlighted in yellow/orange, those that are identical in both texts are yellow, those that are different are orange.[15] The evaluator's task is to express their preference, as well as the strength of their preference, in terms of each quality criterion by moving the slider pointers. Moving the slider to the left means expressing a preference for the text on the left, moving it to the right means preferring the text on the right; the further to the left/right the slider is moved, the stronger the preference. The two criteria of Fluency and Referential Clarity were explained to participants in the introduction with exactly the same wording as described above for GREC-MSR (Section 6.3.1).

In this experiment, unlike in the GREC-MSR experiment, it was not evident to the evaluators that sliders were associated with numerical values. Slider pointers started out in the middle of the scale (no preference). The values associated with the points on the slider ranged from -10.0 to +10.0.

6.4 Human-Based Extrinsic Evaluation of Ease of Comprehension

For GREC-MSR'08, we designed a reading/comprehension experiment in which the task for participants was to read texts one sentence at a time and then to answer three brief multiple-choice comprehension questions after reading each text. The basic idea was that it seemed likely that badly chosen MSR reference chains would adversely affect ease of comprehension, and that this might in turn affect reading speed.

We used a randomly selected subset of 21 texts from GREC-MSR Test Set C, and recruited 21 participants from among the staff, faculty and students of Brighton and Sussex universities. We used a Repeated Latin Squares design in which each combination of text and system was allocated three trials. During the

---

[14] We left out UDel-NEG-1 given our limited resources and the fact that this is a kind of baseline system.

[15] When viewed in black and white, the orange highlights appear slightly darker than the yellow ones.

experiment we recorded *SRTime* (sentence reading time), the time participants took to read sentences (from the point when the sentence appeared on the screen to the point at which the participant requested the next sentence).

We also recorded the speed and accuracy with which participants answered the questions at the end (*Q-Time* and *Q-Acc*). The role of the comprehension questions was to encourage participants to read the texts properly, rather than skimming through them, and we did not necessarily expect any significant results from the associated measures.

The questions were designed to be of varying degrees of difficulty and predictability. There were three questions (each with five possible answers) associated with each text, and questions followed the same pattern across the texts: the first question was always about the subdomain of a text (*The text I just read was about a [city/country/river/person/mountain]*); the second about the geographical location of the main subject (e.g. *The city I just read about is located in [Peshawar/Uttar Pradesh/...]*; *The person I just read about was born in [England/Scotland/...]*); the third question was designed not to be predictable (e.g. *How many hydroelectric power stations are there on this river? [three/five/four/...]*; *This mountain is the location of a neolithic [jadeite quarry/jasper quarry/...]*).

The order of the possible answers was randomised for each question and each participant. The order of texts (with associated questions) was randomised for each participant. We used the DMDX package for presentation of sentences and measuring reading times and question answering accuracy [14]. Subjects did the experiment in a quiet room, under supervision.

7 GREC-MSR'08/09—Participating Systems and Results

7.1 Systems

In this section, we give very brief descriptions of the systems that participated in the two GREC-MSR competitions. Full details can be found in the reports from participating teams in the INLG'08 proceedings (for GREC-MSR'08) and the ENLG'09 proceedings (for GREC-MSR'09).

Base-rand, Base-freq, Base-1st, Base-name: We created four baseline systems. *Base-rand* selects one of the REFEXs at random. *Base-freq* selects the REFEX that is the overall most frequent given the SYNCAT and SEMCAT of the reference. *Base-1st* always selects the REFEX which appears first in the list of REFEXs; and *Base-name* selects the shortest REFEX with attributes REG08-TYPE=name, HEAD=nominal and EMPHATIC=no.[16]

CNTS-Type-g, CNTS-Prop-s (GREC-MSR'08): The CNTS systems are trained using memory-based learning with automatic parameter optimisation. They use

---

[16] Attributes are tried in this order. If for one attribute, the right value is not found, the process ignores that attribute and moves on the next one.

a set of 14 features obtained by various kinds of syntactic preprocessing and named-entity recognition as well as from the corpus annotations: SEMCAT, SYNCAT, position of RE in text, neighbouring words and POS-tags, distance to previous mention, SYNCATs of the three preceding REFEXs, a binary feature indicating whether the most recent named entity was the main subject (entity), and the main verb of the sentence. For *CNTS-Type-g*, a single classifier was trained to predict just the REG08-TYPE property of REFEXs. For *CNTS-Prop-s*, four classifiers were trained, one for each subdomain, to predict all four properties of REFEXs (rather than just REG08-TYPE).

OSU-b-all, OSU-b-nonRE, OSU-n-nonRE (GREC-MSR'08): The *OSU* systems are maximum-entropy classifiers trained on a range of features obtained from the corpus annotations and by preprocessing the text: SEMCAT, SYNCAT, position of RE in text, presence of contrasting discourse entity, distance between current and preceding reference to the entity, string similarity measures between REFEXs and the title of text. *OSU-b-all* and *OSU-b-nonRE* are binary classifiers which give the likelihood of selecting a given REFEX vs. not selecting it, whereas *OSU-n-nonRE* is a 4-class classifier giving the likelihoods of selecting each of the four REG08-TYPEs. *OSU-b-all* also uses the REFEX attributes as features.

IS-G (GREC-MSR'08): The *IS-G* system is a multi-layer perceptron which uses four features obtained by preprocessing texts and from the corpus annotations: SYNCAT, distance between current and preceding reference to the entity, position of RE in text, REG08-TYPE of preceding reference to the entity, feature indicating whether the preceding mention is in the same sentence.

UDel (GREC-MSR'09): The *UDel* system is informed by psycholinguistic research and consists of a preprocessing component performing sentence segmentation and identification of non-referring occurrences of entity names, an RE type selection component (two C5.0 decision trees, one optimised for people and mountains, the other for the other subdomains), and a word string selection component. The RE type selection decision trees use the following features: binary features indicating whether the entity is the subject of the current, preceding and preceding but one sentences, whether the last MSR was in subject position, and whether there are intervening references to other entities between the current and the previous MSR. Other features encode distance to preceding non-referring occurrences of an entity name; sentence and reference IDs; and whether the reference occurred before and after certain words and punctuation marks. Given a selected RE type, the word-string selection component selects (from among those REFEXs that have a matching type) the longest non-emphatic name for the first named reference in an article, and the shortest for subsequent named references; for other types, the first matching word-string is used, backing off to pronoun and name.

ICSI-CRF (GREC-MSR'09): The *ICSI-CRF* system construes the GREC-MSR task as a sequence labelling task and determines the most likely current label given preceding labels, using a Conditional Random Field model trained on the following features for the current, preceding and preceding but one MSR: preceding and following word unigram and bigram; suffix of preceding and following word; preceding and following punctuation; reference ID; and a binary feature encoding whether the current sentence is the beginning of a paragraph. If more than one label remains, the last in the list of possible REs in the GREC-MSR data is selected.

JUNLG (GREC-MSR'09): The *JUNLG* system is based on co-occurrence statistics between REF feature sets and REFEX feature sets as found in the GREC-MSR data. REF feature sets were augmented by a paragraph counter and a within-paragraph REF counter. For each given set of REF features, the system selects the most frequent REFEX feature set (as determined from co-occurrence counts in the training data). If the current set of possible REFEXs does not include a REFEX with the selected feature set, then the second most likely feature set is selected. Several hand-coded default rules override the frequency-based selections, e.g. if the preceding word is a conjunction, and the current SYNCAT is np-subj, then the REG08-Type is empty.

7.2 Results

Table 2 shows mean scores for all evaluation methods in GREC-MSR'08; in the case of the automatically computed metrics, scores were computed against Test Set C-2, which has 3 versions of each text in the test set, each with people REs selected by a (different) human. Statistically significant differences are indicated

Table 2. GREC-MSR'08: REG08-Type Accuracy scores with homogeneous subsets (Tukey HSD, alpha = .05), string-similarity scores, sentence reading time (SRT), Question 1 Accuracy (Q1), and coreference resolver accuracy (CRA), automatic metrics as computed against Test Set C-2

System	REG08-Type Accuracy				WSAcc	BLEU-3	NIST	SE	SRT	Q1	CRA
Corpus	78.58	A			71.18	0.779	7.508	0.723	6548	1.00	43.32
CNTS-Type-g	72.61	A	B		65.61	0.738	6.129	0.884	6436	1.00	48.64
CNTS-Prop-s	71.34		B		65.29	0.676	5.934	0.907	6306	0.98	46.35
IS-G	70.78		B		58.20	0.511	5.610	1.162	6341	0.95	48.05
OSU-n-nonRE	69.82		B		63.85	0.672	5.775	0.967	6423	0.90	51.39
OSU-b-nonRE	58.76			C	51.11	0.496	5.536	1.283	6455	0.97	51.27
OSU-b-all	57.48			C	50.72	0.505	5.606	1.299	6452	0.95	50.87
Base-name	50.00				D	39.41	0.464	5.937	1.518	–	–	52.84
Base-1st	49.28				D	39.09	0.393	5.16	1.645	–	–	53.50
Base-freq	48.17				D	41.32	0.268	3.016	1.543	–	–	41.41
Base-rand	41.24				E	17.99	0.218	2.932	2.322	–	–	35.13

Table 3. GREC-MSR'08: Pearson's correlation coefficients for all evaluation methods in Table 2. **= Correlation is significant at the 0.01 level (2-tailed).

	REG08-T.Acc	WSAcc	BLEU-3	NIST	SE	SRTime	Q1	CRA
REG08-Type Acc.	1	.964 **	.934 **	.795 **	-.937 **	.045	.334	.201
WSAcc	.964 **	1	.934 **	.802 **	-.994 **	.120	.332	.341
BLEU-3	.934 **	.934 **	1	.896 **	-.932 **	.289	.396	.353
NIST	.795 **	.802 **	.896 **	1	-.822 **	.616	.553	.545
SE	-.937 **	-.994 **	-.932 **	-.822 **	1	-.199	-.398	-.390
SRTime	.045	.120	.289	.616	-.199	1	.241	-.140
Q1 Accuracy	.334	.332	.396	.553	-.398	.241	1	-.656
CRA	.201	.341	.353	.545	-.390	-.140	-.656	1

(in the form of homogeneous subsets) only for REG08-Type Accuracy which was the nominated main evaluation method of GREC-MSR'08.[17]

Table 3 shows the corresponding Pearson's correlation coefficients for all evaluation methods in Table 2. The picture is very clear: all corpus-similarity metrics (whether based on string similarity or RE type similarity) are strongly and highly significantly correlated with each other. However they are not correlated significantly with any of the extrinsic methods, and there are also no significant correlations between any of the extrinsic methods.[18]

Table 4 shows analogous results for GREC-MSR'09. This time, the table shows statistical significance for Word String Accuracy, as this was the nominated main evaluation method for GREC-MSR'09. Table 5 shows the corresponding Pearson's correlation coefficients. This time the picture is not quite so simple. Once again, the automatically computed corpus-similarity metrics correlate strongly and highly significantly with each other. Out of the human-assessed intrinsic metrics, Fluency and Coherence correlate strongly with the automatically computed corpus-similarity metrics. However, Clarity only correlates with NIST and (to a lesser extent) with SE. While Fluency and Coherence correlate well with each other, only Coherence is also correlated with Clarity. The correlation between Fluency and Coherence makes sense intuitively, since both could be seen as dimensions of how well a text reads. The weaker correlations with Clarity indicate that the human evaluators were able to consider it to some degree independently from the other criteria, and there must have been some systems that produced texts that were clear but not fluent (or even vice versa). The slightly stronger correlation between Coherence and Clarity also makes sense, since, for example, using pronouns in the right place contributes to both (but not necessarily to Fluency).

---

[17] By 'nominated main evaluation method' we mean that this was the method that participants were told was going to be the main method by which systems would be evaluated. The reason why this is important is that some participants may have optimised their systems for the nominated main evaluation method.

[18] For comparison with a similar lack of correlations between intrinsic and extrinsic methods in the TUNA tasks, see the discussion section below (Section 9).

Table 4. GREC-MSR'09: REG08-Type Accuracy scores, Word String Accuracy with homogeneous subsets (Tukey HSD, alpha = .05), other string-similarity scores, Clarity, Fluency, Coherence scores, and coreference resolver accuracy (CRA), automatic metrics as computed against Test Set C-2

System	REG08-T.Acc.	Word String Accuracy					BLEU-3	NIST	SE	Cla	Flu	Coh	CRA
Corpus	79.30	71.58	A				0.77	5.60	1.04	4.56	4.43	4.40	42.52
UDel	77.71	70.22	A	B			0.74	5.32	1.11	4.35	4.27	4.27	46.19
JUNLG	75.40	64.57		B	C		0.53	4.69	1.34	4.50	4.26	4.33	44.19
ICSI-CRF	75.16	63.69			C		0.54	4.68	1.32	4.45	4.15	4.02	44.47
Base-freq	62.50	57.01				D	0.54	4.30	1.93	4.10	3.33	3.96	63.14
Base-name	51.04	40.21					E	4.76	1.80	4.62	2.84	3.85	65.19
Base-1st	50.32	39.65					E	4.42	1.93	4.27	2.76	3.7	63.77
Base-rand	48.09	26.99						F	0.26	3.02	2.30	3.18	2.15	3.46	42.99

Table 5. GREC-MSR'09: Pearson's correlation coefficients for all evaluation methods in Table 4. **= Correlation is significant at the 0.01 level (2-tailed).

	REG08-T.Acc.	WSA	BLEU-3	NIST	SE	Cla	Flu	Coh	CRA
REG08-T.Acc.	1	.971 **	.862 **	.726 *	-.931 **	.531	.984 **	.922 **	-.609
WSA	.971 **	1	.923 **	.818 *	-.925 **	.645	.983 **	.950 **	-.407
BLEU3	.862 **	.923 **	1	.909 **	-.905 **	.649	.881 **	.909 **	-.293
NIST	.726 *	.818 *	.909 **	1	-.891 **	.880 **	.812 *	.860 **	-.073
SE	-.931 **	-.925 **	-.905 **	-.891 **	1	-.717 *	-.959 **	-.930 **	.488
Clarity	.531	.645	.649	.880 **	-.717 *	1	.670	.714 *	.181
Fluency	.984 **	.983 **	.881 **	.812 *	-.959 **	.670	1	.956 **	-.493
Coherence	.922 **	.950 **	.909 **	.860 **	-.930 **	.714 *	.956 **	1	-.386
CRA	-.609	-.407	-.293	-.073	.488	.181	-.493	-.386	1

8   GREC-NEG'09—Participating Systems and Results

8.1   Systems

In this section, we give very brief descriptions of the systems that participated in the GREC-NEG competition. Individual reports describing participating systems can be found in the proceedings of the UCNLG+SUM workshop.

Base-rand, Base-freq, Base-1st, Base-name: We created four baseline systems each with a different way of selecting a REFEX from those REFEXs in the ALT-REFEX list that have matching entity IDs. *Base-rand* selects a REFEX at random. *Base-1st* selects the first REFEX. *Base-freq* selects the first REFEX with a REG08-TYPE that is the overall most frequent (as determined from the training/development data) given the SYNCAT, SYNFUNC and SEMCAT of the reference. *Base-name* selects the shortest REFEX with attribute REG08-TYPE=name.

UDel-NEG-1, UDel-NEG-2, UDel-NEG-3: The *UDel-NEG-1* system is identical to the *UDel* system that was submitted to the GREC-MSR'09 Task (for a description of that system see Section 7.1 above), except that it was adapted to the different data format of GREC-NEG. *UDel-NEG-2* is identical to *UDel-NEG-1* except that it was retrained on GREC-NEG data and the feature set was extended by entity and mention IDs. *UDel-NEG-3* additionally utilised improved identification of other entities.

ICSI-CRF: see Section 7.1.

WLV-BIAS, WLV-STAND: The *WLV* systems start with sentence splitting and POS tagging. *WLV-STAND* then employs a J48 decision tree classifier to obtain a probability for each REF/REFEX pair that it is a good pair in the current context. The context is represented by the following set of features. Features of the REFEX word string include features encoding whether the string is the longest of the possible REFEXs, the number of words in the string, and all REFEX features supplied in the GREC-NEG data. Features of the REF include features encoding whether the current entity's chain is the first in the text, whether the current mention of the entity is the first, whether the mention is at the beginning of a sentence, and all REF features supplied in GREC-NEG data. Other features include one that encodes whether the current REF is preceded by one of a given set of strings including ", but", "and then" and similar phrases, the distance in sentences to the last mention, the REG08-Types selected for the two preceding REFs, the POS tags of the preceding four words and the following three words, the correlation between SYNFUNC and CASE values, and the size of the chain.

WLV-BIAS is the same except that it is retrained on reweighted training instances. The reweighting scheme assigns a cost of 3 to false negatives and 1 to false positives.

8.2   Results

Table 6 shows mean scores for all evaluation methods used in GREC-NEG'09; in the case of the automatically computed metrics, scores were computed against Test Set 1b, which has 3 versions of each text in the test set, each with people REs selected by a (different) human. Statistically significant differences (in the form of homogeneous subsets) are shown only for REG08-Type Precision and Recall which were the nominated main evaluation methods of GREC-NEG'09.

Table 7 shows the corresponding correlation results (again for Pearson's *r*) at the system level. All intrinsic methods, both automatically computed and human-assessed, correlate well. Unlike in the GREC-MSR correlation results, here, coreference resolver accuracy (CRA) correlates well with all intrinsic methods except Word String Accuracy and string-edit distance. The strongest correlation is with REG08-Type Precision and Recall.

Table 6. GREC-NEG'09: REG08-Type Recall and Precision scores with homogeneous subsets (Tukey HSD, alpha = .05), and all word-string similarity based intrinsic automatic scores. All scores as computed against human topline version of Test Set (1b).

System	REG08-Type Precision	REG08-Type Recall	WSAcc	BLEU-3	NIST	nSE	Cla	Flu	CRA
Corpus	82.67 A	84.01 A	81.90	0.95	7.15	0.25	0	0	59.56
ICSI-CRF	79.33 A B	78.38 B	74.69	0.86	6.36	0.31	-1.45	-0.35	61.28
WLV-BIAS	77.78 B	77.78 B	69.14	0.88	6.18	0.36	-2.44	-2.26	62.64
WLV-STAND	67.51 C	67.51 C	59.84	0.83	5.82	0.45	-4.48	-5.82	51.69
Base-freq	65.38 C	64.37 C	3.24	0.39	2.1	0.90	-8.26	-7.57	55.85
UDel-NEG-2	57.39 D	56.06 D	18.96	0.53	2.42	0.83	-6.67	-7.13	55.9
UDel-NEG-3	57.25 D	55.92 D	18.89	0.53	2.49	0.82	-6.43	-6.26	56.13
Base-name	55.22 D	54.01 D	37.27	0.65	5.57	0.63	-2.58	-4.26	61.11
UDel-NEG-1	53.57 D	52.32 D E	19.25	0.51	2.62	0.82	–	–	54.79
Base-rand	48.46 E	47.75 E	10.45	0.25	1.11	0.89	-8.18	-7.51	34.86
Base-1st	12.54 F	12.54 F	8.65	0.24	1.29	0.92	-9.36	-8.48	26.36

Table 7. Pearson's correlation coefficients for all evaluation methods in Table 6. **= Correlation is significant at the 0.01 level (2-tailed). *= Correlation is significant at the 0.05 level (2-tailed).

	REG08-T. Precision	REG08-T. Recall	WSAcc	BLEU-3	NIST	norm. SE	Clarity	Fluency	CRA
REG08-T. Prec.	1	.999 **	.733 *	.825 **	.749 **	-.760 **	.765 **	.762 *	.844 **
REG08-T. Rec.	.999 **	1	.751 **	.836 **	.763 **	-.777 **	.776 **	.773 **	.831 **
WSAcc	.733 *	.751 **	1	.952 **	.957 **	-.997 **	.933 **	.925 **	.556
BLEU-3	.825 **	.836 **	.952 **	1	.964 **	-.954 **	.934 **	.879 **	.753 **
NIST	.749 **	.763 **	.957 **	.964 **	1	-.968 **	.964 **	.899 **	.678 *
norm. SE	-.760 **	-.777 **	-.997 **	-.954 **	-.968 **	1	-.940 **	-.932 **	-.582
Clarity	.765 **	.776 **	.933 **	.934 **	.964 **	-.940 **	1	.958 **	.734 *
Fluency	.762 *	.773 **	.925 **	.879 **	.899 **	-.932 **	.958 **	1	.664 *
CRA	.844 **	.831 **	.556	.753 **	.678 *	-.582	.734 *	.664 *	1

9   Discussion

9.1   Evaluation Methods

An important research focus in both the TUNA and the GREC shared-task challenges has been the evaluation methods we have used. Our dual aims have been to develop new evaluation methods, and to assess the performance of both existing and new evaluation methods.

One very experimental evaluation metric we developed is Coreference Resolver Accuracy (CRA). The original implementation included one resolver, Java-RAP, that was with hindsight not suitable, because it performs anaphora resolution rather than coreference resolution (and not all references are anaphoric). The removal of JavaRAP may explain the differences in correlation results between GREC-MSR'08 and GREC-MSR'09 (Tables 3 and 5): in the former, correlations tended to go in a positive direction, whereas in the latter, they tended to be in the negative direction. However, the only statistically significant correlation results are from GREC-NEG'09, where correlations (Table 7) with the intrinsic metrics were all in the 'right' direction (i.e. better intrinsic scores also implied better CRA). A clear bias of CRA is that it favours systems that produce a lot of named references (as these are easy for a coreference resolver to identify). CRA remains a highly experimental evaluation metric, but as both coreference resolution methods and named entity generation methods improve, it may become a viable evaluation method.

Another experimental evaluation method we developed is extrinsic evaluation by reading/comprehension experiments. The intuition here was that poorly chosen references would slow down reading speed and interfere with text comprehension. The one time we ran this experiment (GREC-MSR'08), the differences in reading speeds between the participating systems were very small (and no statistically significant differences were found). We did (unexpectedly) find a significant (albeit weak) impact on the comprehension question that asked readers what the article they had just read was about (possible answers were person, city, country, river or mountain).

Of course, we cannot conclude from the lack of statistically significant differences between reading times that there are no differences to be found, and a different experimental design may well reveal such significant differences. For example, a contributing factor to our lack of results may have been that measuring reading time at the sentence level results in measurements on the order of seconds, and there is a lot of variance in such long reading times. We are planning to run this experiment again in the next GREC-NEG evaluation. This time, we will aim to measure reading time on smaller units of text (e.g. individual referential expressions), possibly using eye-tracking; we will also include the baseline systems. For the time being, however, this also remains a very experimental evaluation method.

Much more successful in the immediately term was the preference judgement experiment in GREC-NEG'09. Rating-scale evaluations, where human evaluators assess system outputs by selecting a score on a discrete scale, are the most common form of human-assessed evaluation in NLP, but are problematic for several reasons. Rating scales are unintuitive to use; deciding whether a given text deserves a 5, a 4 or a 3 etc. can be difficult. Furthermore, evaluators may ascribe different meanings to scores and the distances between them. Individual evaluators have different tendencies in using rating scales, e.g. what is known as 'end-aversion' tendency where certain individuals tend to stay away from the extreme ends of scales; other examples are positive skew and acquiescence bias, where individuals make disproportionately many positive or agreeing judgements (see e.g. [8]). It is not surprising then that stable averages of quality judgements, let alone high levels of agreement, are hard to achieve, as has been observed for MT [22,36], text summarisation [35], and language generation [2]. The result of a rating scale experiment is ordinal data (sets of scores selected from the discrete rating scale). The means-based ranks and statistical significance tests that are commonly presented with the results of RSEs are not generally considered appropriate for ordinal data in the statistics literature [33]. At a minimum, "a test on the means imposes the requirement that the measures must be additive, i.e. numerical" [33, p. 14]. Parametric statistics are more powerful than non-parametric alternatives, because they make a number of strong assumptions (including that the data is numerical). If the assumptions are violated then the risks is that the significance of results is overestimated.

In view of these problems, we wanted to try out a preference judgement experiment, where evaluators compare two outputs at a time and simply have to state which of the two they think is better in terms of a given quality criterion. The experiment we designed (described in Section 6.3) added a twist to this in that we used sliders with which participants could express the *strength* of their preference (see Figure 6). In all cases, participants were comparing system outputs to the original Wikipedia texts. Viewed in terms of binary preference decisions, this experiment gave very clear-cut results: all but the top two systems were *dispreferred* almost all of the time; the second ranked system was dispreferred and *neither preferred nor dispreferred* about the same number of times (and almost never preferred); the top ranking system (ICSI-CRF) was *dispreferred* only about a third of the time, and in the case of of Fluency, it was preferred more often than it was dispreferred. This last result indicates that the ICSI-CRF system actually succeeded in improving over the Wikipedia texts.

There was also less variance and more statistically significant differences in the results than in comparable rating-scale experiments (including evaluation experiments that use slider bars to represent rating scales such as in GREC-MSR'09 and TUNA'09). The correlations between Fluency and Clarity on the one hand and the string-similarity metrics on the other were very high (mostly above 0.92). The results are easy to interpret—strength of preference and preference counts are intuitive concepts. On the whole, preference judgements using slider scales with results reported in the above manner are a very promising new evaluation method for NLG, and we are currently conducting follow-up experiments to investigate it further.

In assessing the different evaluation methods we have used, we look at variance, how many statistically significant differences were found, and how intuitive the results are. After GREC-MSR'08, we decided not to use the ROUGE metrics any more (part of the problem was the infeasibly high rank both ROUGE-SU4 and ROUGE-2 assigned to the random baseline). As mentioned above, we continue to consider CRA and reading-comprehension experiments unreliable methods, albeit ones with potential from further development.

However, our main tool is looking at correlation coefficients between sets of system-level evaluation scores. In GREC-MSR'08, there were strong correlations between all pairs of intrinsic evaluation metrics, and none between any intrinsic and extrinsic measures. But the latter result is hard to interpret, because of the highly experimental nature of the extrinsic measures. Furthermore, very few statistically significant results were obtained for the extrinsic measures, so for the time being we do not know how meaningful they are.

In GREC-MSR'09, there were strong correlations between all pairs of intrinsic evaluation metrics (human-assessed and automatically computed) except for Clarity which only correlated significantly with NIST, SE and Coherence. The correlations with CRA were not significant.

Finally, in GREC-NEG'09, virtually all measures correlated strongly with each other; even CRA correlated well with most measures (except for Word String Accuracy and string-edit distance).

While the lack of correlation between extrinsic and intrinsic measures in GREC-MSR'08 and GREC-MSR'09 echoes similar results from the TUNA evaluations, we would caution against drawing any conclusions from this, because of the experimental nature of the extrinsic metrics. The one time we did get plenty of significant differences for CRA (in GREC-NEG'09), we also had good correlation with the intrinsic measures.

Repeated comparative evaluation experiments can reveal patterns in how measures relate to each other. They can also show individual measures to yield inconsistent and/or unintuitive results over time. In the case of the remaining measures—those that do yield consistent and plausible results over time—two measures that consistently correlate highly in different experiments and for different tasks, could conceivably substitute for each other. But what we have avoided is to interpret any one of our measures as a basis for validating other measures. In MT and summarisation evaluation, good correlation with human judgements is often taken as validating an automatic metric (and conversely, lack of correlation as invalidating it). But intrinsic human judgements are simply not consistent and reliable enough to provide an objective meta-evaluation tool.[19] Moreover, all they provide is an insight into what humans (think they) like, not what is best or most useful for them (the two can be two very different matters, as discussed in [4]).

Ultimately, what those evaluation measures that are in themselves consistent, but do not consistently correlate strongly with each other, can provide us with are assessments of two different aspects of system quality. If there is anything we have learned from the numerous evaluation experiments we have carried out for TUNA and GREC, it is that a range of intrinsic and extrinsic methods, human-assessed and automatically computed, need to be applied in order to obtain a comprehensive view of system quality.

9.2   Development of the GREC Tasks

The GREC tasks are designed to build on each other, and to become increasingly complex and application-oriented. Our aims in this have been (i) to create data and define tasks that will enable REG researchers to investigate the REG problem as grounded in discourse context and informed by naturally occurring data; (ii) to build bridges to the summarisation and named-entity recognition communities by designing tasks that should be of specific interest to those researchers; and (iii) to move towards being able to use REG techniques in real application contexts.

Until recently, connections and joint forums between NLG and other fields in which language is automatically generated (notably MT and summarisation) were steadily decreasing as these fields moved away from language generation

---

[19] Agreement among human judges is hard to achieve as has been discussed for MT, summarisation and NLG [2,22,35,36].

techniques and towards data-driven, direct text-to-text mappings. This made it difficult for NLG researchers to engage with these fields effectively and contribute to developing solutions for application tasks.

Grammar-based language generation techniques are making a comeback in MT and summarisation, as stastistical MT researchers have moved towards incorporating syntactic knowledge (e.g. under the heading of syntax-based statistical MT as in work by Knight and colleagues [6,13]), and researchers in extractive summarisation have started to develop post-processing text regeneration techniques to improve the coherence and clarity of summaries [25,27,34].

This means that there is now more common ground than there has been in a long time between NLG and neighbouring fields, providing a good opportunity to bring researchers from the different fields together in working on intersecting tasks such as the GREC tasks.

We have already succeeded in attracting researchers from the Named Entity Recognition (NER) and machine learning communities as participants in GREC-NEG'09. In GREC'10, one of the subtasks is NER and another is an end to end RE regeneration task for extractive summarisation. The latter task will also be the first task to require stand-alone application systems to be developed.

10   Concluding Remarks

Participation in the GREC tasks has so far not reached the high levels of participation seen in the first two years of the TUNA tasks. This may be because the GREC tasks are entirely new research tasks, whereas TUNA involved tasks that many researchers were already working on (the first GIVE challenge also attracted just three teams from outside the team of organisers; see Koller et al. elsewhere in this volume [20]). We are counteracting this to some extent by running each task in two consecutive years.

The GREC research programme started out as an investigation into the REG task as grounded in discourse context and informed by naturally occurring data. We have since created two substantial annotated data resources which between them contain some 21,000 annotated referring expressions. These data resources, along with all automatically computed evaluation methods, become freely available for research purposes as soon as the last competition based on them is concluded.

Acknowledgments. The research reported in this paper was supported under EPSRC (UK) grants EP/E029116/1 (the Prodigy Project), EP/F059760/1 (REG Challenges 2008), and EP/G03995X/1 (Generation Challenges 2009).

References

1. Bagga, A., Baldwin, B.: Algorithms for scoring coreference chains. In: Proceedings of the Linguistic Coreference Workshop at LREC 1998, pp. 563–566 (1998)
2. Belz, A., Reiter, E.: Comparing automatic and human evaluation of NLG systems. In: Proceedings of the 11th Conference of the European Chapter of the Association for Computational Linguistics (EACL 2006), pp. 313–320 (2006)

3. Belz, A., Varges, S.: Generation of repeated references to discourse entities. In: Proceedings of the Eleventh European Workshop on Natural Language Generation (ENLG 2007), pp. 9–16 (2007)
4. Belz, A.: That's nice.. what can you do with it? Computational Linguistics 35(1), 111–118 (2009)
5. Bohnet, B., Dale, R.: Viewing referring expression generation as search. In: Proceedings of the 19th International Joint Conference on Artificial Intelligence (IJCAI 2005), pp. 1004–1009 (2005)
6. Charniak, E., Knight, K., Yamada, K.: Syntax-based language models for machine translation. In: Proceedings of MT Summit IX (2003)
7. Cheng, H., Poesio, M., Henschel, R., Mellish, C.: Corpus-based np modifier generation. In: Proceedings of The Second Meeting of the North American Chapter of the Association for Computational Linguistics, NAACL 2001 (2001)
8. Choi, B., Pak, A.: A catalog of biases in questionnaires. Preventing Chronic Disease 2(1) (2005)
9. Dale, R.: Cooking up referring expressions. In: Proceedings of the 27th Annual Meeting of the Association for Computational Linguistics, ACL 1989 (1989)
10. Dale, R., Reiter, E.: Computational interpretations of the Gricean maxims in the generation of referring expressions. Cognitive Science 19(2), 233–263 (1995)
11. van Deemter, K.: Generating referring expressions: Boolean extensions of the Incremental Algorithm. Computational Linguistics 28(1), 37–52 (2002)
12. van Deemter, K., van der Sluis, I., Gatt, A.: Building a semantically transparent corpus for the generation of referring expressions. In: Proceedings of the 4th International Conference on Natural Language Generation (INLG 2006), Sydney, Australia, pp. 130–132 (2006)
13. Deneefe, S., Knight, K.: Synchronous tree adjoining machine translation. In: Proceedings of the 2009 Conference on Empirical Methods in Natural Language Processing, EMNLP 2009 (2009)
14. Forster, K.I., Forster, J.C.: DMDX: A windows display program with millisecond accuracy. Behavior Research Methods, Instruments, & Computers 35(1), 116–124 (2003)
15. Gatt, A., Belz, A.: Introducing Shared Tasks to NLG: The TUNA Shared Task Evaluation Challenges. In: Krahmer, E., Theune, M. (eds.) Empirical Methods in NLG. LNCS (LNAI), vol. 5790, pp. 264–293. Springer, Heidelberg (2010)
16. Gupta, S., Stent, A.: Automatic evaluation of referring expression generation using corpora. In: Proceedings of the 1st Workshop on Using Copora in Natural Language Generation, pp. 1–6 (2005)
17. Huddleston, R., Pullum, G.: The Cambridge Grammar of the English Language. Cambridge University Press, Cambridge (2002)
18. Jordan, P.W.: Contextual influences on attribute selection for repeated descriptions. In: van Deemter, K., Kibble, R. (eds.) Information Sharing: Reference and Presupposition in Language Generation and Interpretation. CSLI Publications, Stanford (2002)
19. Jordan, P.W., Walker, M.: Learning attribute selections for non-pronominal expressions. In: Proceedings of the 38th Annual Meeting of the Association for Computational Linguistics, ACL 2000 (2000)
20. Koller, A., Striegnitz, K., Byron, D., Cassell, J., Dale, R., Moore, J., Oberlander, J.: The first challenge on generating instructions in virtual environments. In: Krahmer, E., Theune, M. (eds.) Empirical Methods in NLG. LNCS (LNAI), vol. 5790, pp. 329–353. Springer, Heidelberg (2010)

21. Krahmer, E., Theune, M.: Efficient context-sensitive generation of referring expressions. In: van Deemter, K., Kibble, R. (eds.) Information Sharing: Reference and Presupposition in Language Generation and Interpretation, pp. 223–264. CSLI Publications, Stanford (2002)
22. Lin, C.Y., Och, F.J.: ORANGE: A method for evaluating automatic evaluation metrics for machine translation. In: Proceedings of the 20th International Conference on Computational Linguistics (COLING 2004), Geneva, pp. 501–507 (2004)
23. Luo, X.: On coreference resolution performance metrics. In: Proceedings of HLT-EMNLP, pp. 25–32 (2005)
24. Morton, T.: Using Semantic Relations to Improve Information Retrieval. Ph.D. thesis, University of Pensylvania (2005)
25. Nenkova, A.: Entity-driven rewrite for multi-document summarization. In: Proceedings of the Third International Joint Conference on Natural Language Processing, IJCNLP 2008 (2008)
26. Nenkova, A.: Understanding the process of multi-document summarization: content selection, rewrite and evaluation. Ph.D. thesis, Columbia University (2006)
27. Otterbacher, J., Radev, D., Luo, A.: Revisions that improve cohesion in multi-document summaries: a preliminary study. In: Proceedings of the ACL 2002 Workshop on Automatic Summarization, pp. 27–36 (2002)
28. Poesio, M.: Annotating a corpus to develop and evaluate discourse entity realization algorithms: issues and preliminary results. In: Proceedings of the Second International Conference on Language Resources and Evaluation, LREC 2000 (2000)
29. Poesio, M.: Discourse annotation and semantic annotation in the GNOME corpus. In: Proceedings of the ACL 2004 Discourse Annotation Workshop (2004)
30. Qiu, L., Kan, M., Chua, T.S.: A public reference implementation of the rap anaphora resolution algorithm. In: Proceedings of the Fourth International Conference on Language Resources and Evaluation (LREC 2004), pp. 291–294 (2004)
31. Reiter, E., Dale, R.: A fast algorithm for the generation of referring expressions. In: Proceedings of the 14th International Conference on Computational Linguistics (ACL 1992), Nantes, France, pp. 232–238 (1992)
32. Siddharthan, A., Copestake, A.: Generating referring expressions in open domains. In: Proceedings of the 42nd Annual Meeting of the Association for Computational Linguistics (ACL 2004), Barcelona, Spain (2004)
33. Siegel, S.: Non-parametric statistics. The American Statistician 11(3), 13–19 (1957)
34. Steinberger, J., Poesio, M., Kabadjov, M., Jezek, K.: Two uses of anaphora resolution in summarization. Information Processing and Management: Special issue on Summarization 43(6), 1663–1680 (2007)
35. Trang Dang, H.: DUC 2005: Evaluation of question-focused summarization systems. In: Proceedings of the COLING-ACL 2006 Workshop on Task-Focused Summarization and Question Answering, Prague, pp. 48–55 (2006)
36. Turian, J., Shen, L., Melamed, I.D.: Evaluation of machine translation and its evaluation. In: Proceedings of MT Summmit IX, New Orleans, pp. 386–393 (2003)
37. Viethen, J., Dale, R.: Towards the evaluation of referring expression generation. In: Proceedings of the 4th Australasian Language Technology Workshop (ALTW 2006), pp. 115–122 (2006)
38. Vilain, M., Burger, J., Aberdeen, J., Connolly, D., Hirschman, L.: A model-theoretic coreference scoring scheme. In: Proceedings of the 6th Conference on Message Understanding (MUC-6), pp. 45–52 (1995)

The First Challenge on Generating Instructions in Virtual Environments

Alexander Koller[1], Kristina Striegnitz[2], Donna Byron[3], Justine Cassell[4],
Robert Dale[5], Johanna Moore[6], and Jon Oberlander[6]

[1] Saarland University, Saarbrücken, Germany
[2] Union College, Schenectady, NY, USA
[3] Northeastern University, Boston, MA, USA
[4] Northwestern University, Evanston, IL, USA
[5] Macquarie University, Sydney, Australia
[6] University of Edinburgh, Edinburgh, UK

Abstract. This paper describes the First Challenge on Generating Instructions in Virtual Environments (GIVE-1). GIVE is a shared task for generation systems which give real-time natural-language instructions to users in a virtual 3D world. These systems are evaluated by connecting users and NLG systems over the Internet. We describe the design and results of GIVE-1 as well as the participating NLG systems, and validate the experimental methodology by comparing the results collected over the Internet with results from a more traditional laboratory-based experiment.

1 Introduction

The ability to evaluate and compare our algorithms, techniques and systems is fundamental for the progress of research in computational linguistics. Challenges such as the TREC Question-Answering competition[1] and the NIST machine translation competition[2] have helped spawn tremendous interest in their respective subfields. More recently, the Recognizing Textual Entailment challenge[3] has revived broad interest in computational semantics. By making systems comparable and progress measurable, these challenges have improved systems, contributed to our scientific understanding of the research issues, and helped create research communities in their respective areas.

The field of natural language generation (NLG) lags behind other areas in terms of its ability to·evaluate system performance, because NLG systems are inherently hard to evaluate. On the one hand, evaluations based on annotated corpora (e.g. [2]) will misjudge some systems because in NLG a mismatch between the gold standard and a system's output does not necessarily indicate that the system's output is inferior (e.g. [4,16,24] and also see [5,6,9,18] in this

---

[1] http://trec.nist.gov/
[2] http://www.nist.gov/speech/tests/mt/
[3] http://pascallin.ecs.soton.ac.uk/Challenges/

E. Krahmer, M. Theune (Eds.): Empirical Methods in NLG, LNAI 5790, pp. 328–352, 2010.

volume). On the other hand, evaluation studies where human subjects interact with a system or compare the outputs of competing NLG systems are time-consuming and expensive to run, and may be completely infeasible outside of large well-established laboratories due to lack of funding or an insufficient pool of potential subjects.

There has recently been a growing consensus that a convincing evaluation method for NLG is needed [3,13]. But all recent NLG-specific shared tasks that we know about [6,18] (this volume) have been limited to a very specific NLG subtask, that of generating referring expressions. Conversely, the large-scale DARPA Communicator challenge [28] and its (telephone-based) evaluation methodology were tied to evaluating end-to-end spoken dialogue systems and too coarse-grained for a convenient evaluation of just the NLG components. Thus the question of what a generic evaluation method for NLG systems should look like is still largely unanswered.

In this paper, we report on the results of the First Challenge on Generating Instructions in Virtual Environments (GIVE-1). In this shared task, an NLG system must generate natural-language instructions which guide a user towards performing some task in a virtual 3D environment. This makes it possible to explore situated communication in a simulated environment; but perhaps even more importantly, the GIVE Challenge opens up a novel approach to NLG system evaluation, in which human experimental subjects are connected to NLG systems over the Internet. In GIVE-1, this allowed us to collect data from 1143 separate interactions with NLG systems, making GIVE, to our knowledge, the largest ever NLG evaluation effort in terms of the number of experimental subjects taking part.

The paper is structured as follows. We start by reviewing the state of the art in NLG system evaluation, and present the Internet-based evaluation strategy used by GIVE, in Section 2. We then present the specific setup of GIVE-1 in Section 3, and summarize the five NLG systems that participated in GIVE-1 in Section 4. In Section 5, we present the results of GIVE-1. Finally, we validate the Internet-based evaluation methodology we used in GIVE-1 in Section 6, by comparing the results of the Internet evaluation with those obtained in a more traditional laboratory-based setting. Section 7 provides an outlook towards GIVE-2 and concludes.

2   Evaluating NLG Systems

The evaluation of NLG systems has recently been the subject of much discussion. We review some of the main trends, and then introduce a new methodology on which the GIVE Challenge is based: evaluating NLG systems over the Internet.

2.1   Previous Work

Evaluation efforts for NLG systems typically fall into one of three classes: similarity with respect to a gold standard, effectiveness in terms of task performance, and ratings by human judges. The first class aims at judging the output of an

NLG system by comparing it against one or more gold standards produced by human annotators, using comparison metrics such as those proposed by [2]. The advantage of this approach is that evaluations are cheap and easily repeatable once the gold-standard corpus has been built. This is why the first major shared tasks for NLG, the TUNA [18] and GREC [6] challenges on generating referring expressions, both focused on this approach and only added very small studies involving human subjects.

Unfortunately, evaluation against a gold standard is a much less natural evaluation method for NLG than it is, for example, in parsing. One problem is that the same meaning can usually be expressed equally well in many different ways, and a gold standard cannot capture all possible variations. Furthermore, it has been shown on the TUNA data that those gold-standard-based metrics that are currently available do not correlate with task-based evaluation measures [4,18]. Similarly, gold-standard-based metrics have been found to not always correlate well with the ratings of human judges [7,10]. In our opinion, gold-standard evaluations may be useful for tracking improvements in a single NLG system, but are less suitable for large-scale evaluation efforts.

Alternatively, human subjects can be asked to to perform some task that depends on the system output (such as identifying the target referent [4]). Their task performance is a measure of the NLG system's effectiveness for this task. The problem with this approach is that such experiments, which involve getting large numbers of subjects into a laboratory, are expensive and time-consuming. As a consequence, the number of subjects tends to be relatively low (for instance, the TUNA 2009 evaluation used sixteen subjects [17]), which can limit the evaluation's ability to detect significant differences. We know of only one evaluation effort that tested systems involving NLG on a task-based evaluation with human subjects on a large scale: the DARPA Communicator evaluation [28]. However, this was an end-to-end evaluation effort for spoken dialogue systems, which makes it too coarse-grained to tell us much about the specific performance of the NLG components.

The third alternative is to present the system outputs to human judges and ask them to rate their quality. Similarly to task-based evaluations, such evaluations involve significant costs and time requirements. These may still be lower than those of a task-based evaluation since judges are typically not naive, but have some training in linguistics and are aware of the study and its purpose, so that they can work independently and can handle larger amounts of data. However, ratings by judges do not produce a direct measure for how "real" users would interact with an NLG system. Interestingly, the TUNA 2009 evaluation [17] found a correlation between task-based evaluation measures and ratings by judges, but since this is, to the best of our knowledge, the only work so far that compares these two approaches, further research is needed.

2.2   Internet-Based Evaluation of NLG Systems

In this paper, we propose a new approach to evaluating NLG systems which is meant to provide results that are as meaningful as those of a laboratory-based evaluation with human subjects, but at a much lower cost. The key idea is to provide a software infrastructure that allows us to physically separate the experimental subject and the NLG system; the subject runs a client program on their own computer, which connects to the NLG system over the Internet. Access to subjects is obtained by couching the subject's task in a game-like environment, and making it easy to start the client from a public website.

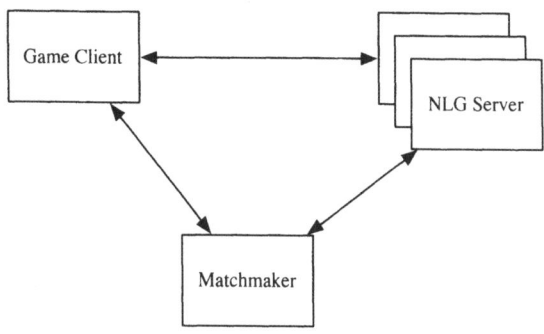

Fig. 1. The architecture of the evaluation software

More concretely, the software architecture we implemented for the GIVE Challenge involves three different software components (see Figure 1):

1. the *client*, which implements the task which the users must perform, and displays the NLG system's outputs;
2. the *NLG servers*, which generate the natural-language instructions; and
3. the *Matchmaker*, which establishes connections between clients and NLG servers and enters game logs into a database for future analysis.

These three components run on different machines. The client is downloaded by users from a central website and run on their local machine; each NLG server is run on a server at the institution that implemented it; and the Matchmaker runs on a central server provided by the evaluation organizers. When a user starts the client, it connects to the Matchmaker and is randomly assigned an NLG server and a task instance (e.g., one particular 3D world). The client and NLG server then communicate over the course of one task execution. At the end of this run, the client displays a questionnaire to the user, and the log of the user's actions, the NLG system's utterances, and the questionnaire data are uploaded to the Matchmaker and stored in a database.

The Java implementation of our framework is available as an open-source project.[4] While the current implementation is targeted specifically at the GIVE task, it should be relatively straightforward to adapt it to other NLG tasks.

---

[4] See http://www.give-challenge.org/research/page.php?id=software

2.3 Access to Subjects

One huge advantage of a web-based evaluation is that it provides easy access to the entire population of the Internet as a pool of potential experimental subjects. This resource has been successfully exploited in the past for such tasks as image labelling [1], script learning [23], and corpus creation [11], as well as in psychological and psycholinguistic "web experiments" [20].

From the perspective of the experimental subject, they simply perform a task in which they are guided by natural-language instructions; in principle, they don't even need to know that the instructions come from an automated system. They can be recruited by any means that will get them to visit the website from which they can download the client. As we will see below, even relatively conservative methods of recruiting subjects already provide quite satisfactory numbers.

One complication that must be addressed in web-based evaluations is the lack of control over the subject pool. As we will show below, several demographic factors including language proficiency and gender had measurable effects on the results of the evaluation. We mitigated this concern by logging the IP addresses from which the clients connected (which can be resolved to countries) and asking users to specify their gender and self-rate their language proficiency in a questionnaire. We also made each player complete a tutorial before they started on the game world proper, to ensure at least a basic familiarity with controlling a 3D game.

2.4 Evaluation

Because the complete log of the user's actions and the NLG system's output is stored in the database, it is straightforward to extract a number of objective evaluation measures from the data. This includes measures such as the users' task success rate for each NLG system and each task instance, and the task completion times for successful runs. This data can be collected completely unobtrusively, without requiring any user intervention at all. In addition, experimental subjects can be asked to fill in a questionnaire to provide subjective data such as whether they found the generated texts helpful. Unfortunately, we are not aware of a reasonable way over the Web to force a user to actually provide meaningful answers to all items in a questionnaire.

Based on the individual data points with the objective and subjective measures for each subject, it is possible to compute aggregate evaluation measures for each NLG system and determine which differences between these aggregate values are statistically significant. However, more fine-grained analyses are also possible, especially because each NLG system can leave behind timestamped messages in the log at any point, which can later be analyzed by its developers in detail.

Note that the approach to evaluation we have just presented focuses on *gathering* the data on which the evaluation should be based; we are not arguing for a specific set of appropriate measures for evaluating NLG systems, except that it must be possible to collect data for these measures online. In this way, our proposal is orthogonal to frameworks like PARADISE [27], which provide concrete evaluation measures and show how to combine them.

3   The GIVE Challenge

We will now describe how we implemented this data-gathering framework in the GIVE Challenge. In the GIVE scenario, subjects try to solve a treasure hunt in a virtual 3D world that they have not seen before. The NLG system has access to a complete symbolic representation of the virtual world. The challenge is to generate, in real time, natural-language instructions that will guide the user to the successful completion of their task.

Users participating in the GIVE evaluation start the 3D game from our website at www.give-challenge.org. They then see a 3D game window as in Figure 2, which displays instructions and allows them to move around in the world and manipulate objects. The first room they visit is a tutorial room where users learn how to interact with the system; they then enter one of three evaluation worlds, where instructions for solving the treasure hunt are generated by an NLG system. Users can either finish a game successfully, lose it by triggering an alarm, or cancel the game. This result is stored in a database for later analysis, along with a complete log of the game.

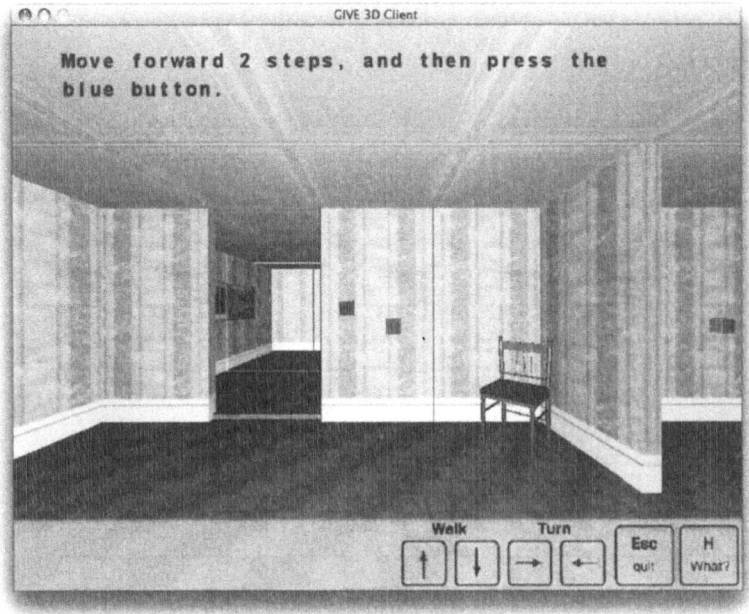

Fig. 2. What the user sees when playing the GIVE game

3.1 GIVE as a Special Case of the Web-Based Method

In its reliance on connecting users and NLG systems over the Web, GIVE is a special case of the framework we presented in Section 2. GIVE specializes this general approach in two ways.

First, GIVE is specifically about generating *instructions* in *real time*. This differentiates GIVE from NLG research which focuses on the generation of descriptions or narratives. It also differentiates GIVE from related work which focuses on "batch" instructions, in which an entire discourse of instructions is presented at once, with the result that the instruction giver cannot react dynamically to the instruction follower's behavior.

Second, GIVE focuses on *situated* natural language generation. This makes the NLG task quite different to that found in other NLG challenges. For example, experiments have shown that human instruction givers make the instruction follower move to a different location in order to use a simpler referring expression [25]. That is, referring expression generation becomes a very different problem than the classical non-situated Dale and Reiter style referring expression generation [14], which focuses on generating referring expressions that are single noun phrases in the context of an unchanging world.

Both of these characteristics make GIVE a task that fits well with web-based evaluation: It is easy to judge whether and how easily a human user has followed real-time instructions, and by embedding the task in a virtual environment, it becomes possible to present it as an online game. On the other hand, the GIVE task is still open-ended enough to potentially be of interest to a wide range of NLG researchers. This is most obvious for research in sentence planning (i.e., issues such as referring expression generation, aggregation, and lexical choice) and realization, where the real-time nature of the task imposes high demands on the system's efficiency. But there are various ways of extending the task (e.g., to two-way dialog, speech output, or a different scenario) such that it also involves issues of prosody generation (i.e., research on text/concept-to-speech generation) and discourse generation. Finally, the game world can be designed to focus on specific issues in NLG, such as the generation of referring expressions or the generation of navigation instructions.

3.2 Game Worlds

Each GIVE game run takes place in a *game world*, consisting of the map of a virtual environment along with descriptions of the objects in the world with their 3D positions and information about relationships between the objects: for example, the fact that pressing a certain button will open a certain door. The teams participating in GIVE-1 were given a *development world* early on in the challenge, against which they tested their systems.

The data gathering then took place on three new *evaluation worlds*. We made these worlds available to the NLG system developers one week before starting the data-gathering phase to ensure the compatibility of all systems with these

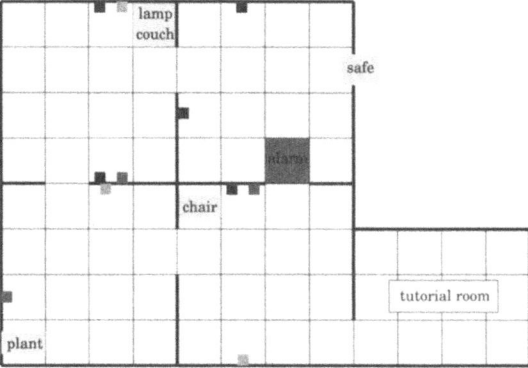

Fig. 3. World 1

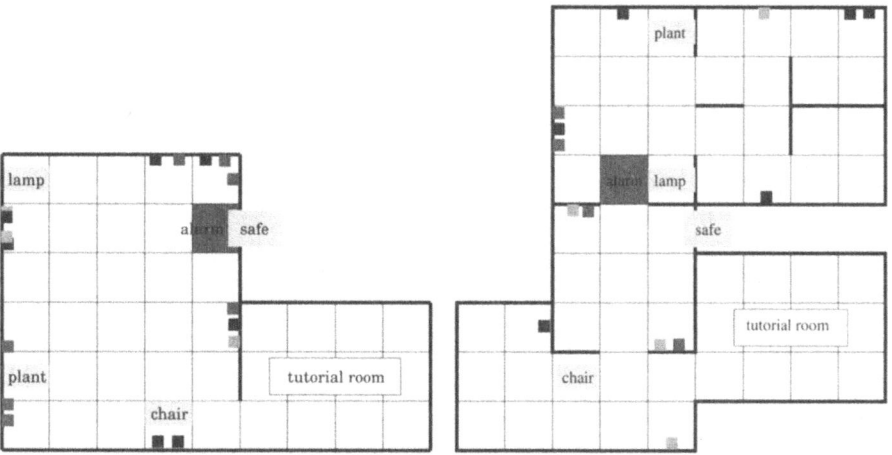

Fig. 4. World 2   **Fig. 5.** World 3

worlds, but asked the developers not to specifically adapt their systems to these worlds. Figures 3–5 show the layout of the three evaluation worlds. The worlds were intended to provide varying levels of difficulty for the direction-giving systems and to focus on different aspects of the problem. World 1 is at a level of complexity similar to that of the development world. World 2 was intended to focus on referring expressions: the world has only one room which is full of objects and buttons, many of which cannot be distinguished by simple descriptions. World 3, on the other hand, puts more emphasis on navigation directions, as the world has many interconnected rooms and hallways.

All GIVE-1 worlds were split up into square tiles, and only permitted the user to jump from the center of one tile to the center of the next, and to turn in discrete 90-degree increments. The game client enforced the requirement that only such discrete movements could be made.

```
abstract class NlgSystem:
    void connectionEstablished();
    void connectionDisconnected();
    void handleAction(Atom actionInstance, List⟨Formula⟩ updates);
    void handleMoveTurnAction(Direction direction);
    void handleDidNotUnderstand();
    void handleStatusInformation(Position playerPosition,
                 Orientation playerOrientation, List⟨String⟩ visibleObjects);
    ...
```

Fig. 6. The interface of an NLG system

3.3 Software Infrastructure

The NLG system developers also had access to the GIVE software described in Section 2.2, and used it to develop their own systems.

The client we use in GIVE is responsible for displaying the virtual world, allowing the user to interact with it, and exchanging data with the NLG server over the Internet. To maximize portability, the client, like all other components of the GIVE software, is implemented in Java; users start it directly from the website using Java Web Start. Although the low-level details of drawing OpenGL graphics are handled by the jMonkeyEngine, a free 3D game library for Java, we still spent the majority of our development effort on the 3D graphics. We could have reduced this effort by building upon an existing virtual 3D world system such as Second Life. However, the effort needed to adapt such a system to our needs would have been at least as high (in particular, we would have had to ensure that the user could only move according to the rules of the GIVE game and to instrument the virtual world to obtain real-time updates about events), and the result would have been less extensible to future installments of the challenge.

To simplify the job of the NLG system developers, we implemented a scaffold for the NLG system servers that handled all the necessary networking. This allowed system developers to focus on the development of their NLG systems. Specifically, they implemented concrete subclasses of the class `NlgSystem`, shown in Figure 6. This involved overriding the six abstract callback methods in this class with concrete implementations in which the NLG system reacts to specific events. The methods `connectionEstablished` and `connectionDisconnected` are called when users enter the game world and when they disconnect from the game. The method `handleAction` gets called whenever the user performs some physical action, such as pushing a button, and specifies what has changed in the world as a consequence of this action; `handleMoveTurnAction` gets called whenever the user moves; `handleDidNotUnderstand` gets called whenever the user presses the H key to signal that they didn't understand the previous instruction; and `handleStatusInformation` gets called once per second and after each user action to inform the server of the player's position and orientation, and the objects which are visible from that position. Ultimately, each of these method calls gets triggered by a message that the client sends over the network in reaction to some event; but this is completely hidden from the NLG system developer.

The NLG system can use the method **send** to send a string to the client to be displayed. It also has access to various methods for querying the state of the game world and to an interface to an external planner which can compute a sequence of actions leading to the goal. The planner that worked best in the GIVE planning domain, and which we provided in Linux and MacOS versions for GIVE-1, was SGPLAN 5.2.2 [19,21].

3.4   Timeline

After the GIVE Challenge was publicized in March 2008, eight research teams signed up for participation. We distributed an initial version of the GIVE software and a development world to these teams. In the end, four teams submitted NLG systems. These were connected to a central Matchmaker instance that ran for about three months, from 7 November 2008 to 5 February 2009.

During this time, we advertised the GIVE Challenge to the public in order to encourage people to play the GIVE game and serve as experimental subjects. Subjects were recruited via online press releases (in English and German), postings to email lists, and online gaming forums. As an incentive, one Amazon voucher was given away to a randomly-chosen subject each month.

4   Systems Participating in GIVE-1

The four participating research teams submitted the following five systems to the evaluation:

System	Research Team
Austin [12]	University of Texas at Austin
Madrid [15]	Universidad Complutense de Madrid
Twente [8]	University of Twente
Union [26]	Union College
Warm-Cold [8]	University of Twente

We provide here a brief comparative overview of the systems; for more details see the papers cited in the above table.

The *Warm-Cold* system stands out as the only system that does not purely focus on generating instructions that are easy to understand and follow, but tries to create a more game-like and entertaining atmosphere. Instead of navigation instructions, it only provides the users with some feedback on whether they are getting closer to the next button they needed to press ("warmer") or are moving away from it ("colder").

The *Austin* team's primary focus was on improving the plans produced by the off-the-shelf planner provided by the organizers, which sometimes lead the user on unintuitive detours. Their system only retains the object manipulation instructions from the plan and uses A* search for path planning. Otherwise, the system uses a relatively simple generation strategy that maps plan actions to instructions almost step-by-step. The only aggregation that is done combines sequences of actions of the same type into one instruction: for example, it generates *move forward three steps* to combine three movement actions, or *turn around* to combine two turn left or turn right actions.

The other three systems, *Madrid*, *Twente*, and *Union*, all have the ability to switch between different levels or modes of instruction giving, which allow the systems, in certain situations, to use higher-level instructions that do not explicitly mention every single action in the plan.

The *Madrid* system directs the user to a door or button by describing the appearance and location of these objects, but without prescribing the individual steps for how to get to there. If this strategy is not possible—for example, because the object is not visible—the system guides the user from reorientation point to reorientation point until it becomes visible. This system produces relatively complex object descriptions. It uses concepts (such as spatial regions like rooms and corners) that are not available in the world representation provided by the GIVE framework, but are computed by the NLG system by analyzing the world map. Another aspect that distinguishes the *Madrid* system is that it pro-actively produces alerts or warnings to keep the user from doing potentially dangerous things.

The *Twente* team focused on making their system adapt to the user's behavior. The system distinguishes between three levels of direction giving: at the first level all instructions contain only one action type (like the *Austin* system), at the second level instructions can combine forward movements with another action (e.g., *walk forward 3 steps and then press the button*), and at the third level, once users can see the object that needs to be manipulated next, they are instructed to do so without prescribing the path to that object (similar to *Madrid*'s strategy). The system starts out in level two and then continuously (re-)estimates the user's success by counting how many actions they perform in 5 seconds. If they perform many actions, they are assumed to be doing well, and the instruction level goes up. If they carry out few actions, they are assumed to have problems and the level goes down. The system also switches to a lower level if the user explicitly asks for help by pressing "H".

Unfortunately, the referring expression generation module of the *Twente* system had a bug which sometimes resulted in the production of referring expressions where left and right were switched; this is fatal in evaluation worlds 2 and 3.

The *Union* team focused on producing landmark-based directions. Their system switches between a landmark mode and a path-based mode. The path-based mode is similar to *Twente*'s second level, in that instructions can combine movement actions with another action type. In the landmark mode, the system checks whether the object that needs to be manipulated next is visible. If so, it is described and the user is instructed to manipulate it without receiving any further instructions on how to reach it. If that object is not visible, the system tries to find another visible object along the way that can be used as a landmark. The path-based mode is only used if the landmark mode is not possible because there are no visible objects that can be used as landmarks, or if users are deemed to have problems because they press "H" or because they are not making enough progress toward their target.

5   Results

We now report on the results of GIVE-1. We start with some basic demographics; then we discuss objective and subjective evaluation measures.

Notice that some of our evaluation measures are in tension with each other: For instance, a system which gives very low-level instructions (*move forward*; *ok, now move forward*; *ok, now turn left*) will lead the user to complete the task in a minimum number of steps; but it will require more instructions than a system that aggregates these. This tension is intentional, and emphasizes both the exploratory character of GIVE-1 and our desire to make GIVE a friendly comparative challenge rather than a competition with a clear winner. Our goal is to provide as many useful measures as possible in order to establish a framework in which research teams can evaluate and compare their systems along a variety of dimensions.

5.1   Demographics

Over the course of three months, we collected 1143 valid games. A game counted as valid if the game client didn't crash, the game wasn't marked as a test game by the developers, and the player completed the tutorial.

Of these games, 80.1% were played by males and 9.9% by females; a further 10% didn't specify their gender. The players were widely distributed over countries: 37% connected from an IP address in the US, 33% from an IP address in Germany, and 17% from China; Canada, the UK, and Austria also accounted for more than 2% of the participants each, and the remaining 2% of participants connected from a further 42 countries. This imbalance stems from very successful press releases that were issued in Germany and the US and which were further picked up by blogs, including one in China. Despite this geographical spread, over 90% of the participants who answered this question self-rated their English proficiency as "good" or better. About 75% of users connected with a client running on Windows, with the rest split about evenly between Linux and Mac OS X.

The effect of the press releases is also plainly visible if we look at the distribution of the valid games over the days from November 7, 2008 to February 5, 2009 (Figure 7). There are huge peaks at the very beginning of the evaluation period, coinciding with press releases through Saarland University in Germany and Northwestern University in the US, which were picked up by science and technology blogs on the Web. The US peak contains a smaller peak of connections from China, which were sparked by coverage in a Chinese blog. By comparison, posting an invitation to connect to the GIVE game to the mailing list of

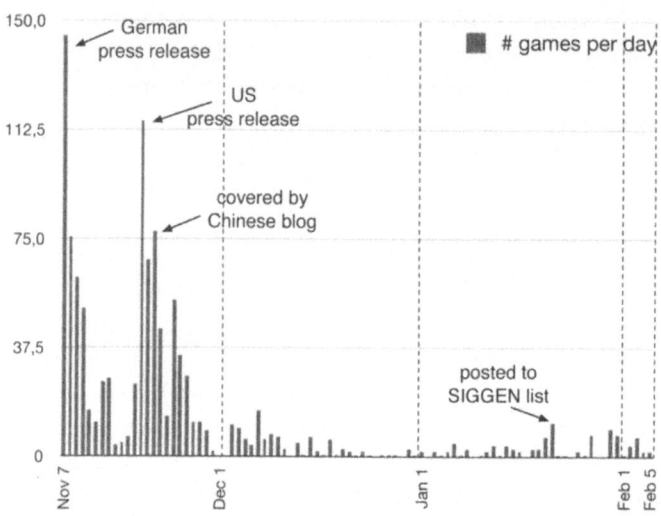

Fig. 7. Histogram of the connections per day

SIGGEN (the ACL's Special Interest Group on NLG) yielded a much weaker response. This illustrates the potential for recruiting experimental subjects over the Internet, compared to recruitment within the scientific community.

There is a risk that a user who plays twice will perform better the second time, because they may be more familiar with the task. We did not actively control this because we wanted to keep access to the online game as simple as possible. In the end, users from about 20% of the participating IP addresses connected multiple times. We believe this an acceptably low number, particularly because the duplicates are distributed evenly over the NLG systems.

5.2 Objective Measures

We extracted objective and subjective measurements from the valid games. The objective measures are summarized in Figure 8. For each system and game world, we measured the percentage of games which were completed successfully. Furthermore, for each game we counted the number of instructions the system sent

- task success (Did the player get the trophy?)
- instructions (Number of instructions produced by the NLG system.*)
- steps (Number of all player actions.*)
- actions (Number of object manipulation actions.*)
- second (Time in seconds.*)

* Measured from the end of the tutorial until the end of the game.

Fig. 8. Objective measurements

	Austin	Madrid	Twente	Union	Warm-Cold
task success	40% B	71% A	35% B	73% A	18% C
instructions	83.2 B	58.3 A	121.2 C	80.3 B	214.1 D
steps	103.6 A	124.3 B	160.9 C	117.5 A B	307.4 D
actions	11.2 B	8.7 A	14.8 C	9.0 A	15.1 C
seconds	129.3 A	174.8 B	207.0 C	175.2 B	320.7 D

Fig. 9. *Objective* measures by system. Task success is reported as the percentage of successfully completed games. The other measures are reported as the mean number of instructions/steps/actions/seconds, respectively. Letters group indistinguishable systems; systems that don't share a letter were found to be significantly different with $p < 0.05$.

to the user, measured the time until task completion, and counted the number of low-level steps executed by the user (any key press, to either move or manipulate an object) as well as the number of task-relevant actions (such as pushing a button to open a door). To ensure comparability, we only counted successfully completed games for all these measures, and only started counting when the user left the tutorial room. Crucially, all objective measures were collected completely unobtrusively, without requiring any action on the user's part.

Figure 9 shows the results of these objective measures. This figure assigns systems to groups A, B, etc. for each evaluation measure. Systems in group A are better than systems in group B, etc.; if two systems don't share the same letter, the difference between these two systems is significant with $p < 0.05$. Significance was tested using a χ^2 test for task success and ANOVAs for instructions, steps, actions, and seconds. These were followed by post hoc tests (pairwise χ^2 and Tukey) to compare the NLG systems pairwise.

Overall, there is a top group consisting of the *Austin*, *Madrid*, and *Union* systems: while *Madrid* and *Union* outperform *Austin* on task success (with 70–80% of successfully completed games, depending on the world), *Austin* significantly outperforms all other systems in terms of task completion time. As expected, the *Warm-Cold* system performs significantly worse than all others in almost all categories. This confirms the ability of the approach described here to distinguish the performances of different systems.

5.3   Subjective Measures

The subjective measures, which were obtained by asking the users to fill in a questionnaire after each game, are shown in Figure 10. All of the questions were answered on 5-point Likert scales, with the exception of "overall", which used a 7-point scale, and the "informativity" and "timing" questions, which had nominal answers. For each question, the user could choose not to answer.

7-point scale items:

overall: What is your overall evaluation of the quality of the direction-giving system? (very bad 1 ... 7 very good)

5-point scale items:

task difficulty: How easy or difficult was the task for you to solve? (very difficult 1 2 3 4 5 very easy)

goal clarity: How easy was it to understand what you were supposed to do? (very difficult 1 2 3 4 5 very easy)

play again: Would you want to play this game again? (no way! 1 2 3 4 5 yes please!)

instruction clarity: How clear were the directions? (totally unclear 1 2 3 4 5 very clear)

instruction helpfulness: How effective were the directions at helping you complete the task? (not effective 1 2 3 4 5 very effective)

choice of words: How easy to understand was the system's choice of wording in its directions to you? (totally unclear 1 2 3 4 5 very clear)

referring expressions: How easy was it to pick out which object in the world the system was referring to? (very hard 1 2 3 4 5 very easy)

navigation instructions: How easy was it to navigate to a particular spot, based on the system's directions? (very hard 1 2 3 4 5 very easy)

friendliness: How would you rate the friendliness of the system? (very unfriendly 1 2 3 4 5 very friendly)

Nominal items:

informativity: Did you feel the amount of information you were given was: too little / just right / too much

timing: Did the directions come ... too early / just at the right time / too late

Fig. 10. Questionnaire items

	Austin	Madrid	Twente	Union	Warm-Cold
overall	4.9 A	4.9 A	4.3	4.6 A	3.6
			B	B	
					C
task difficulty	4.3 A	4.3 A	4.0 A	4.3 A	3.5
					B
goal clarity	4.0 A	3.7 A	3.9 A	3.7 A	3.3
					B
play again	2.8 A	2.6 A	2.4 A	2.9 A	2.5 A
instruction clarity	4.0 A	3.6 A	3.8 A	3.6	3.0
		B	B	B	
					C
instruction helpfulness	3.8 A	3.9 A	3.6 A	3.7 A	2.9
					B
choice of words	4.2 A	3.8	4.1 A	3.7	3.5
		B	B		
		C		C	C
referring expressions	3.4	3.9 A	3.7 A	3.7 A	3.5
	B		B	B	B
navigation instructions	4.6 A	4.0	4.0	3.7	3.2
		B	B	B	
					C
friendliness	3.4 A	3.8 A	3.1	3.6 A	3.1
	B		B		B
informativity	46%	68% A	51%	56%	51%
	B		B	B	B
timing	78% A	62%	60%	62%	49%
		B	B	B	
			C		C

Fig. 11. *Subjective* measures by system. Informativity and timing are reported as the percentage of successfully completed games in which users chose "just right". The other measures are the mean ratings reported by the players. Letters group indistinguishable systems; systems that don't share a letter were found to be significantly different with $p < 0.05$.

The results of the subjective measurements are summarized in Figure 11, in the same format as for the objective measurements. We ran χ^2 tests for the nominal variables informativity and timing, and ANOVAs for the scale data. Again, we used post hoc pairwise χ^2 and Tukey tests to compare the NLG systems to each other one by one.

Here there are fewer significant differences between different groups than for the objective measures: For the "play again" category, there is no significant difference at all. Nevertheless, *Austin* is shown to be particularly good at navigation instructions and timing, whereas *Madrid* outperforms the rest of the field in "informativity". In the overall subjective evaluation, the earlier top group of *Austin*, *Madrid*, and *Union* is confirmed, although the difference between *Union* and *Twente* is not significant. However, *Warm-Cold* again performs significantly worse than all other systems in most measures. Furthermore, although most systems perform similarly on "informativity" and "timing" in terms of the number of users who judged them as "just right", there are differences in the tendencies: *Twente* and *Union* tend to be overinformative, whereas *Austin* and *Warm-Cold* tend to be underinformative; *Twente* and *Union* tend to give their instructions too late, whereas *Madrid* and *Warm-Cold* tend to give them too early.

5.4   Further Analysis

In addition to the differences between NLG systems, there may be other factors which also influence the outcome of our objective and subjective measures. We tested the following five factors: evaluation world, gender, age, computer expertise, and English proficiency (as reported by the users on the questionnaire). We found that there is a significant difference in task success rate for different evaluation worlds and between users with different levels of English proficiency.

The interaction graphs in Figures 12 and 13 also suggest that the NLG systems differ in their robustness with respect to these factors. χ^2 tests that compare the success rate of each system in the three evaluation worlds show that while the instructions of *Union* and *Madrid* seem to work equally well in all three worlds, the performance of the other three systems differs dramatically between the different worlds. World 2 was especially challenging for some systems as it required relational object descriptions, such as *the blue button on the left of another blue button*.

The players' English skills also affected the systems in different ways. *Union* and *Twente* seem to communicate well with players on all levels of proficiency (χ^2 tests do not find a significant difference). *Austin*, *Madrid* and *Warm-Cold*, on the other hand, don't manage to lead players with only basic English skills to success as often as other players. However, if we remove the players with the lowest level of English proficiency, language skills no longer have an effect on the task success rate for any of the systems.

We also asked the participants to rate their computer expertise and how many hours of video games they played per week. Neither of these showed an effect on our measures.

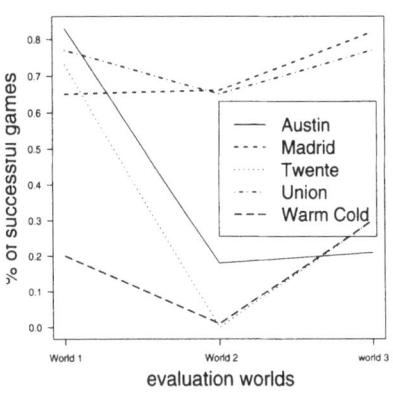
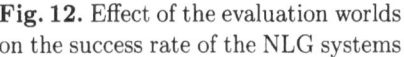
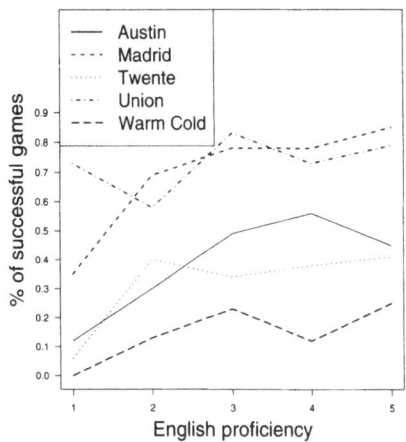

Fig. 12. Effect of the evaluation worlds on the success rate of the NLG systems

Fig. 13. Effect of the players' English skills on the success rate of the NLG systems

6   Validating the Experimental Approach

We have just seen that the web-based data gathering method provided informative results for the GIVE Challenge in that we managed to access a large number of experimental subjects and detect a number of significant differences between the different NLG systems. However, one could argue that, because of the lack of control we have over the selection of subjects over the Internet, these results may be artificially skewed or less precise than the results we would have obtained with a more traditional laboratory-based experiment. To allay such concerns, we repeated part of the GIVE-1 Challenge evaluation in a laboratory setting and compared the results to those of the web-based setting.

6.1   The Laboratory Experiment

For this laboratory-based evaluation, we recruited 91 participants from a college campus. Each participant played the GIVE game once with each of the five NLG systems, in different orders. To avoid learning effects, we only used the first game run from each subject in the comparison with the web experiment; as a consequence, subjects were distributed evenly over the NLG systems. To accommodate for the much lower number of participants, the laboratory experiment only used a single game world: World 1, which was known from the online version to be the easiest world.

Among this group of subjects, 93% self-rated their English proficiency as "expert" or better; 81% were native speakers. In contrast to the online experiment, 31% of participants were male and 65% were female (4% did not specify their gender).

6.2 Results

Figure 14 shows the results of the laboratory experiment for the objective measures collected. The table also includes the results for the Internet experiment on World 1; this makes the data comparable between the two experiments, and is in contrast to the data in Figure 9, which reports aggregate data for all three worlds. The task success rate is only evaluated on games that were completed successfully or lost, not cancelled, as laboratory subjects were asked not to cancel. This brings the number of Internet subjects to 322 for the success rate, and to 227 (only successful games) for the other measures. Significance was tested and is reported as in Section 5.2.

The results for the subjective measures are summarized in Figure 15, in the same format and using the same analysis method as for the objective measures. Also as above, the table is based only on games in World 1, and on games that were completed successfully. We justify this latter choice below. One

	Austin	Madrid	Twente	Union	Warm-Cold	Austin	Madrid	Twente	Union	Warm-Cold
task success	91% A	76% B	85% A B	93% A B	24% C	100% A	95% A	93% A	100% A	17% B
instructions	83.4 B	68.1 A	97.8 C	99.8 C	159.7 D	78.2 A B	66.3 A	107.2 C	88.8 B C	134.5 D
steps	99.8 A	145.1 B	142.1 B	142.6 B	256.0 C	93.4 A	141.8 B	134.6 B	128.8 B	213.5 C
actions	9.4 A	10.0 A B	9.7 A B	10.3 B	9.6 A B	9.9 A	10.5 A	9.6 A	9.8 A	10.0 A
seconds	123.9 A	195.4 B C	174.4 B	194.0 B C	234.1 C	143.9 A	211.8 B	205.6 B	195.1 A B	252.5 B
	Web experiment					**Laboratory experiment**				

Fig. 14. Comparison of the five NLG systems in terms of the *objective* measures collected in the *web* (left) and the *laboratory* (right) experiments on World 1.

consequence, however, is that the results for the *Warm-Cold* system in the lab experiment are based on only two subjects (all others lost when working with *Warm-Cold*) and may therefore not be meaningful.

6.3 Discussion

The primary question that interests us in a comparative evaluation is which NLG systems performed significantly better or worse on any given evaluation measure. In the experiments above, we find that of the 170 possible significant differences (= 17 measures × 10 pairs of NLG systems), the laboratory experiment only found six that the web-based experiment didn't find. Conversely, there are 26 significant differences that only the Internet-based experiment found. But even more importantly, all pairwise rankings are consistent across the two evaluations: Where both systems found a significant difference between two systems, they always ranked them in the same order. We conclude that the Internet experiment provides significance judgments that are comparable to, and in fact more precise than, the laboratory experiment.

Nevertheless, there are important differences between the laboratory and Internet-based results. For instance, the success rates in the laboratory tend to be higher than on the Internet, but so are the completion times. We believe that these differences can be attributed to the demographic characteristics of the participants in the two experiments. To substantiate this claim, we looked in some detail at three of these differences: gender, language proficiency, and questionnaire response rates.

First, the gender distribution differed greatly between the Internet experiment (10% female) and the laboratory experiment (65% female). This is relevant because gender had a significant effect on task completion time (women took longer) and on six subjective measures including "overall evaluation" in the laboratory. We speculate that the difference in task completion time may be related to reported gender differences in processing navigation instructions [22].

Second, the two experiments collected data from subjects with different language proficiencies. While 93% of the participants in the laboratory experiment self-rated their English proficiency as "expert" or better, only 62% of the Internet participants did. This partially explains the lower task success rates on the Internet, as Internet subjects with English proficiencies of 3–5 performed significantly better on "task success" than the group with proficiencies 1–2. If we only look at the results of high-English-proficiency subjects on the Internet, the success rates for all NLG systems except *Warm-Cold* rise to at least 86%, and are thus close to the laboratory results.

Finally, the Internet data are skewed by the tendency of unsuccessful participants to not fill in the questionnaire. Figure 16 summarizes some data about the "overall evaluation" question. Users who didn't complete the task successfully tended to judge the systems much lower than successful users, but at the same time tended not to answer the question at all. This skew causes the mean subjective judgments across all Internet subjects to be artificially high. It is to make the data more comparable with respect to this that Figure 15 in-

	Austin	Madrid	Twente	Union	Warm-Cold	Austin	Madrid	Twente	Union	Warm-Cold
overall	5.6	5.0	4.5	4.5	4.8	5.7	5.4	4.9	5.7	5.0
	A	A			A	A	A	A	A	A
		B	B	B	B					
task difficulty	4.9	4.3	4.3	4.4	4.0	4.9	4.0	4.4	4.3	4.0
	A					A		A	A	A
		B	B	B	B		B	B	B	B
goal clarity	4.9	4.1	4.2	4.0	4.2	4.9	3.7	4.3	4.6	3.0
	A				A	A		A	A	A
		B	B	B	B		B	B	B	B
play again	2.2	2.7	2.2	2.8	3.3	2.9	2.7	1.9	1.9	1.0
	A	A	A	A	A	A	A	A	A	A
instruction clarity	4.9	3.9	4.1	3.9	3.5	4.8	3.8	4.4	4.8	2.0
	A					A		A	A	
		B	B	B	B		B	B		
							C			C
instruction helpfulness	4.6	4.1	3.8	3.6	3.24	5.0	4.4	3.3	4.5	4.0
	A	A				A	A		A	A
		B	B	B	B			B		B
choice of words	4.7	3.8	4.4	4.0	3.8	4.7	3.8	4.5	4.7	4.5
	A		A			A		A	A	A
		B	B	B	B		B	B		B
referring expressions	4.7	4.0	4.3	4.0	4.2	4.8	4.3	4.4	4.3	4.0
	A		A		A	A	A	A	A	A
		B	B	B	B					
navigation instructions	4.6	4.0	4.1	3.8	3.4	4.7	3.7	4.1	4.3	4.0
	A		A			A	A	A	A	A
		B	B	B	B					
friendliness	3.8	3.9	3.3	3.7	3.5	3.9	4.3	3.5	4.1	3.0
	A	A	A	A	A	A	A	A	A	A
informativity	63%	67%	48%	62%	59%	77%	84%	43%	75%	100%
	A	A	A	A	A	A	A	A	A	A
timing	81%	70%	73%	51%	50%	92%	95%	64%	100%	100%
	A	A	A			A	A	A	A	
		B	B		B	B	B	B		B
		C		C	C					

Web experiment   **Laboratory experiment**

Fig. 15. Comparison of the five NLG systems in terms of *subjective* measures collected in the *web* (left) and the *laboratory* (right) experiments on World 1

Web

	# of games	reported	mean
success	227 = 61%	93%	4.9
lost	92 = 24%	48%	3.4
cancelled	55 = 15%	16%	3.3

Lab

	# of games	reported	mean
success	73 = 80%	100%	5.4
lost	18 = 20%	94%	3.3
cancelled	0	–	–

Fig. 16. Skewed results for "overall evaluation". "Reported" is the percentage of subjects who answered the "overall evaluation" question after having succeeded/lost/cancelled the game. "Mean" is the mean "overall evaluation" score.

cludes only judgments from successful games in both the laboratory and Internet experiments.

In summary, we find that while the two experiments made consistent significance judgments, and the Internet-based evaluation methodology thus produces meaningful results, the absolute values they find for the individual evaluation measures differ due to the demographic characteristics of the participants in the two studies. This could be taken as a possible disadvantage of the Internet-based evaluation. However, we believe the opposite to be the case. In many ways, an online user is in a much more natural communicative situation than a laboratory subject who is being discouraged from cancelling a frustrating task. In addition, every experiment, whether in the laboratory or on the Web, suffers from some skew in the subject population due to sampling bias; for instance, one could argue that an evaluation that is based almost exclusively on native speakers in universities leads to overly benign judgments about the quality of NLG systems.

One advantage of the Internet-based approach to data collection over the laboratory-based one is that, due to the sheer number of subjects, we can detect such skews and deal with them appropriately. For instance, we might decide that we are only interested in the results from proficient English speakers and ignore the rest of the data; but we retain the option to run the analysis over all participants, and to analyze how much each system relies on the user's language proficiency. The amount of data also means that we can obtain much more fine-grained comparisons between NLG systems. For instance, the second and third evaluation worlds specifically exercised an NLG system's abilities in generating referring expressions and navigation instructions respectively, and there were significant differences in the performance of some systems across different worlds. Such data, which is highly valuable for pinpointing specific weaknesses of a system, would have been prohibitively costly and time-consuming to collect using laboratory subjects.

7 Conclusion

In this paper, we have described GIVE-1, the first installment of the GIVE Challenge. GIVE uses a novel evaluation methodology for NLG systems: It connects NLG systems to human subjects over the Internet and evaluates them according to several objective measures related to task success, as well as a variety of subjective measures in a questionnaire. GIVE-1 collected data from 1143 valid games over a period of three months, and found a number of significant differences between the five NLG systems under evaluation. We established that these results are comparable to, but more precise than, those gained from a laboratory-based version of the evaluation, and thus validated the web-based data-gathering strategy for NLG systems.

We are currently running GIVE-2, which differs from GIVE-1 in that it allows users to move and turn freely, as opposed to the discrete moves and turns they could execute in the GIVE-1 client. This makes the NLG task much harder. For instance, the *Austin* system exclusively used navigation instructions of the form *walk three steps forward* and *turn left*; but *three steps* is not a distance measure that a GIVE-2 user will be able to interpret, and it is an open research question how far a user will turn after the instruction *turn left*.

Beyond this, there are many directions in which one could take GIVE in the future. In particular, it would be interesting to try GIVE with spoken rather than written language generation; to extend GIVE to a dialogue challenge by allowing the instruction follower to speak; and to reverse GIVE into a challenge for instruction *understanding* systems. More generally, we plan to generalize the GIVE software into a generic platform for the Internet-based evaluation of NLG systems. This would then allow evaluators to replace the specific implementations of the GIVE client and NLG system interface by tools that are suitable to their domains, while retaining the networking, database, and world management backbone that the GIVE software provides.

Acknowledgments. We are grateful to the participants of the 2007 NSF/SIGGEN Workshop on Shared Tasks and Evaluation in NLG and the ENLG 2009 workshop, and to many other colleagues for fruitful discussions while we were designing the GIVE Challenge, and to the organizers of Generation Challenges 2009 and ENLG 2009 for their support and the opportunity to present the results at ENLG. We also thank the four participating research teams for their contributions and their patience while we were working out bugs in the GIVE software. We are indebted to Sara Dalzel-Job for running the laboratory-based experiment at the University of Edinburgh. Particular thanks go to Mariët Theune for her insightful and detailed comments on this paper. The creation of the GIVE infrastructure was supported in part by a Small Projects grant from the University of Edinburgh and by the Cluster of Excellence "Multimodal Computing and Interaction" at Saarland University.

References

1. von Ahn, L., Dabbish, L.: Labeling images with a computer game. In: Proceedings of the ACM CHI Conference (2004)
2. Bangalore, S., Rambow, O., Whittaker, S.: Evaluation metrics for generation. In: Proceedings of the First International Natural Language Generation Conference (INLG 2000), Mitzpe Ramon, pp. 1–8 (2000)
3. Belz, A.: That's nice... what can you do with it? Computational Linguistics 35(1), 111–118 (2009)
4. Belz, A., Gatt, A.: Intrinsic vs. extrinsic evaluation measures for referring expression generation. In: Proceedings of ACL 2008: HLT, Short Papers, Columbus, Ohio, pp. 197–200 (2008)
5. Belz, A., Kow, E.: Assessing the trade-off between system building cost and output quality in data-to-text generation. In: Krahmer, E., Theune, M. (eds.) Empirical Methods in NLG. LNCS (LNAI), vol. 5790, pp. 180–200. Springer, Heidelberg (2010)
6. Belz, A., Kow, E., Viethen, J., Gatt, A.: Generating referring expressions in context: The GREC task evaluation challenges. In: Krahmer, E., Theune, M. (eds.) Empirical Methods in NLG. LNCS (LNAI), vol. 5790, pp. 294–328. Springer, Heidelberg (2010)
7. Belz, A., Reiter, E.: Comparing automatic and human evaluation of NLG systems. In: Proceedings of EACL 2006, Trento, Italy, pp. 249–256 (2006)
8. Boer Rookhuiszen, R., Obbink, M., Theune, M.: Two approaches to GIVE: dynamic level adaptation versus playfulness. In: Proceedings of the First NLG Challenge on Generating Instructions in Virtual Environments (2009), http://www.give-challenge.org/research
9. Cahill, A., Forst, M.: Human Evaluation of a German Surface Realisation Ranker. In: Krahmer, E., Theune, M. (eds.) Empirical Methods in NLG. LNCS (LNAI), vol. 5790, pp. 201–221. Springer, Heidelberg (2010)
10. Callison-Burch, C., Osborne, M., Koehn, P.: Re-evaluating the role of Bleu in machine translation research. In: Proceedings of the 11th Conference of the European Chapter of the Association for Computational Linguistics (EACL 2006), Trento, Italy, pp. 249–256 (2006)
11. Chamberlain, J., Poesio, M., Kruschwitz, U.: Addressing the resource bottleneck to create large-scale annotated texts. In: Bos, J., Delmonte, R. (eds.) Proceedings of the Symposium on Semantics in Text Processing (STEP), pp. 375–380 (2008)
12. Chen, D., Karpov, I.: The GIVE-1 Austin system. In: Proceedings of the First NLG Challenge on Generating Instructions in Virtual Environments (2009), http://www.give-challenge.org/research
13. Dale, R., White, M. (eds.): Proceedings of the NSF/SIGGEN Workshop for Shared Tasks and Comparative Evaluation in NLG, Arlington, VA (2007)
14. Dale, R., Reiter, E.: Computational interpretations of the Gricean maxims in the generation of referring expressions. Cognitive Science 19(2), 233–263 (1995)
15. Dionne, D., de la Puente, S., León, C., Hervás, R., Gervás, P.: Guide. In: Proceedings of the First NLG Challenge on Generating Instructions in Virtual Environments (2009), http://www.give-challenge.org/research
16. Foster, M.E.: Automated metrics that agree with human judgements on generated output for an embodied conversational agent. In: Proceedings of the Fifth International Natural Language Generation Conference (INLG 2008), Salt Fork, OH, pp. 95–103 (2008)

17. Gatt, A., Belz, A., Kow, E.: The TUNA-REG challenge 2009: Overview and evaluation results. In: Proceedings of the 12th European Workshop on Natural Language Generation (ENLG 2009), pp. 174–182 (2009)
18. Gatt, A., Belz, A.: Introducing Shared Tasks to NLG: The TUNA Shared Task Evaluation Challenges. In: Krahmer, E., Theune, M. (eds.) Empirical Methods in NLG. LNCS (LNAI), vol. 5790, pp. 264–293. Springer, Heidelberg (2010)
19. Hsu, C.W., Wah, B.W., Huang, R., Chen, Y.X.: New features in SGPlan for handling soft constraints and goal preferences in PDDL 3.0. In: Proceedings of the Fifth International Planning Competition, 16th International Conference on Automated Planning and Scheduling, pp. 39–41 (2006)
20. Keller, F., Gunasekharan, S., Mayo, N., Corley, M.: Timing accuracy of web experiments: A case study using the WebExp software package. Behavior Research Methods 41(1), 1–12 (2009)
21. Koller, A., Petrick, R.: Experiences with planning for natural language generation. In: Proceedings of SPARK 2008: The ICAPS 2008 Scheduling and Planning Applications Workshop, Sydney, Australia (2008)
22. Moffat, S., Hampson, E., Hatzipantelis, M.: Navigation in a "virtual" maze: Sex differences and correlation with psychometric measures of spatial ability in humans. Evolution and Human Behavior 19(2), 73–87 (1998)
23. Orkin, J., Roy, D.: The restaurant game: Learning social behavior and language from thousands of players online. Journal of Game Development 3(1), 39–60 (2007)
24. Stent, A., Marge, M., Singhai, M.: Evaluating evaluation methods for generation in the presence of variation. In: Gelbukh, A. (ed.) CICLing 2005. LNCS, vol. 3406, pp. 341–351. Springer, Heidelberg (2005)
25. Stoia, L., Shockley, D.M., Byron, D.K., Fosler-Lussier, E.: Noun phrase generation for situated dialogs. In: Proceedings of the Fourth International Natural Language Generation Conference (INLG 2006), Sydney (2006)
26. Striegnitz, K., Majda, F.: Landmarks in navigation instructions for a virtual environment. In: Proceedings of the First NLG Challenge on Generating Instructions in Virtual Environments (2009), http://www.give-challenge.org/research
27. Walker, M., Litman, D., Kamm, C., Abella, A.: PARADISE: A framework for evaluating spoken dialogue agents. In: Proceedings of ACL 1997, Madrid, Spain, pp. 271–280 (1997)
28. Walker, M., Rudnicky, A., Prasad, R., Aberdeen, J., Bratt, E.O., Garofolo, J., Hastie, H., Le, A., Pellom, B., Potamianos, A., Passonneau, R., Roukos, S., Sanders, G., Seneff, S., Stallard, D.: DARPA communicator: Cross-system results for the 2001 evaluation. In: ICSLP 2002: Inter. Conf. on Spoken Language Processing, Denver, CO USA, vol. 1, pp. 273–276 (2002)

Author Index